Sustainable
Development Handbook

By Stephen A. Roosa, Ph.D.

THE FAIRMONT PRESS, INC.

CRC Press
Taylor & Francis Group

Library of Congress Cataloging-in-Publication Data

Roosa, Stephen A., 1952-
  Sustainable development handbook / by Stephen A. Roosa.
    p. cm.
  Includes bibliographical references and index.
  ISBN 0-88173-565-5 (alk. paper) -- ISBN 0-88173-566-3 (electronic) -- ISBN
1-4200-7382-6 (Taylor & Francis : alk. paper)
    1. Sustainable development--Handbooks, manuals, etc. 2. Environmental
protection--Government policy--Handbooks, manuals, etc. 3. Energy
conservation--Government policy--Handbooks, manuals, etc. 4. City
planning--Handbooks, manuals, etc. I. Title.

  HC79.E5R652 2007
  338.9'27--dc22

                                                                    2007037898

Published by The Fairmont Press, Inc.
700 Indian Trail
Lilburn, GA 30047
tel: 770-925-9388; fax: 770-381-9865
http://www.fairmontpress.com

Distributed by Taylor & Francis Ltd.
6000 Broken Sound Parkway NW, Suite 300
Boca Raton, FL 33487, USA
E-mail: orders@crcpress.com

Distributed by Taylor & Francis Ltd.
23-25 Blades Court
Deodar Road
London SW15 2NU, UK
E-mail: uk.tandf@thomsonpublishingservices.co.uk

Printed in the United States of America
10 9 8 7 6 5 4 3 2 1

0-88173-565-5 (The Fairmont Press, Inc.)
1-4200-7382-6 (Taylor & Francis Ltd.)

Table of Contents

Foreword

"Sustainable"… Take a few quiet moments.
Think what this term means to you—then read this book!

Preface

Stephen Roosa's *Sustainable Development Handbook* is one of the field's most important contributions. Energy, how we use it, and the types of energy we use, are vitally important to achieving a sustainable world. There is a growing recognition by society in general and our communities in particular that energy engineering solutions are closely linked to sustainability. This insightful new book offers with refined clarity an understanding of how sustainability can be achieved. It provides ways we can address the growing energy and environmental crisis and offers methodologies that can lead us to a sustainable future.

This book should be used in classrooms, read by state and local governmental leaders, and used as a source book for architects, energy engineers, sustainability professionals, planners and journalists. There is a tremendous sustainability crisis—one with international implications—and it is the number one issue confronting us. We must educate ourselves and act before it is too late. The most up-to-date information on this subject is found in this book.

There are two approaches we can take. Madonna told her audience during the world-wide coverage of the Live Earth concert in 2007 that you should "jump up and down" to the beat of her music to end global warming. Others said we should change our light bulbs on the front porch. These actions alone will not solve the looming problems we face—reading this book, becoming informed and acting in the political, energy and environmental arenas is far more powerful. The *Sustainable Development Handbook* provides one of the best assessments of our sustainability crisis and the mitigation actions that need to be taken next. This book needs to be widely read, discussed and debated. It's an important contribution to the energy and environmental issues that we face. It profiles the steps we must take so that our children can live in a world with a future—a sustainable one that be enjoyed by many generations after us. This book shows how cities, like New York, Las Vegas, Amsterdam, Havana or Venice, can make a difference using a mixture of public and private programs and policies.

We can become more sustainable by redesigning and rebuilding our cities house by house, block by block, and neighborhood by neighborhood. If Lagos, Nigeria, ranks as among the worst cities in the world in terms of environmental blight, Amsterdam, The Netherlands, would be among the most sustainable in the world. People are tired of long commutes to and from home and work. They want other transportation options. People want to do something about global warming and the impacts of the greenhouse effect, and they are choosing to reduce carbon dioxide emissions.

People also desire a more active lifestyle, one offering a greater opportunity to walk or ride a bike to work. "Green" lifestyle choices are not only better for the environment, but they provide the added benefit of helping people be healthier. For example, in Amsterdam the lifespan of those who walk or ride a bike to work is 2-1/2 to 4 years longer than those who don't. Health care costs are substantially reduced. Cities that have taken the lead on sustainability create win-win situations that reduce healthcare costs and help us become better citizens for the world community. Cities that fail to embrace sustainability face economic and moral decline.

Another reason for the shift is the "back to the city" movement, where people are finding affordable, historic housing in neighborhoods near downtown. These older neighborhoods will continue to improve if they take the following actions. First, new planning initiatives such as traffic calming can make neighborhoods more family friendly. Second, we need bike lanes in these neighborhoods for safer commutes from work to home. Third, mixed-use developments that include housing, restaurants, commercial, and cultural activities need to be embraced—and not hindered by our planning decisions. Fourth, stronger historic preservation ordinances are needed. Fifth, private-public partnerships need to be created to champion solutions to our looming sustainability problems. Sixth, we need to embrace alternative energy in our homes and businesses along with energy efficient design. Most importantly, we must understand the relationship between energy use and sustainability, and redesign infrastructure accordingly. The *Sustainable Development Handbook* shows us ways these issues can be addressed and resolved.

There is a greener future for our buildings, neighborhoods, downtowns and cities. The *Sustainable Development Handbook* details the progressive actions we can take and what actions need to be avoided. This book is the definitive work on sustainability. Read it and share it with others.

John I. Gilderbloom, Ph.D.

Dr. John I. Gilderbloom is a professor in the Urban and Public Affairs Graduate Program at the University of Louisville, where he also directs the Center for Sustainable Urban Neighborhoods.

Acknowledgements

During the course of my career in energy, I have been truly blessed to have had the opportunity to learn and share experiences with people who have helped me understand the salient aspects and importance of sustainable development. My work has been dedicated to exploring the linkage between energy and sustainability and implementing policies, programs and technologies that are sustainable.

Early in my career, two professors at the University of Kentucky helped shape and support my belief that societies were not on the right track when it came to energy. Dr. Leonard Ravitz, with whom I and Herb Burns co-taught a course in Educational Psychology, helped me understand the long-term implications of decisions we make. Sarah Tate, a former architecture instructor at the University of Kentucky, helped me understand the importance of architectural design principles that can shape structures in ways that will solve environmental needs. I would like to thank Al Thumann, Executive Director of the Association of Energy Engineers, and Jerry Taylor, a noted Kentucky Architect and Engineer, for their friendship, insights and support.

Professors and counterparts from the days when I worked my way though the doctoral program at the University of Louisville helped me by reviewing the content and logic found in Chapters 7, 9, and 10. These gentlemen included Dr. Hank Savitch, Dr. Thomas Lyons, and Dr. Charles Ziegler. All assisted me with the development of my Theory of Divergence (found in Chapter 9). In addition, Dr. Wayne Usui, a statistics guru if ever one existed, assisted by helping to structure the statistical analysis in Chapter 9, and to sort out key variables that defined the basis of the sustainability indexing methodology. Without their aid at the time when many of these ideas were just beginning to congeal, this project would have ended in a false start.

Researching the topic of sustainable development—truly an international phenomenon—quickly became a travel-intensive endeavor. I thank Dr. Albin Zsebig, from the Budapest University of Science and Technology, who invited me to Hungary on several occasions where I had the opportunity to co-chair international conferences that he had organize. The many comments of the participants about the novelty of an American coming to Eastern Europe to present a report on sustainability gave me the incentive to research the subject internationally. Also, Dr. John Gilder-bloom, who invited me to co-chair conferences in Honolulu and Havana and to travel with him and his students to Amsterdam, provided me with opportunities to study sustainability issues in these cities. I thank him for his encouragement and friendship. I also thank, April Li and other members of the organizing committee of the Electrical and Mechanical Safety and Energy Efficiency Symposium 2007 in Hong Kong, who had invited me as keynote speaker, and provided me with the opportunity to research sustainability issues in China.

Without the invaluable efforts of the book's contributors, sustainable development could not have been so rigorously addressed. My sincere thanks go to:

Sieglinde Kinne, author of Chapter 5, and contributor to Chapter 3, who relished the opportunity to participate on this project.

Matt Hanka, a former student of mine at the University of Louisville who coauthored Chapter 3 and provided perspective and valuable insights regarding the discussions of the Kyoto and Montreal Protocols.

Kristen Cooper-Carter, from California State University, Chico, who provided rich insight and important research on sustainability initiatives in Butte County California and who contributed descriptions of the initiatives at CSU in Chapter 12.

I also thank my many former Environmental Policy students, graduate and undergraduate, at the University of Louisville. It was in those classes that many of the ideas in this book were tested, re-evaluated, reinvented and tested again. I wish them well and know that by their efforts and others around the world like them, our future is in good hands.

It was only fitting that Bill Payne, the editor for my first published journal article about 20 years ago, should be asked to serve as editor on this project. From our first discussion, his enthusiasm about this topic inspired me. A finer man you will never encounter. I thank him for his outstanding work and contributions to this book.

There are friends, professors, and associates who took the time to listen to my many ideas about sustainability (some for endless hours) and provided important insights into to the topic by offering their thoughts suggestions, and support. Among them are Carl Salas, Dr. Doug Kelbaugh, Dr. Ron Vogel, Nailya Kutzhanova, Lin Ye, and Dr. Peter Myers. I have been moved and motivated by the investment my friends, family, and professional counterparts have made in me and I thank them for their support.

Introduction

Make no mistake—these are interesting and challenging times. We are at yet another crossroads in human history. The decisions we are making today will profoundly shape the course of future events. In the last few decades, there has been a radical shift in thinking—we now see energy and environmental issues among the most important challenges we face. This importance cannot be understated. This simple change in perspective is redefining values, changing policies and changing the rules by which we live. We are rolling the dice on our planet's future and the odds of winning have not yet been posted. The stakes are high. We sense that "business as usual" is no longer an option. We feel that the costs of delaying action are increasing. The risks have simply become too great. Ecosystems are in flux and man's activities are to blame. Energy—the types of energy we use and how we choose to use them—once part of the solution are now part of the problem.

Suddenly, energy and environmental problems seem very complicated and their complexity overwhelming. Developmental problems seem interconnected—effort to solve one creates others. Our economically-based, decision-making processes seem to be lacking—and in many cases have failed to support a rationale that leads to correct decisions. Many of the solutions being proposed—ways offered to solve energy problems, solve environmental problems and create sustainability—rapidly devolve into a new set of unmanageable dilemmas. It is like we awoke one morning only to be surprised by the dawning reality that our quick fix solutions were no longer working.

Our best hope is found in a new movement that is gaining momentum on the world stage—one that is redefining the policies of businesses, corporations and governments. This movement is still in its infancy. This movement is causing cities and governments to re-evaluate and rewrite their policies to deal with a wide range of common issues and developing problems. This movement is changing the way we view our future, impacting the decisions that we make and in the process, causing us to re-think the formulation of policies, the design of our processes and products, and the configuration of our cities. This movement is called *sustainable development*. The genesis of this new paradigm is upon us. Like a tidal wave looming on the horizon, it is mounting in strength. What is it about this explosive social evolution called sustainable development that makes it so enticing and appealing? If you happen to unfamiliar with it—you are about to find out.

Sustainable development is an overarching set of integrative principles that involves energy, urban management, environmental ecosystems, economic development, social equity, policy integration, and the idea that effective solutions can be achieved in a cooperative manner. It asks that we consider the future repercussions of decisions we make today—not all of which are favorable. Sustainable development offers opportunities and challenges—but most importantly—solutions. This *Handbook* clarifies the issues regarding sustainable development, explains how it can be achieved, and defines the opportunities that it offers. Technical solutions are leading the way to sustainability.

While it offers hope, the definitions and goals of global sustainable development are both misunderstood and variously applied. Sustainable development is hindered by a lack of consensus as to its definition and by a lack of agreement as to how it can be implemented. This confusion sometimes clouds our judgment and creates justification for inaction. While there are a wide range of defined objectives, empirical examples of sustainable development often fail to be comprehensive.

The ways we incorporate sustainable development initiatives into manageable policies are relevant. Policy initiatives that respond to the idea of sustainable development vary in scope and application. There are indeed a range of choices and means to customize sustainability initiatives. These choices include using energy and environmental policies to reduce urban energy use and improve environmental conditions.

There are strong links between energy usage and sustainable development. The types of energy being used, the ways energy is used and the impacts resulting from energy use can directly contribute to ecosystem disruption. One means of achieving sustainability is to develop and implement local policies that simultaneously impact energy use and the configuration of urban areas. Entities and organizations that have the means of implementing sustainable solutions include institutions, businesses, corporations, local planning agencies, governmental entities—and each of us.

The purpose of this book is to present the history, policies and agenda that comprise the principal components of sustainable development. To this end, this book provides a detailed discussion about sustainable development, demystifies the history of sustainability, and discusses comparative issues concerning sustainability. Energy and sustainability are uniquely related. Sustainability cannot be achieved without energy—using too much of the wrong kinds of energy in improper ways can make a sustainable future an impossibility. It is encouraging that policies have evolved upon which broad based-efforts toward sustainability can be mounted. Changing energy consumption patterns plays an important role in the solution.

In the *Sustainable Development Handbook*, engineering and architectural approaches are detailed that offer solutions to meet the goals of sustainable development. With concerted efforts, the initiatives of governments, institutions, and corporations to promote energy efficiency, energy conservation, and alternatives to carbon-based energy consumption will contribute to enhanced sustainability.

Programming and planning initiatives also play an important role. To clarify this role, a comparative assessment of how urban communities are currently incorporating sustainable development agendas will be provided. Operational policy actions that have been undertaken in cities in order to achieve sustainability goals will be assessed, providing further definition to how polices can be successful. Examples of sustainability programs will be provided. A set of characteristic policies will be used to assess sustainable activities in cities. In addition, a set of quantitative indicators will gauge and measure the variables that show how much impact these policies are achieving. The technologies we choose and the ways we choose to deploy them are also a part of the solution—and these technologies will be described.

Chapter 1 introduces the idea of sustainability and sustainable development. These concepts may be our generation's most important contribution. Sustainable development is an exciting new policy alternative. The various theories and beliefs that concern sustainability are considered as is the concept that it provides a new vision for future development. Energy is gauged to be the most critical resource. The underlying causes influencing sustainable development—population growth, urban development and increasing urban energy use—will be discussed at length. The underlying effects—environmental impacts, urban dislocation, and changes in urban infrastructure—are detailed. In addition, a wide range of public policy alternatives and available technologies are explored. Policies are in flux. Sustainability and energy are closely linked.

Chapter 2 considers the goals of sustainable development and the idea that it is a new social force. Its goal structures offer hope for a new future. Operational definitions of "sustainability" and "sustainable development" are provided. This chapter explores the early development of the idea of sustainability and details how it evolved to became the new international buzzword. Insight is given as to how the concepts of energy and sustainability are interrelated. Chapter 2 also discusses the Rio Conference and Agenda 21 to identify what sustainability means today. The crucial role energy systems play in regard to sustainability is discussed. Cities, the largest users of energy, have experienced growth due to the low cost and availability of energy. Alternative energy, energy conservation and energy efficiency are proposed as ways to meet the goals of sustainability policies.

Sustainable development evokes strong emotions in many and can be threatening to some. To explain why, the political views on sustainable development are discussed as are the arguments for and against sustainability. In order to achieve a sustainable future, a long-term view and holistic perspective is necessary.

Chapter 3 discusses some of the key environmental issues that are linked to sustainability. These include waste management, air quality, water quality, pollution prevention and global warming. Fresh water resources are under assault. Greenhouse gas generation, the culprit contributing to global warming, has become a central sustainability issue. Ozone layer depletion is another. Despite these looming environmental problems, natural capital must be preserved.

They are drivers that generate a need for change. Environmental catastrophes—caused my mankind's interventions—have occurred in the past and more will occur in the future. Indeed, clear air and water are no longer viewed as free goods but as necessities, basic requirements of life that everyone has a right to enjoy. Minimizing production, construction and consumer wastefulness can reduce product costs and at the same time reduce negative environmental impacts. These initiatives can reduce raw material requirements at the front end of production processes and also promote post-consumption recycling of waste products. The process of mining and extracting energy resources, such as coal and oil pollute the environment. There are alternatives. Efforts to prevent and minimize pollution can mitigate adverse environmental impacts.

There are interrelated environmental effects. Ecosystems are in flux. Climates are changing. Species are on the move, expanding their ranges and invading new territories. Many are adapting to changes in climate, others are in decline and some are becoming extinct. All of these are linked to the sustainability of our planet.

Chapter 4 discusses exciting developments in sustainable buildings and related building policies. Buildings consume lots of energy. The types of energy they consume and how they use energy are important to their sustainability. Sustainable construction practices are a part of the solution and offer hope for an alternative future. This chapter considers the architectural and engineering tools that are available to achieve sustainable development in the man-made environments. New construction standards and new materials are being used in the construction industry. Energy efficiency standards and codes (e.g., Energy Star and the International Energy Conservation Code) for buildings provide engineering and architectural solutions intended to reduce energy consumption in buildings and thus their impact on the environment. Green construction technologies are being used in new buildings. Chapter 4 considers the pros and cons of the LEED building rating system, which is an example of how criteria-based sustainable construction is becoming more commonplace. Measurement and Verification provides a means of evaluating and quantifying savings from green construction procedures, and ensures that buildings and process meet energy efficiency requirements. Green construction practices are being used successfully in commercial construction including new schools.

Chapter 5 discusses technology solutions such as energy conservation and renewable energy as a means of achieving sustainability. Technologies are evolving and improving. There is great potential to improve the energy efficiency of existing facilities and equipment in residential, commercial and industrial buildings. Energy efficiency, energy conservation and alternative energy will play an important role in changing the economics of energy production. Renewable energy solutions, including biomass energy, geothermal energy, wind energy, solar energy, and others are forms of sustainable energy production—they are substitutes for carbon-based energy production. Other alternatives including landfill gas production and atomic power will have a role in achieving sustainability.

Chapter 6 considers how corporations are dealing with the concept sustainability, and how they develop, implement and promote their sustainability programs. Corporations are changing the ways they do business and instituting programs that address social responsibility. Increasingly, corporate programs promoting sustainability are in vogue. Approaches vary widely—from marginalizing the idea of sustainability to having very elaborate sustainability programs. Many result from their internal sustainability programs. In Chapter 6, the features and attributes of corporate sustainability programs are discussed in detail.

Corporations tend to combine programs into an agenda and by implementing their agendas move their companies toward sustainability. Agendas form the basis for corporate policies and programs are developed to implement them. Some corporations may have internal recycling programs while others have environmental management programs. Still others tailor their purchasing processes or product designs to become more sustainable.

Industries are improving management systems, replacing central plants, and becoming more environmentally friendly. They are changing how they do business and using sustainability as the justification to implement new ideas. They are concerned about their resource consumption and the impact of manufacturing processes on the environment. Energy efficiency is being used a basis to improve, update and modify industrial processes to achieve enhanced sustainability. The types of materials being used in products are being scrutinized and re-evaluated.

"Green" power production is becoming commonplace. Google has installed the largest rooftop solar-electric generation system at its new corporate headquarters in California. Utility companies are providing energy conservation incentives, upgrading production and transmission facilities, and investing in alternative energy production. Corporations are proud of their sustainability programs and projects. They often advertise their approaches to becoming sustainable.

The process of developing and implementing corporate sustainability programs also is outlined. The evidence is clear—corporations are changing their agendas in an effort to be seen by their stakeholders as supporting internal sustainability programs.

Chapter 7 focuses on local sustainable development programs—the policies and practices employed by cities and their governments. It is inevitable that local governments respond and change their agendas. This chapter sheds light on local efforts to pursue sustainable development agendas. In the U.S., city and county governments have been instrumental in developing and implementing sustainability plans—generally without the leadership and support of the federal government. Their sustainability programs may result from combining programs within their government to form an agenda—then modifying them to meet new sustainability criterion.

Sustainability plans reflect geographic, political and economic concerns of the local governments. Chapter 7 considers sustainability programs for metropolitan governments in the U.S. Local programs are customized and creative. City and local governments are successfully providing an experimental proving ground for sustainability programs. The idea is spreading that cities have the ability to select policies that impact their energy use while implementing programs that align with their choices—all in an effort to achieve sustainability. These may include establishing recycling programs, air quality requirements, local planning codes, transportation standards, green building statutes, and other initiatives. In many southwestern states, water conservation, energy management, and green building standards are the focus of their agendas. In New Jersey and California, there are incentives to expand the use of solar energy—tailored to the needs and concerns of their localities. To incorporate policies that foster sustainability, local governments are modifying their intermediate and long range plans.

The cities in the U.S. Sunbelt are of particular interest due to the ways they grow and develop. Growth in the past has been fueled by the diversion of water resources and the availabilitly of inexpensive energy supplies. This is changing as energy resources become more expensive. Assessments comparing the results of sustainable policies in 25 of the largest Sunbelt cities are provided. Their approaches are assessed by analyzing ten policies.

Chapter 8 details the importance of planning mechanisms for sustainable development. Planning provides a structure for solutions. Planning solutions and energy use are linked. Planning deals with the decisions that are made regarding the placement and design of communities, the facilities constructed in communities, the design of transportation conduits, and site components involved. Considerations such as geography, and landscapes and their inter-relationships are fundamental to the decision making process. The ideal of sustainability intersects the planning process and overlays a fresh set of criteria to be incorporated and implemented. This overlay considers environmental impact, energy consumption, materials, economic impacts and other components. These are keys to achieving sustainability.

Planning alternatives include urban homesteading, revitalizing historic neighborhoods, creating amenities in downtown areas, redeveloping abandoned commercial and industrial sites, and encouraging mixed-use developments. The New Urbanism offers a fresh approach to planning developments and communities. This chapter provides rich examples of the types of sustainable planning solutions in the U.S. and Europe that have been implemented.

Chapter 9 is about how sustainability is measured. It explores ways to create indexes and ranking systems to measure sustainability within peer groups. Sustainability has traditionally been evaluated at the national scale, using macroeconomic attributes. More recently, ways to compare the sustainability of local governments have been developed. Based on the types and nature of individual programs, these can be used to compare the relative success of cities in their efforts to become more sustainable.

In a competitive environment, where corporations, cities and counties feel it is important to be considered sustainable, comparisons within their peer group can be important. This is also true for local governments. Rating systems use key indicators of sustainability to devise a means of weighting the indicators. A tabulation of policies for a sample of major Sunbelt cities is included in chapter 9 which ranks them in terms of the number of policies they have in place.

Next, formulas are used to provide values for rankings, combining both qualitative and quantitative indicators of sustainability. A new sustainability index is developed, and cities are ranked based on this index. Finally, this chapter identifies two types of Sunbelt cities, each with a fundamentally different set of characteristics. It is determined that some cities are more energy efficient than others and that these are leading the way to sustainability. The results are surprising—the major cities in California are aggressively pursuing sustainable development programs.

A new *Theory of Divergence* is proposed that helps us understand how cities are on diverging paths in their efforts to achieve sustainability. This new theory is important—some cities are on the path to sustainability and others are floundering. This theory provides a way to explain the forces causing this divergence.

Chapter 10 offers a specific local governmental case analysis. While most studies focus on cities such as New York and Los Angeles, Las Vegas is selected as a case study for an in depth examination of sustainability issues. Why Las Vegas? Las Vegas is truly unique. This chapter describes how sustainability evolves as a concern in a rapidly-growing metropolitan city. This case study shows how the development of Las Vegas evolved, the environmental and sustainability issues encountered, and which policies were selected as the process unfolded.

The discussion about Las Vegas delves into its history, population growth, urban development, environmental impacts, energy consumption, and water resources. This case study considers the dimensions of qualitative policies and the quantitative measures which were discussed in previous chapters. The types of polices adopted by Las Vegas are elaborated upon in depth. Quantitative measures are used to describe how Las Vegas compares to other cities. Las Vegas is typical of most cities in that it has yet to adopt a comprehensive sustainability policy. Regardless, like many U.S. cities, there are programs in place and projects are underway, that focus on sustainability. With a style all its own, Las Vegas faces challenges on its path to a sustainable future—and these challenges are identified and discussed.

Chapter 11 considers international efforts to achieve sustainability. Sustainable development is a worldwide phenomenon replete with variations on a theme. In European countries, programs are often driven from the top down. They start at the European Union level—then move to the states and their regions and cities. In European countries, cities are often *pulled* into the need to develop local agendas and plans for sustainability, rather then being motivated to develop them on their own. Without guidance from the federal government, U.S. agendas are often developed and tested at the local level; programs that have broad appeal are then *pushed* to the state level.

Holland is an EU country with a historical need to protect itself from the North Sea. Concerns regarding sustainability have historic roots. Amsterdam, once one of the major ports of Europe, offers a glimpse of how Holland is creatively devising alternatives for transportation systems that are both characteristic of the Dutch and a model for the rest of the developed world. Alternatively, Nigeria has allowed its oil resources to be exploited, yielding political upheaval, environmental degradation, and unevenly distributed economic benefits.

Havana, Cuba, the Jewel of the Caribbean, has unique sustainability issues. In the countryside, there are efforts to re-establish nature preserves in areas that were once stripped of jungles to create plantations. Havana has a socialist economy yet is finding ways to revive tourism and redevelop its historic district. Venice, Italy, the Jewel of the Adriatic, is a world-asset and tourist destination, yet it suffers from sinking buildings and rising seas that threaten its existence. It is a city that is challenged with loss of population and a need to preserve its rich cultural heritage and architecture. The in-depth analysis of these cities considers characteristics such as their population growth and environmental issues. Energy plays a key role in the sustainability of both cities. These cities demonstrate how sustainable development is an international phenomenon.

Chapter 12 considers the exciting future of sustainability and sustainable development. The greatest gift of sustainable development is the hope that it offers. This chapter explores what the future holds for sustainability programs. It considers some of the new tools in the sustainable development tool kit, looks at trends that may impact the ability to achieve results and provides predictions on how the sustainable programs of tomorrow will evolve. Technologies are improving and many that will be used in the future are available today. There are differences and similarities between programs at the local and national levels. The most challenging aspects of implementing sustainability might well be the need to create policies and programs that are integrated among local, state, federal and international governments.

This chapter considers how polices will be shaped in the future and how companies, cities and local governments will achieve sustainability. It explores how buildings of the future will manage energy and water usage, how new technologies will be used and how products will be changed. Partnerships among companies, governments, educational institutions and their utilities will form to achieve local objectives and a sustainable future.

The idea of merging common sustainability policies offers hope that programs and solutions can be implemented more rapidly. Creating new policies will be necessary if sustainable programs are to be successfully implemented. In addition, technological developments—especially in the fields of energy and energy efficiency—are occurring internationally at an explosive pace.

We are at a crossroads. Sustainable development polices offer hope for an alternative future—a bright one for humankind. Yet there is much work to be done, and most of it requires tough decisions and involves heavy lifting. The *Sustainable Development Handbook* delves deeply into the policies, programs, and technologies that are necessary to achieve the goals of sustainability.

Contributors

Matthew J. Hanka

Mr. Matthew Hanka is a doctoral candidate (ABD) in the Urban and Public Affairs program at the University of Louisville with a focus in urban policy and administration. Mr. Hanka earned a B.A. in history and world politics from The Catholic University of America in Washington, DC, and earned an M.A. in political science from the University of Louisville. He currently is the graduate research assistant for the Center for Sustainable Urban Neighborhoods (SUN) in Louisville. His interests include housing policy, urban politics and policy, community development, comparative urban development, historic preservation, and economic development.

Sieglinde Kinne

Sieglinde Kinne is a mechanical engineer and renewable energy specialist. She has been working in renewable energy since in 1989. She graduated from Red Rocks Community College with an Associates Degree in active solar technology in 1993 and from Colorado State University with a Bachelor of Science in mechanical engineering in 2003. In the mid-1990s she worked in Colorado repairing and installing solar thermal systems. During her mechanical engineering program, she was fortunate to work with the National Renewable Energy Lab (National Wind Testing Facility) and the Industrial Assessment Center program. From 2004 to 2007 she worked at the Kentucky Pollution Prevention Center, where she provided technical assistance in energy efficiency and renewable energy for state and federal grants. She currently is working as an energy engineer for Siemens Energy and Environmental Solutions in Louisville, Kentucky.

Chapter 1

What is Sustainable Development?

"For most of human history there have been only a few million people alive at any one time. Levels have probably fluctuated, but reached perhaps 300 million some 2,000 years ago and about 600 million 500 years ago. Numbers have increased steadily since the late 17th century. In 1987 the world's population exceeded 5 billion. The rate of increase now appears to be exponential. The population took 1,500 years from the year 0 to 1500 AD to double to 600 million, just 150 years from 1750 to 1900 to double to 1.7 billion and only 30 years from 1950 to 1980 to double to 4.8 billion... At current rates if growth, there will be standing room only by about 2500. Global resource flow potential and pollution absorption capacity would probably be exceeded before that point."

CLAYTON AND RADCLIFFE (1996:75)

The evidence is undisputable—sustainable development is generating significant attention throughout the world. Sustainability is an ideal that has evolved to become the buzzword for a new era. It is an important concept, worthy of the attention it is generating. Understanding what the concepts of "sustainable development" and "sustainability" really mean in practice is important and has a wide range of implications. Sustainability manifests itself as a set of policies, programs, and initiatives, each with its own implications. Does sustainability represent a new vision for future development? If so, how did this concept develop and what are its underlying causes and effects?

While there are a number of factors and events driving the interest in sustainability, this chapter focuses primarily on population growth, environmental considerations, developmental issues, and increasing energy usage as its underlying causes. Effects include adverse environmental impacts, dislocation of populations, and changes in urban infrastructure. The broad issues concerning cities and the manner in which their policies directly and indirectly impact urban energy use will be explored.

Sustainable development is a theoretical construct. After all, sustainability within the time frame of the human experience on Earth, may be difficult if not impossible to achieve. Therefore, using the concept of sustainability as a catch-all solution for all global problems is fraught with peril. According to Harken (1993:211), "Most global problems cannot be solved globally because they are global symptoms of local problems with roots in reductionist thinking that goes back to the scientific revolution and the beginnings of industrialism."

This reality does nothing to diminish the importance of sustainable development. Many aspects of the concept of sustainability have problem-solving implications for industry, institutions, corporations, and governments. Most often, sustainability becomes defined more by the policies that are deployed and the agendas that these policies establish. To complicate the difficulties in implementing the agenda, understand that sustainability has far-reaching and universal implications. Sustainability is about resources, management policies, energy, social concerns, planning, economics, environmental impacts, construction practices, and much more. Responding to its agenda has caused institutions to rethink basic processes, with the potential of yielding fresh and creative solutions to current problems.

Understanding sustainability is a complex task. Berke believes that sustainability may be the next paradigm or "framework to dramatically shift the practice of local participation from dominance by narrow special interests toward a more holistic and inclusive view" (Berke 2002:23). To achieve a holistic and inclusive view, sustainability is considered to be an overarching principle that has a corresponding set of guidelines. One strength of the agenda is the recognition that the decisions and investments we make today can have serious implications for our future and the future generations to come. The long-term impact of our present decisions demands serious consideration. We must temper present actions with caution. One weakness of a sustainability agenda is that it is often misinterpreted—at times by intent. *A sustainability agenda can also be politicized.*

Sustainability clothes itself in a systems analysis approach that considers how processes are redesigned and managed, with the hope of yielding better long-term outcomes. More favorable outcomes are those that best meet the goals of the agenda after trade-offs are considered. While "sustainability" hopefully occurs when the agenda's guidelines are successfully implemented, "sustainable development" can be thought of as physical outcomes that occur when the guidelines are followed.

Is "sustainable development" a policy, a set of policies, a management philosophy, an agenda, a new set of solutions, a new set of problems, or all of these? How did the agenda evolve and what are the components of the agenda? Can sustainability, an abstraction, be defined and how could it possibly be measured? What solutions, if any, does it offer? Is sustainability worth pursuing? Before such questions are addressed, let's look at what drives a sustainability agenda.

HOW SUSTAINABILITY EVOLVED

There are many forces responsible for the concept of sustainability. These include social issues, economic concerns, resource allocation, environmental damage, population growth, access to potable water, health and energy usage, among others. Several of these will be discussed in detail to demonstrate how sustainability evolved. There are a number of causes and effects that seem destined to make sustainable development an important priority of a new world agenda. Underlying causes of specific global problems (such as urban development, population growth, and urban energy use) and their effects (such as pollution and changes in urban infrastructure) will be considered. Specific policies supporting sustainable development will be discussed to examine the nature of its scope. This chapter will consider the types of policy objectives available that can be implemented in response to the increased demands for energy.

MAJOR UNDERLYING CAUSE: POPULATION GROWTH

Population growth has been a driving force increasing the demand for all types of resources. The expansion of urban populations has long been recognized. Over half of the world's population lives in urban areas. These areas have gained over one billion in population in the last 30 years. By 1990, the World Bank noted that the world urban population had grown to nearly equal country-side settlements and that urban populations were growing at a rate of 4.5% per year (Drakakis-Smith 2000:8). This population growth rate meant that the world urban populations were projected to double in just 16 to 20 years.

By 2005, the estimated world urban population was 3.18 billion of a total world population of 6.46 billion, meaning that 49% of the total population of the world is living in urban areas (Worldwatch Institute 200:4). To put this in perspective, the world's urban population in 1950 was about 730 million, or less than a quarter of present estimates. In less than 60 years we have witnessed the urbanization of the world's population. While London, England was the first city since ancient Rome to have a population greater than one million, today there are 414 (many happen to be in China) with over 1,000 projected to exist within 35 years (Rifkin 2006:8). *Is population growth at such astonishing rates sustainable? Are we planning for unprecedented urban growth on such a grand scale?*

The U.S. reached a population of 300 million in October of 2006 and with the population increasing at the rate of 1.8 million annually, is expected to hit 400 million by 2043 (Brown 2006:1). Increases in U.S. urban population are the result of population redistribution, migration, population growth and changes in urban density. As Stephens and Wilstrom (2000:16) noted, "Ninety-nine percent of the 153 million increase in (U.S.) population between 1930 and the estimate for year 2000 has occurred in the nation's metropolitan areas." By 1970 in the U.S., more people lived in the suburbs than in urban or rural areas (DiGaetano and Klemanski 1999:45). The U.S. is now predominately a suburban nation.

Figure 1-1. Honolulu, HI

In addition, populations have been migrating from the "Frostbelt" to the "Sunbelt," and from older industrial (Rustbelt) regions, to locales that are developing information-based service economies. According to Platt (2000:2), "between 1990 and 1996, the 25 fastest-growing metro areas were in Southern and Western Sunbelt states, and 20 of the 25 fastest-declining were in the Northeast and upper Middle West."

Dramatic population increases have indeed occurred in places such as Honolulu, Hawaii, Los Angeles, California, Las Vegas, Nevada, Austin, Texas and Mesa, Arizona. By the beginning of 2007, the city of Los Angeles had a population over 4 million (a population greater than almost half of the U.S. states), increasing from roughly 100,000 in 1900. Migration from Louisiana to other locations due to Hurricane Katrina is partially responsible for the decline in the population of New Orleans (-22.2%) and Gulfport (-7.4%) from 2000 to 2006 while the population increased in Atlanta (+21%), Houston (+17.5%), and Dallas (+16.3) during the same period (Ohlemacher 2007:A3). In-migration is contributing to population increases in many cities. The New York metro area added 1 million immigrants to its population from 2000 to 2006 and without them the region would have declined in population by 600,000.[1]

North American examples are often dwarfed by the population increases in many third world cities (e.g., Calcutta, Mexico City, Rio de Janeiro, and Sao Paolo). Currently, world urban populations are increasing by 2.3% annually. The rapid development and industrialization of many third world cities is astonishing. The once quiet fishing village of Shenzhen, China, provides an explosive example of rapid urbanization: its population increased from 100,000 to 3 million in only 15 years (Rogers 1997:2-41). It has continued to grow to become an industrial metropolis of 7 million inhabitants today and boasts the highest per capita gross domestic product in China. The number of "mega-cities," or those with populations of over 10 million, has increased from 5 in 1960 to 14 in 1995 and is expected to reach 26 by 2015.[2] Despite regulations on the number of births, as China's economy continues to strengthen and its people become more prosperous, they are likely to desire larger families. This may cause China's population to grow more rapidly.

Interestingly, urban expansion can be accommodated by increasing population densities, by increasing the size of metro areas, or by a combination of both. Houston, for example, is characterized by its low density, with only 9 persons per hectare (22.2 per acre) (Synergy 1997:2). Cities with higher population densities require less land to accommodate their residents than do cities with lower population densities.

Using year 2000 U.S. census data as a guide, over 90% of the land areas of the states of Alabama, California, Florida, and Nevada are now included in a named Metropolitan Statistical Area (MSA), despite the existence of large underdeveloped areas. In Los Angeles County, population density increased from 385 persons per square kilometer (1,023 per square mile) in 1950, to 711 persons per square kilometer (1,842 per square mile) in 1980, and to 905 persons per square kilometer (2,345 per square mile) in 2000.

Many urban areas of the northeastern U.S. that were initially developed prior to the advent of the automobile (e.g., Baltimore, Boston, New York, and Philadelphia) tend to exhibit higher population densities than those cities that developed at a time when the automobile was the principle means of transportation (e.g., Los Angeles, Las Vegas, and Houston). Even older cities in the Midwest and West, such as Chicago and San Francisco, that developed when locomotives, cable cars, and automobiles were available, typically have higher population densities than most major cities in the South or Southwest.

Major cities in the more recently developed Sunbelt tend to be less densely populated than most cities in other regions of the country. Miami, the Sunbelt city with the highest urban density, is much less densely populated than New York, Newark, San Francisco, Chicago, Boston, or Philadelphia, and is similar in density to Washington, D.C. Using 2000 population data (from the U.S. Census Bureau, Census 2000), Table 1-1 compares the U.S. cities that have the greatest population densities:

U.S. population densities of about 31 persons per square kilometer (80 persons per square mile) are relatively low compared to other industrialized countries such as France with 108 persons per km^2 (280 per square mile), or Britain with 239 persons per km^2 (620 per square mile) (Desai 2002:140). Population densities in areas of non-western cities are almost unimaginable: In Calcutta's Metropolitan District the population density in 1991 was 23,801 per km^2 (61,644 per square mile); In Cairo the population density in 1996 was 32,005 per km^2 (82,893 per square mile); and in Hong Kong's Kowloon district, the population density in 1999 was 45,474 per km^2 (117,776 per square mile).[3] The process of increasing population densities in cities is called densification.

Where will we find the resources to support the growth of our urban populations? Sustainable development advocates often suggest increasing urban densities and reducing (as opposed to enlarging) the urban "foot-

Table 1-1. U.S. Cities Ranked by Urban Population Densities (2000)

	Total Population	*Population Per Sq. KM*	*Population Per Sq. Mile*
Sunbelt Cities			
Miami, Florida	362,470	3,889	10,072
Long Beach, California	461,522	3,316	8,588
Los Angeles, California	3,694,820	2,868	7,428
Fresno, California	427,652	1,380	3,574
San Diego, California	1,223,400	1,323	3,426
Las Vegas, Nevada	478,434	1,197	3,100
Houston, Texas	1,953,631	1,166	3,020
Atlanta, Georgia	416,474	1,154	2,989
Other Cities			
New York, New York	8,008,278	10,010	25,925
San Francisco, California	776,733	6,422	16,632
Chicago, Illinois	2,896,016	4,922	12,747
Boston, Massachusetts	589,141	4,700	12,172
Newark, New Jersey	273,546	4,438	11,494
Philadelphia, Pennsylvania	1,517,550	4,337	11,233
Washington, D.C.	572,059	3,597	9,317
Baltimore, Maryland	651,154	3,112	8,059

1-2. Hong Kong, China - High Rise Development

print" as a means of improving sustainability. Techniques such as using improved infrastructure, infill construction, preserving existing structures, brownfield redevelopment, high rise developments, and more creative design of physical environments provide logical alternatives that often yield increases in urban density. The examples of New York and Hong Kong certainly offer proof that much higher population densities are possible and are a potential means of accommodating growth in population.

A SECOND MAJOR UNDERLYING CAUSE: URBAN DEVELOPMENT

Growing populations create impetus for changes in urban development. Urban development enlarges the urban service areas as populations migrate to them in search of employment and other amenities that cities can provide. Cities specialize in their types of development and the types of employment they offer. Coastal cities such as Miami typically have greater development and higher population densities near their beaches. Miami is

1-3. Beachfront Development in Miami, FL

also a major port and center for a growing tourist and cruise ship industry.

New buildings, transportation systems, utility infrastructure, and distribution systems are constructed to meet the demands of larger populations. The rate of development can be astounding. In Shanghai during the year of 2005, the amount of building space constructed was greater than all of the commercial office that exists in New York City (Worldwatch 2007:93). Spirn (1984) believes that the process of urbanization is associated with declining environmental quality and environmental degradation. In the cities of highly industrialized countries, vast areas are required for housing, marketplaces, schools, hospitals, commercial centers, and parking.

The components of urban sprawl have been variously identified. Sources of sprawl in the U.S. have been linked to measures such as total population and per capita land consumption—with mixed results (Kolankiewics and Beck 2000:24-25). Sprawl has been linked primarily to per-capita land consumption in cities such as New York and Philadelphia, and primarily to population growth in Los Angeles, Dallas-Fort Worth, Houston, and Phoenix (Kolankiewics and Beck 2000:26).

Some observers categorize urban sprawl into five primary compo-

1-4. Cruise Ships in Port

nents: Housing Subdivisions, Shopping Centers, Office Parks, Civic Institutions, and Roadways (Duany, Plater-Zyberk and Speck 2000:6). Another contributor to sprawl is the fact that developers are often required to design their sites based on the standard of the largest local fire trucks, which requires ever-larger turning radiuses. In order to improve safety, roads are widened to accommodate ever-larger vehicles. However, widening roads enables increases in speed, often making roadways less safe, thus creating an interesting paradox. Roads are made wider to be safer—while vehicular traffic and speeds increase, creating a greater threat to life safety (Duany, Plater-Zyberk and Speck 2000:67). In the U.S., automobiles have killed more people than all the wars in America's history and continue to kill over one million wild animals per week (Harkin *et al.* 1999:22-23).

Other forces cause us to consume land as our cities grow. To create larger facilities while using less-expensive land, manufacturing facilities tend to move to the urban perimeter. New roads are constructed. "Leapfrog" development requires costly extension of transportation networks. Utility infrastructure must expand—but these costs are often subsidized.

Due to the internationalization of industries and lower shipping costs, manufacturing now leapfrogs national borders, and even continents, seeking relatively inexpensive labor, relaxed environmental restrictions, green fields for development, and access to materials. In the process, infrastructure that existed in the original locations is often abandoned. People then migrate to areas where employment is available. Shopping areas are constructed nearby to serve them. Central business districts eventually fall into decline and additional taxpayer dollars are dedicated their revitalization. This cycle creates ever-greater demands on land use. However, costs of energy and the impacts of energy use are rarely considered in the decision-making processes.

Cities are occupying more space due to both in-migration and mergers. Dramatic increases in urban area in U.S. cities began to occur after 1950. Examples include:

1)   Oklahoma City, Oklahoma—which grew from 129 to 1,572 square kilometers (50 to 607 square miles) from 1950 to 1970;

2)   Tulsa, Oklahoma—which grew from 70 to 505 square kilometers (27 to 195 square miles) from 1950 to 1970;

3)   Atlanta, Georgia—which grew from 96 to 342 square kilometers (37 to 132 square miles) from 1950 to 1970;

4)   Kansas City, Missouri—which grew from 220 to 818 square kilometers (85 to 316 square miles) from 1950 to 1960 (Stephens and Wilkstrom 2000:46).

Urban development occurs in both the urban cores and suburban areas as cities adapt to population changes. Kolankiewicz and Beck (2000:46) assessed changes in land use patterns for the cities of El Paso and Phoenix among others, seeking to understand relationships between their central cores and their corresponding suburbs. They found that from 1970 to 1990, El Paso's central core grew from 306 to 421 square kilometers (118.3 to 162.7 square miles) while the suburbs grew from 3 to 149 square kilometers (1.1 to 57.6 square miles). In Phoenix, they found that the central core grew from 642 to 1,223 square kilometers (247.9 to 472.1 square miles) while the suburbs grew from 362 to 697 square kilometers (139.6 to 269.0 square miles).

Cities use various approaches to foster expansion. While in San Antonio, annexation was a primary strategy for sustaining growth, Metro Louisville and Nashville-Davidson County are among those cities that expanded by merging with their surrounding county governments (Cisneros 1995:5).

As cities increase in geographic area, commuting typically increases. "Lower density suburbs feature less public transit infrastructure and thus, unlike city residents, suburbanites must use private vehicles to travel" (Kahn 2000:570). While suburban development has increased time wasted in commuting (Harrigan and Vogel 2000:303-305), a more significant concern is that almost all of the grid-locked vehicles are consuming fossil fuels.

It is paradoxical that in the U.S., there are now more registered automobiles than licensed drivers. Interestingly, the fleet fuel consumption averages of newly manufactured vehicles has been declining in recent years as multi-use vehicles (MUVs), sport-utility vehicles (SUVs) and Hummer-like urban assault vehicles (UAVs) become the new standards for urban mobility. Studies indicate that if the urban structure of Houston and Phoenix were comparable to Boston or Washington, their gasoline consumption would be as much as 30% lower (Newman and Kenworthy 1989:55;1999). Others suggest that even more dramatic reductions are possible in special cases. In fact, it is considered possible to lower energy demands for transportation by as much as 90% by substantially modifying the physical design characteristics of neighborhoods (O'Meara 1999).

Urban development often means enlarging the urban service areas. One reason for the movement to suburban areas is that additional land is required due to the need for single-story buildings. These are considered a better means of supporting current production and material handling processes. Flat land is used for worker parking to eliminate the need to construct multilevel parking structures.

1-5. Trucking Distribution Center in Nashville, TN

The increasing shift from railroads to trucks as a means of transporting freight is also a factor. The trucking industry lobbies for improved interstate roadways to move its goods and requires extensive facilities for docks and distribution centers. Rather than being linked by light rail systems, outer suburbs in the U.S. almost exclusively depend on automobiles. This dependence can dramatically increase energy usage. Dr. Kentworthy of Murdock University reports that people in the outermost U.S. suburbs consume an estimated "five times more fuel per capita for transport than their inner city counterparts" (Synergy 1997:1).

In addition to driving more miles, suburbanites have larger residences. It is paradoxical that as residences in the U.S. grow ever-larger, the

number of people living in them continues to decline. The average household size of owner-occupied housing declined from 3.45 to 2.71 members
from 1980 to 2000, a decline of 21%. Table 1.2 indicates that the decline was
particularly precipitous between 1980 and 1990.

Table 1.2 - Size of U.S. Households (1980-2000)

	1980	1990	2000
Average household size of owner-occupied unit	3.45	2.75	2.71
Average household size of renter-occupied unit	3.11	2.42	2.36

Source: U.S. Census Bureau

There are a growing number of single-occupant households in U.S.
urban areas. In 2005, married couples were a minority in the U.S. An estimated 51% of women were living without a spouse, an increase from 35%
in 1950.[4] In 2006, there were 55 million people 18 and older who had never been married, an increase of 10 million over ten years.[5] While in many
cases some are living with other adults, others are simply choosing to live
alone, thus creating more households.

Despite the movement toward smaller households, the physical
size of newly constructed houses increased between 1982 and 2002. In
only 20 years, new single family residences in the U.S. increased from a
median size of 141.2 m^2 (1,520 ft^2) to 196.4 m^2 (2,114 ft^2), a 39% increase
(Clements 2003). Using 1980 and 2000 data for household size with 1982
and 2002 data for owner-occupied residences, this equates to a change
from 40.9 m^2 (440 ft^2) per resident to 72.5 m^2 (780 ft^2) per resident. This
is an astounding increase of 77% in personal space for new owner-occupied residences during a 20 year period. If not carefully designed, larger
homes not only use more land area and require more raw materials, they
also consume more energy.

Most population growth is occurring in the suburbs. Most new
homes are being constructed there as well. "An unintended consequence
of suburban growth is greater resource consumption leading to greater environmental damage than if households stayed in the city" (Kahn
2000:584). U.S. households in suburbia consume 31% more resources and
more than twice as much land as their central city counterparts (Kahn
2000:584). The new and growing suburban developments kindle a sort of
urban pathos as central city neighborhoods often lose their vitality and
fall into decline.

The availability of inexpensive energy supplies facilitates urban development. While rapid population growth on a worldwide scale has increasingly placed a growing burden on Earth's resources in general, physical urban development is creating ever-greater demands for all forms of energy.

A THIRD UNDERLYING CAUSE: INCREASING ENERGY USAGE

Energy is a key to the concept of sustainable development. According to Andrews (1999:295), "No sector of human activity impacts the environment more pervasively than the production and use of energy." Energy conservation, use of alternative energy, and improved energy efficiency fill a significant role by enhancing the potential for sustainable development. Inefficient and environmentally insensitive energy use causes a decline in the sustainability of urban areas. This is due to the extensive environmental impacts resulting from energy use. Energy use also creates economic concerns.

Abundant (yet exhaustible) energy resources are used to meet growing urban demands. Today, cities are responsible for consuming roughly three-quarters of the world's energy (Rogers 1997:227). Taking a long-term view, energy usage has increased substantially, from about 136 kg (300 lbs.) of equivalent annual usage of coal per person in 1860 to more than 1,814 kg (4,000 lbs.) in 1984 (Perhac 1989:41-44). This increase continues. Global reserve-production ratios of coal, natural gas and oil are estimated at 221, 61 and 40 years, respectively (Saha 2003:1053). In the U.S, energy usage continues to increase unabated. Current projections are that U.S. energy usage will grow by more than one-third by 2030, with electricity use increasing by 40% (U.S. Energy Information Administration 2006).

ENERGY: SOCIETY'S MOST CRITICAL RESOURCE

Examples of the benefits of having widely available, commercially usable, and affordable energy are found everywhere. Energy is required to operate every means of transportation, all of our appliances, and all types of equipment. Energy creates usable and comfortable interior environments. Indeed, the benefits of having abundant energy are endless. Without readily accessible energy supplies, modern societies would be seriously hampered, and our cities rendered useless. Energy is society's most critical resource. At over $4 trillion annually, energy is now the world's largest business.

Energy conversion is clearly the engine propelling improvements in the world's living standards, the quality of lifestyles and the progress of civilization. Access to energy is considered a birthright in the industrialized world, to be taken for granted unless disrupted. Yet, elsewhere in the world, approximately 2 billion people lack access to modern energy services (Saha 2003:1056).

This gap in standards provides yet another interesting paradox. While space heating and hot water may not be available in many areas of certain third world countries, these same services are often considered absolute necessities in the industrialized countries of Europe, North America, and elsewhere. Lack of either can be cause for legal action by local governments against home and building owners, resulting in condemnation, hefty fines and in extreme cases, imprisonment of the offender.

Reliable energy is an important infrastructure component of the world economy and the prime mover of all else. When electrical power disruptions occur for extended periods in our urban cities, the events are termed disasters by the media and often require intervention by civil authorities. In North America, electrical transmission systems are regionally interconnected. During the summer of 2003, an electrical system failure (due to a tree limb falling on a power line) in rural Ohio precipitated a chain of events that within minutes turned out the lights in Boston and New York. Building occupants fled their high rises into sweltering streets. Such outages disrupt urban areas, causing widespread transportation system congestion, loss of refrigeration, interruption of medical services, substantial economic and financial losses, and even loss of life. U.S. transmission system reliability is perceived to be declining. A 2007 survey of energy engineers and managers asked if they believed that the reliability of the national electric transmission grid had declined in the previous year— 71% answered yes (Thumann *et al.* 2007).

The developmental issues associated with energy consumption have global implications. According to Harken (1993:211), "It is not surprising that this energy-driven growth is producing cities around the world much like our own (U.S.) urban areas, with comparable slums, crowding, pollution and crime."

Energy usage in our built environment has been increasing. Globalization has created opportunities to export industries and their physical plants to countries with more lax environmental regulations and where energy costs are perceived to be lower. Rapid increases in population and the resulting demands for conditioned space are major causes of increased energy use. The need for highly conditioned space has given rise to new standards for human comfort, especially in workplaces where productivity can be increased.

Interestingly, energy usage tends to be highly decentralized while energy production tends to be relatively centralized. External costs include those not only associated with water and air pollution, but also capital availability, and social equity. Today, businesses are increasingly finding themselves in situations where they are required to internalize such costs.

More efficient use of energy in the built environment can have a significant impact on reducing direct economic costs. These include the ability to provide for urban expansion without constructing additional power generating facilities while simultaneously mitigating environmental impact. Technologies are available to provide more efficient use of energy. On a world scale, indexes expressing energy production and transmission efficiency (final energy consumption divided by the total primary energy supplied) yield a surprisingly consistent result of approximately 70 percent. Electrical system efficiencies, included in this average, are much lower. This suggests that 30% of the energy being used is ideally recoverable, indicating that additional energy efficiency improvements are certainly possible (Klevas and Minkstimas 2004:313).

The categories of energy are divided unevenly into nonrenewable sources (typically carbon-based energy sources such as coal, oil, and oil shale), and renewable sources, such as wind, solar, geothermal, and gravitational water sources (e.g., hydroelectric, tidal, and water currents). Technically speaking, most carbon-based energy forms are fossilized biomass and therefore renewable. However, the geophysical and biochemical processes involved require hundreds of millions of years for the renewal process to be effective. Much of the solar energy captured and stored in the Earth by fossilized hydrocarbon processes that are being extracted and consumed today, date from the Paleozoic period, roughly 600 million years ago, predating mankind's existence. As a result, renewable energy sources can be further categorized as *sustainable*, while most nonrenewable energy sources are *potentially unsustainable*. While there is movement toward greater use of renewable energy, hydrocarbon energy consump-

1-6. Cogeneration Facility Using Landfill Gas in Johnson City, TN

tion is also increasing.

Though energy usage is increasing worldwide, most of it is carbon-based. The United States (the world's largest energy consumer), provides a heuristic example. In the U.S. from 1970 to 1996, total energy consumption grew from 17.7 quadrillion kilocalories (67.9 quadrillion Btus) to 23.8 quadrillion kilocalories (92.2 quadrillion Btus), an increase of 34%.[6] Energy from renewable sources grew from 0.7 quadrillion kilocalories (2.7 quadrillion Btus), or 4% of the total, to 1.8 quadrillion kilocalories (7.2 quadrillion Btus), 7.7% of the total.[7] In 1999, U.S. energy usage totaled 24.3 quadrillion kilocalories (96.6 quadrillion Btus) with transportation fuels totaling 6.5 quadrillion kilocalories (25.9 quadrillion Btus) or 26.8% of the total (U.S. Census Bureau 2000:583). By 2004, energy consumption increased to 25.1 quadrillion kilocalories (99.5 quadrillion Btus).[8]

Up to the 1950s, the U.S. was the world's leading producer of oil. As U.S. energy consumption increases, the concern is all about oil. In 1990, over 100,000 Iraqi troops entered Kuwait in what might be described as the first war fought over oil. It was an international coalition, led by the U.S., which reclaimed Kuwait and reinstalled its former government. The

Iraqis departed leaving the Kuwaiti oil fields ablaze. It can be argued that the civil war in the bi-cultural (Arab and Black African) country of Sudan, which has displaced millions in its southern regions, may indeed be a grab by the north to control southern oil reserves.

Today the U.S. remains deeply involved in a war in Iraq and remains politically supportive of Saudi Arabia, a non-democratic monarchy which functionally endorses the Qur'an as its constitution. In his 2007 State of the Union Address, U.S. President George W. Bush stated that the U.S. "has been dependent on foreign oil... this dependence leaves us more vulnerable to hostile regimes, and to terrorists—who could cause huge disruptions of oil shipments, and raise the price of oil, and do great harm to our economy." Others believe that by purchasing Mid-Eastern oil, "the U.S. is actually indirectly subsidizing terrorist activities with their oil purchases" (Steffes 2002:5). One example is the funding of the U.S.-designated terrorist group Hezbollah in Lebanon, estimated by Western analysts to be receiving $200 million annually from Iran.[9] Regardless, we are witnessing the beginnings of an international and inter-generational food fight over control of oil and the potential wealth that it can bring.

Despite this volatile political climate, the U.S. is increasing oil imports to sustain its economy. How much oil is actually being imported? According to the U.S. Census Bureau, in 1973, the U.S. imported an average 3,244,000 barrels of crude oil per day. Recall that 1973 was the beginning of the "oil crisis" when members of the Organization of Arab Petroleum Exporting Countries (OAPEC) decided during the Yom Kipper War that they would not ship oil to countries (including the U.S.) that supported Israel in the conflict. Despite periodic calls for energy independence during the interim, by 1998, total imports had risen to an average of 8,706,000 barrels per day, an increase of 168% (U.S. Census Bureau 2000). Meanwhile domestic production declined 36%, from an average of 9,208,000 barrels per day to 5,925,000 barrels per day. Oil imported from the Organization of Petroleum Exporting Countries (OPEC) rose nearly three-fold, from 258 million barrels in 1973 to 746 million barrels in 1998 (U.S. Census Bureau 2000). Over the twenty-five year period from 1973 to 1998, U.S. importation of oil increased three-fold, yet there was no longer a perceived oil crisis.

In 2004, the U.S. was importing oil at rate of roughly 12.4 million barrels of oil per day. OPEC, other oil exporting countries, and the "Seven Sisters" oil conglomerates were awash in cash while consumers in oil-importing countries were becoming cash poor. By 2005, the U.S was consuming oil at the rate of 20.7 million barrels per day[10] –still not an apparent crises, but simply getting more costly. In 2006, most oil companies were reporting record profits while getting more tax breaks. ExxonMobil recorded the largest annual profit in history by a U.S. company—$39.5 billion on revenues of $377.6 billion.[11] Royal Dutch Shell, Marathon Oil and Valero Energy combined for record profits of and additional $75.6 billion for the year.[12] *However, efforts to implement energy efficiency improvements that could provide insurance against potential energy supply disruptions were comparatively minimal.*

Power production and power use have urban and regional impact. Questions immediately arise. Where will the next power plant be constructed? Will strip mining and mountain-top removals be allowed in Appalachia? Will a dam be constructed that could disrupt natural water flows and eliminate yet another recreation resource? Will an oil tank farm be allowed to be built next to a river? How much longer will a 30-year-old nuclear power plant be allowed to operate? How many commercial and industrial zones can be developed—given the energy supply available? Is a new cross-country gas pipeline really necessary? How can coal-fired electrical generating plants be modified to reduce pollution? Will we open more federal lands to oil drilling? What, if anything, needs to be done about CO_2 levels in the atmosphere? The list of relevant questions seems endless. What these questions have in common is that all are related to sustainability.

TECHNICAL SOLUTIONS: READY, NOW!

From a technological capability standpoint, we now have flexibility to select from multiple energy sources to satisfy a given requirement, allowing the most appropriate energy sources to be used. In addition, we now have the technology to design and build extremely efficient new facilities.

Are we doing this? Or can we suggest that the U.S. and many other nations are experiencing on-going energy shortages that we must simply learn to cope with? Is improving energy efficiency unimportant?

Conventional wisdom suggests that our buildings today are more energy-efficient than ones constructed in earlier times. While this is certainly possible, many of our buildings waste energy needlessly and often incorporate few conservation or alternative energy features (Meckler

1994:41). It is interesting to note that despite improved architectural design standards, the availability of more efficient equipment, and the advent of digital environmental control systems, many newer buildings actually use far more energy than older ones. This is due to their size, design, location, improved standards, the construction technologies employed, and the activities and processes within them.

Buildings today are often larger in order to meet higher standards for personal space, access, egress, accessibility, spaces for mechanical equipment, and amenities. Causes for the increases in energy use include more widespread use of central air conditioning systems, air filtration requirements, and improved fresh air standards for occupied spaces. Increasing ventilation air means greater energy costs in order to produce conditioned occupancy air (filtered, cooled or heated, humidified or dehumidified) from unconditioned outside air. Widespread use of electrical motors in buildings not only increases electricity demand requirements but also creates the need for still more electricity to exhaust the heat which these motors generate.

Though service economies at the macroeconomic level are generally thought to consume less energy than industrially based economies, yet another paradox is that the demand for energy appears to be surging in the service and high-tech sectors while it has not dropped appreciably in other sectors (Sioshansi 2001:64-70). Consider the new high-tech companies that are prized by cities competing to attract growing service industries. Highly automated manufacturing plants with few actual employees often condition the air in their entire facilities as if they were fully occupied by humans. High-tech companies such as Cisco Systems, Oracle, Sun Microsystems, HP and Intel are major energy consumers of electricity. Oracle's complex in Silicon Valley uses an estimated 13 MW and Sun's campus requires over 26 MW (Sioshansi 2001:64-70). Our computer age requires electrical energy, often of the highest quality, to operate our computer networks and energy-intensive "data-bahn" systems that allow information to link any number of countries and their economies.

There are many renewable and non-renewable energy sources available. Conventional non-renewable energy sources include coal, oil, natural gas and propane. Coal is used to generate heat and provide electricity. Oil is a fuel and lubricant source for all manner of vehicles and is popular in the northeastern U.S. as a residential heating source. Both natural gas and propane gas are widely used for both water and space heating. These, along with coal and oil, are in a subset of energy sources commonly referred to as carbon-based "fossil fuels."

The category of energy sources which are often referred to as "alternative fuels" or "renewable" energy sources include: 1) solar power used directly for space heating, with heat exchangers for water heating, or for direct electrical energy generation using photovoltaic cells; 2) wind power used for water pumping and direct conversion of electricity, 3) biomass fuels for heat production and cogeneration; 4) fuel cells for various uses; and 5) hydropower for electrical generation.

1-7. Wind-Turbine Generator in Central Hungary

Improved energy efficiency and energy conservation are considered by some to be "alternative fuels" in the sense that they reduce the recurring need for fossil-fuels. Since the 1990s, they have combined to provide a greater amount of new "energy" than any other fuel. It is clear that there are many energy options available to meet urban energy needs.

The complexity of energy problems is best represented by the following question: How can urban areas provide completely safe, totally dependable, readily transportable, seemingly inexhaustible, efficiently convertible, easily containable, environmentally sensitive, relatively inexpensive, convenient energy to our citizens? Further: these goals must be achieved without using too much land, causing injuries, making too much noise, stopping traffic, having to buy lots of hardware, demanding excessive maintenance, needing too much infrastructure, all while avoiding environmental externalities and without creating any other associated problems.

ONE UNDERLYING EFFECT:
ENVIRONMENTAL DETERIORATION

Sustainability deals with the environmental impacts of development. These include global climate changes such as deforestation, loss of wetlands, damage to the ozone layer, and the greenhouse effect. Regional environmental impacts include air- and water-pollution, acid gas emissions resulting in acid rain, destruction of microclimates by expanding urban areas, and the localized health impacts of uncontained environmental toxins. A preponderance of literature focuses on various aspects of these issues (Brown 2001; Burgess and Jenks 2000; Choucri 1993; Crenson 1971; Dincer 1999). Local solutions have limited impact, but examples include strategies to dispose of wastes without causing additional environmental damage.

Energy use impacts the environment in various ways (Rosen 2004:27). The "central element in urban environmental sustainability is the adoption of appropriate energy policies, since most environmental externalities are directly or indirectly related to energy use" (Nijkamp and Pepping 1998:1481).

Externalities are subsidized costs that are not accounted for. The idea of externalities implies that benefits can accrue—while the resulting costs, sometimes hidden and often difficult to quantify, have a corresponding negative impact. There are specific externalities related to environmental pollution that must be considered. The U.S. generates 1.4 billion tons of wastes annually, most of which is shipped to landfills (Durst 2007:78). The costs of transportation, disposal, and landfill development are externalities that are not included in the costs of the products.

Urban development has long been associated with reduced environmental quality and environmental degradation (Spirn 1984). According to the United Nations Commission on Sustainable Development, "air and water pollution in urban areas are associated with excess morbidity and mortality... Environmental pollution as a result of energy production, transportation, industry or lifestyle choices adversely affects health. This would include such factors as ambient and indoor air pollution, water pollution, inadequate waste management, noise, pesticides and radiation" (United Nations 2001:38). It has been suggested that the earth's capacity to "absorb energy-related pollution will be exceeded before energy shortages become a problem" (Chasek 2000:430).

Environmental issues raise an additional set of concerns. Pollution generated in urban areas raises equity issues. In the U.S., issues evolved regarding levels of income and pollution, the idea being that the poor are more likely to be exposed to toxic environments than the wealthy. This is not limited to the U.S. A study of emissions in Birmingham, England indicated a "striking relationship between modeled emissions and poverty indicators and ethnicity" (Brainard *et al.* 2002:695-716).

In addition to social concerns, what happens when systems fail? Oil tankers can and do run aground on reefs and beaches (Prudhoe Bay, Alaska). Despite publicized remediation efforts, most oil spills are not mitigated. Coal-sludge dikes can break, causing environmental damage to rivers (Big Sandy River, Kentucky). Nuclear power plants occasionally have major accidents (Three Mile Island, Pennsylvania and Chernobyl, Ukraine)—and minor accidents happen almost daily. Air and water has been contaminated due to sulfur and carbon dioxides, which damage lakes and marine life. This has been experienced in the northeastern United States and elsewhere.

New development in urban areas creates the need for more paved areas (asphalt is a petroleum-based byproduct). Solid roof areas and thermal heat from equipment create a microclimatological condition known as the "urban heat island effect." While there is significant debate over the extent of environmental impacts caused by increasing energy usage, there are known, verified effects.

Dealing with environmental impacts has yielded various policy responses. These include "ignore the problem or else legitimize it" as issues emerge, "react and attempt to cure" when a publicized event occurs, and "exert command-and-control" styles of governance to contain the problems. These sometimes, yet rarely, evolve to a more proactive "anticipate and prevent" stance.

Pathologies on a global scale gradually become difficult to ignore. In the last 50 years, most global warming has been due to human activities (IPPC 2001). A root cause is the global appetite for hydrocarbon fuels. The United Nations (1992:11) in its *Guide to Agenda 21* states:

> *"Curbing the global appetite for fossil fuels is the most important single action that must be taken to reduce adverse impacts to the atmosphere. This requires shifts in consumer and industrial practices that are deeply entrenched and movement towards a pattern of energy production and consumption that relies more on efficiency and environmentally sound energy systems, particularly renewables... Energy is at the heart of some of the principal environmental problems we face. The unprecedented economic growth that has occurred mainly in industrialized countries in this century has depended to a very great extent on availability of low-cost energy, principally in the form of fossil fuels."*

Strategic objectives to implement sustainability often involve changing how energy is used, developing more appropriate housing solutions, improving transportation systems, creating higher-order employment, providing creative alternate-energy solutions, and reducing environmental impact. Housing that is more energy-efficient provides the opportunity to lower the recurring costs of renting or owning a home. Public transportation systems generally reduce the *per capita* use of transportation fuels and offer broader mobility options to the public at large.

There exists an unfounded myth in the U.S that addressing environmental concerns and supporting alternative energy solutions creates employment losses. Contrary to conventional wisdom, there is evidence that improved environmental management actually creates jobs and does not involve major sacrifices (Bezdek, Welding and Jones 1989:247-279; Nemtzow 2003:20; Meadows 1989:M1). This is also true of decentralized forms of energy production. Wind power, as one example of an alternative energy source, creates 27% more employment than coal and 66% more employment than natural gas, per kilowatt-hour produced (Sanghi 1992).

Improved energy efficiency and environmental management can create employment within the U.S. (Bezdek, Wendling and Jones 1989:274-279). A regional study by the Environmental Law and Policy Center considered ten Midwestern U.S. states and painted a clearer picture. According to their study, re-powering the U.S. Midwest with a "clean energy" plan by promoting energy-efficient technologies and developing renewable energy resources has the potential to provide 200,000 new jobs in the region by 2020—while creating up to $20 billion in increased economic activity (Hewings and Moshe 2002:2).

While the U.S. falters, other countries are less short-sighted. Both Japan and Germany view their advances in alternative energy and environmental improvement technologies as having the potential to develop into major export markets for their service and equipment industries.

Regardless, with less than 5% of the world's population residing in the U.S., the country is responsible for over one-fifth of global annual carbon emissions (Benfield *et al.* 2001:49) and roughly a quarter of the world's primary energy use.

That energy usage is a major contributor to environmental pollution is incontrovertible and its human impacts have been documented. Among many other examples, air pollution in Houston has been citied as the major reason why lung cancer rates are twice the national average (Capek and Gilderbloom 1992:215). Resource conservation can provide for urban expansion without constructing as many additional power generating facilities, while simultaneously mitigating environmental impact. Targeted policies and concerted action can reduce emissions of greenhouse gases (but may fall short of resolving all associated problems).

In addition to energy conservation, another solution is to build more power plants that can generate more electricity. Increasing electricity production by using atomic power is being revisited as a plausible option in the U.S and elsewhere as electrical demand continues to grow. Atomic power generation, unlike generation by coal fired power plants, emits no greenhouse gases. However, the problem of disposing of nuclear wastes generated by atomic power production remains unresolved. Radioactive wastes require careful handling, storage and management. The French solution at their COGEMA La Hague facility is to reprocess nuclear waste materials—separate them into usable elements such as uranium which are then converted into usable fuel. Reprocessing substantially reduces the amount of material that requires disposal.

Today, with $2 billion in tax incentives available for the nuclear power industry from the U.S. Federal Energy Policy Act of 2005, more than 20 nuclear power plants are being planned in addition to the 104 that are presently operative. However, the dearth of employment opportunity in the atomic power industry from the 1980s to 2005, when no atomic power generating facilities were being constructed, has led to a shortage of skilled nuclear engineers. The industry faces a "severe shortage of qualified workers."[13] In the nuclear power industry, by the year 2010, roughly 15,600 workers will be eligible for retirement and there will be yet another 7,600 jobs available through turnover (Aston 2007:78). So... where are the skills to safely design, build and maintain these plants? The U.S. will soon need many trained nuclear engineers who understand how to construct and operate safe atomic reactors.

A SECOND UNDERLYING EFFECT: URBAN DISLOCATION

The impacts of urban dislocation due to environmental changes are only now beginning to be observed. Ocean levels are rising as a result of the warming climate. Overall, the average temperature in the Artic has increased an average of 2.2° C (4° F) in the last 30 years. Warming has been observed in Alaska, where the average temperature has spiked 5.5° C (10° F) since 1971, causing glaciers to reduce in thickness and recede by as much as 15% (Verrengia 2002c:A13). As noted by Verrengia (2002c:A13), "Alaskan melt-water accounts for half of the worldwide sea level rise of 7.8 inches (20 centimeters) in the past 100 years." Shishmaref, Alaska, a coastal Eskimo community and ancestral home of the Inupiat, is being forced to relocate five miles inland (at a cost to U.S. taxpayers of over $100 million) due to rising ocean waters (Verrengia 2002c:A13). As a result, the village of Shishmaref has the distinction of becoming the first place in the U.S. to be dislocated by the effects of global warming. Rising sea levels, in part a result of global warming, has resulted in the loss of lowlands and marshlands in Louisiana.

Dislocation takes on local forms as well. Relocation of entire urban neighborhoods has occurred due to highway construction, point source pollution, airport noise, and energy-related disasters. Environmental waste has caused certain locales to become uninhabitable, reducing property values and eventually contributing to judicial backlogs.

Interesting international legal questions come to mind. For example, if it is proved that the U.S. is one-quarter responsible for a rise in the world's oceans, should the U.S. be required to bear one-quarter of the costs of restoring the ocean's water levels and paying claimants for remediation? Venice, Italy, a city having difficulty with its primary urban environmental problem (flooding from the rising Adriatic) may be among the initial claimants along with The Netherlands and several island nations in the Pacific, to seek financial relief.

A THIRD UNDERLYING EFFECT: CHANGES IN URBAN INFRASTRUCTURE

Without doubt, "cities are as much a part of nature as rainforests and tidal marshes" (Kelbaugh 2002:9). Urban development increases demand for resources. As cities grow, this usually means constructing new infrastructure or modifying existing infrastructure, enlarging the urban service areas, and expanding suburban areas. For the U.S. in general and the Sunbelt in particular, green field developments are the norm. New buildings, transportation systems, and energy distribution systems are required to meet increases in urban demands for shelter and services. As new facilities are constructed, energy usage can be overlooked in the planning process. Infrastructure planning for supplying energy to cities is usually performed by utility companies in concert with local governments. Planning for energy consumption is performed by the facility owners. Without planning, it is assumed that new buildings can simply be plugged into the utility system, not unlike a kitchen appliance being plugged into an electrical receptacle.

Cities experiencing rapid development will often support creative alternatives in order to adapt to changes. Examples include reusing existing structures, redeveloping brownfield sites and promoting new construction in vacant areas. Adaptive reuse of existing structures often results in lower total energy use as renovated structures represent an embedded energy content already "invested" during their original construction.

Inner cities experiencing population losses have historically focused on "urban renewal" efforts, often demolishing structures of significance in the name of eliminating urban blight. Such changes in urban infrastructure cannot always be classified as improvements. While there can be a lack of funds to support restoration and revitalization in areas where infrastructure is already in place, funds for new and expansive suburban infrastructure are somehow found.

The tendency to provide for new construction while eliminating older structures is particularly evident in North American cities.[14] This impacts low-income housing. The city of Houston, for example, demolished thousands of rental units during a period in the 1980s when there was a shortage of housing. Thousands of citizens were made homeless (Gilderbloom and Capek 1992:245).

PUBLIC POLICIES IN THE UNITED STATES:
SUSTAINABILITY, ENERGY AND CITIES

In the United States, public policy is defined at the federal, state and local levels. The policies of local governments concerning sustainable development and energy can be examined in order to bring light as to how policies in the U.S. are implemented and how policy constraints filter to urban governments. Each level of government has different motivations, legislative processes and controls, enforcement powers, means of taxation and budgetary approaches with which to define and implement policies. As a result, actual policies result from not only what is stated in them, but also by how they are implemented, interpreted, enforced and funded. Workable policies are often reduced to folklore when funds for implementation are not budgeted or when allocated funds are withdrawn. Lack of enforcement can neutralize policies. On the other hand, enforcing inappropriate policies can be equally dysfunctional.

Policies concerning energy and sustainable development are clearly important. In the U.S., local and state governments use policy instruments such as regulation, legislation, training, education, and fiscal and financial incentives to promote energy conservation (Kablan 2003:1). In addition, analytical approaches are available to provide decision support to promote energy conservation (Kablan 2003:1).

Central governmental policies can be categorized as external and internal. External policies are those that the government promotes for the country as a whole while internal policies are those that it enforces within its own organization and agencies. The *raison d'etre* of appropriate policies is that they advance changes that provide our society with benefits (such as reducing use of foreign oil or improving efficiency) while moving away from changes that are less desirable (such as increasing dependency on foreign oil or decreasing efficiency). In order to maximize effectiveness "energy conservation programs in any country should encourage the different enterprises, utilities and individuals to employ efficient processes, techniques, equipment and materials" (Kablan 2003:1).

At the present time, the U.S. federal government lacks a legislated agenda for sustainable development. The U.S. Environmental Protection Agency and the Interior Department are among those agencies that enforce environmental policies. To confound widespread policy implementation in the U.S., environmental policy is shared with the 50 states, with U.S. territories and with over 80,000 local governments (which include cities, counties, merged governments and parishes).

The prospects for future U.S. energy use remain sobering. Currently, the central government anticipates increased needs over the next two decades for natural gas, an increase in oil imports (from a 55% share to a 65% share of total oil consumption), and increased electrical power generation which will require approximately 60 to 70 new power plants annually. With the difficulties of bringing atomic and coal-fired generation on line, increasing natural gas usage is one of the fuel policies of choice for new electrical generation. Though technical solutions exist and funding is available, building power plants is not a straightforward process. Construction for an estimated 30% of the new electrical plants that were scheduled to come on line during 2003-2004 have been indefinitely postponed or cancelled (Tauzin 2003:33-34).

One might hope that the central government has a strategy in place to provide incentives to address these needs. Actually, unlike most European countries, the U.S. lacks an effective national policy. U.S. energy policy over the last 25 years has been driven not by environmental concerns but by "worries about energy shortages or the national security implications of U.S. dependency on foreign sources of petroleum" (Chertow and Esty 1997:218). The National Energy Policy Act (1992) was an attempt to develop a broad energy strategy. It resulted in policies that dealt with energy management within the federal government, a farsighted effort led by Senator John Glenn of Ohio. A decade later, the U.S. administration's 2001 National Energy Plan declared that the "Persian Gulf will be a primary focus of U.S. international energy policy."

The balance of the policy has recently been broadly criticized as having either a neutral impact or being patently ineffective. Representative Tauzin (2003:32), Chair of the House of Representatives Energy and Commerce Committee, recently stated, "For the past decade, our nation has gone without a viable national energy policy… we cannot afford to wait any longer to take action."

In the spring of 2003, the U.S. House and Senate each developed proposals (H.R. 6 and S. 14 respectively) for the "National Energy Policy Act of 2003"—but enactment did not occur until late 2005. Through its agencies, the federal government advocates and funds the development and use of nuclear energy, provides targeted research funding, manages the Clean Cities Program to promote the use of alternative fuel vehicles (ATVs), and advocates the use of energy-efficient equipment by means of the Department of Energy's Energy Star™ Program—most of which was in place prior to the Act.

The U.S. central government has had greater success with internal policies, focusing on incremental improvements. As the world's single largest energy user, the federal government expends huge sums annually in direct energy expenses for over 500,000 buildings. Using executive orders and various federal energy management programs, energy use in federally owned buildings declined by over 20% between 1985 and 2000 (based on energy use per unit area of occupied space). This has been accomplished by performing surveys to identify opportunities, by establishing energy reduction targets for all departments and agencies, and by implementing Facility and equipment improvements using budgeting authority and energy-savings performance contracts.

As Koven and Lyons (2003:4) observe, the U.S. government "often supplies the funding and defines the rules, but it relies upon the state and local levels to implement the policy." At other times, they simple establish unfunded mandates. State governments throughout the U.S. have long histories of delivering specialized legislation for a wide range of energy concerns including edicts concerning transmission, production, sale of energy, performance contracting, conservation, demand side management, etc. Through their state energy offices, many states have programs in place to encourage energy conservation efforts in their facilities, apply for federal research grants, fund energy studies, institute demonstration projects, support energy management programs in educational facilities, and promote public outreach and education.

Examples of state initiatives in Texas include a 1999 legislated requirement to establish statewide goals for renewable energy and energy efficiency, create a 10% goal for energy efficiency improvements in transmission and distribution utilities, and require a minimum of 2,000 MW of renewable electrical energy capacity by 2009 (Zarnikou 2003:1671). Legislative support of such goals lends credence to their viability. Peterson (1981:41) identified three primary forms of urban public policies:

- *Developmental policies* deal with the economic position of a city.
- *Redistributive policies* focus on benefiting low-income residents that have the potential of negatively affecting the local economy.
- *Allocational policies* are those that have neutral impact.

While sustainable development policies have both redistributive and allocational effect, they fit primarily in the category of developmental policies since they tend to enhance the economic position and vitality of the community. The long-term benefits of pursuing sustainable development have potential to exceed their costs. Policies that reduce energy costs may also be effective and exhibit a high benefit/cost ratio. In addition, lowered energy costs can have redistributive benefits: low-income families can reduce the proportion of their total income which they must devote to energy.

SUMMARY AND CONCLUSIONS

This chapter introduced the concept of sustainability, and asked if it indeed forms a new vision for the growth of our societies. The forces that drive policies relative to energy and sustainable development were discussed. The chapter considered the broader issues and relationships involving energy and sustainability.

The underlying causes and effects of factors influencing sustainable development were reviewed. These include population growth, urban development and increasing energy usage. Population growth in urban areas was found to be a driver that increases demands for resources. Cities adjust to population changes with changes in urban density. Urban development increases the demand for services and facilities, creating the need for expansion of urban boundaries. Urbanization requires energy to fuel expansion. The demand for energy is increasing, especially in the U.S.

Underlying effects that were identified include environmental impacts, urban dislocation, and changes in urban infrastructure. Environmental impacts have both regional and worldwide impacts. Energy is a principle contributor to ecosystem disruptions. Climate and other ecosystem changes cause urban dislocation. Urban redevelopment efforts often attempt to inappropriately apply suburban requirements to inner cities.

Policy alternatives were explored. In the U.S., cities are influenced by federal and state governments. While both follow their own agendas, many constraints filter down to local governments. Since they have the potential to enhance the economic position of the community in the long term, sustainable development policies are identified as environmental and developmental policies. Attitudes are changing as "there is emerging a constituency that believes in approaching global environmental issues by appropriately managing resources at the local level" (Naisbitt 1994:207).

The recent debates concerning sustainable development create a timely opportunity to explore the concept in greater detail. Concerns about urban sustainability appear to center upon strains on the natural environment due to population growth, development policies, energy usage and environmental impacts. Undeveloped land, rural areas, and growing urban and suburban areas are affected. According to Dincer and Rosen (2004:4), since "much environmental impact is associated with energy use, sustainable development requires the use of energy resources which cause as little environmental impact as possible."

The background information provided in this chapter suggests that:

1) The causes and effects driving the need for sustainability are real and unlikely to be resolved in the near future;
2) Environmental and economic impacts suggest that sustainability and energy are closely linked;
3) The sustainability policies of cities are likely variable and subject to influence by state and central government decisions; and
4) Public policies are in a state of flux, changing and evolving in response to the challenges each city encounters.

The use of energy is central to the ways urban development proceeds and how development impacts the urban environment. Energy and sustainability are interrelated. The next chapter explores how sustainable development has become a new social concept.

Endnotes

1. Additional examples from Ohlemacher 2007: Without the influx of immigrants, population declines would have occurred in Los Angeles (-200,000), San Francisco (-188,000), and Boston (-101,000) from 2000-2006.
2. See www.overpopulation.org/human.html, accessed 7 April 2007.
3. Population Density: Selected International Urban Areas and Components. www.demographica.com/db-dense-nhd.htm, accessed 2 June 2007.
4. Roberts, Sam (2007, 17 January). More than half of women are living without spouses. *New York Times*.

5. Jayson, S. (2007, 12 April). Being single has its benefits. *USA Today*. p. 4D.

6. Data from the Energy Information Administration: www.eia.doe.gov/emeu/aer/txt/ptb0105.html

7. According to Verrengia (2002a, August), renewable energy contributes 14% of the total worldwide energy use. Since the U.S. consumes roughly one fourth of all world energy, one can estimate that the world's reliance on renewable energy, excluding the U.S. is roughly 16% of total energy usage.

8. Gibbon, G. (2004) *U.S. Energy Sources and Consumption*. www.sc-2.psc.edu/news/USEnergy.ppt#309,1,U.S., accessed 18 April 2007.

9. Wilson, S. (2004, 20 December). Lebanese Wary of a Rising Hezbollah. *The Washington Post*. p. A17.

10. Energy Information Administration. www.eia.doe.gov/emeu/cbs/topworldtables3_4.html, accessed 18 April 2007.

11. Porretto, J. (2007, 2 February) Exxon posts record U.S. annual profit. *The Courier-Journal*. p. D3.

12. Ibid.

13. Quote by Andy White, President of GE Nuclear Energy, Inc. in *Business Week*, 22 January 2007.

14. Consider the experience of Louisville, Kentucky in the 1950s through the 1990s. To protect the historic neighborhoods that had not been devastated by urban renewal policies, resident organizations were formed to directly oppose city government policies. While many historic buildings were lost, and entire neighborhoods were demolished in order to provide for highway, urban center and airport expansion programs, neighborhood organizations were often successful in obtaining concessions from a monolithic urban regime that ignored citizen desires and was viewed as unsympathetic to needs of its citizens. These efforts prevented the city government from condemning structures in historic neighborhoods, allowing time for certain neighborhoods (e.g., Old Louisville) to obtain protection.

Chapter 2

Sustainable Development— A New Social Concept

"States shall cooperate in a spirit of global partnership to conserve, protect and restore the health and integrity of the Earth's ecosystem. In view of the different contributions to global environmental degradation, States have common but differentiated responsibilities. The developed countries acknowledge the responsibility that they bear in the international pursuit of sustainable development in view of the pressures their societies place on the global environment and of the technologies and financial resources they command."

AGENDA 21, THE RIO DECLARATION ON ENVIRONMENT AND
DEVELOPMENT, PRINCIPLE 7 (1992).

Sustainable development is like a tidal wave building beyond the horizon, almost ready to come ashore. Sustainability is evolving and growing in strength, changing how we think, changing our agendas, changing how we design buildings and infrastructure, changing the processes we use and the changing the technological solutions we implement. Our views on the environment will be redefined. The process is rippling through our corporations, governments and institutions, changing cities and regions. The wave will soon be upon us and come ashore. As it washes over us, it forces us to redesign our future and the future of our descendants.

This chapter considers the history, definition and application of sustainable development, explains its origins, reviews the pertinent academic information available concerning the concept, and explores the view that sustainability may indeed offer an enlightened vision for the future. Yet another goal of this chapter is to study the intellectual terrain regarding sustainable development and to frame sustainability as an overarching strategy for developing social, political, economic, and technological policy frameworks. Since the concept of sustainability tends to be variously interpreted and sometimes politicized, the categorical views of the interpretations are explored.

SUSTAINABILITY EMERGES AS A NEW SOCIAL FORCE

The theory of sustainability has it roots in the 1960s environmental movement when "problems such as overpopulation, resource depletion, decreasing water supplies, air pollution and the spread of chemicals and heavy metals in nature came into focus" (Low *et al.* 2000:37). Warning of the dire consequences of pesticide use, Rachel Carson's descriptions of vanishing species of birds in *Silent Spring*, plus fear about overpopulation leading to resource disruptions (Ehrlich 1969), and the potential of food shortages (Commoner 1971), summarized the concerns of the day. These events contributed to the theory that there may be *Limits to Growth* (Meadows *et al.* 1972).

"Sustainability" rapidly evolved to become a buzzword for the dawning of the new century. Being sustainable has become the socially preferable approach to almost everything. There have been references to sustainable policies, sustainable communities, sustainable agriculture, sustainable horticulture, sustainable use of the oceans, sustainable ecosystems, sustainable housing, not to mention sustainable businesses, sustainable practices, sustainable business practices and sustainable *ad nauseam*. Sustainable development as currently used traces its origins to the 1987 Bruntland Report of the World Commission on Development and the Environment. Expressed simply, sustainable development is commonly considered to be "development that meets the needs of the present without compromising the ability of future generations to meet their needs" (Holland *et al.* 2000:10). *Sustainability is an integrative concept.*

Ultimately, the coining of sustainability has caused the term to become nearly ubiquitous. At other times, it seems purposefully politicized, such as the use of "sustainability" in the U.S. Federal Agriculture Improvement and Reform Act (1996), which defined sustainability as a way "to continue primary emphasis on large-scale industrial agriculture for competitive production in a global export economy" (Andrews 1999:307). However, few understand what is meant by sustainable development and fewer still have a clear idea of how its promise might be fulfilled.

The concepts of sustainability and sustainable development evoke a broad range of questions. What is sustainability and how does it relate to urban development patterns and policies? What aspects of the "sustainable development" agendas have implications for urban areas? What are "sustainable development" policies and are they worth pursing? To comprehend answers to such questions, one must initially consider how the concept of sustainable development came about.

Sustainability evolved from multiple sources. Among these are theories from the sciences that suggested that many problems and potential solutions, first thought to be individually manageable and resolvable, might be interrelated across systems. Einstein was among those who led the way by suggesting that all "phenomenon of nature, all the laws of nature, are the same for all systems that move uniformly relative to one another" (Barnett 1957:46). He also believed that everything indicated that the universe was ultimately progressing to "darkness and decay" (Barnett 1957:105), suggesting that disorder and entropy would eventually result. The idea that models of various systems may have common, interrelated characteristics was proposed by Bertalanffy (1933) in the Theory *of General Systems.* Much later, the broader implications and relevance of Bertalanffy's theories were applied across sciences previously considered to be unrelated (Bertalanffy 1968; Laszlo 1972).

Theories of ordering systems into hierarchies soon evolved (Pattee 1973). The "systems idea" was very quickly adapted to the planning and design processes for large-scale systems such as cities (Ferguson 1975:55-73). Miller's inspired work, *Living Systems,* resulted in the merger of hierarchy theory and systems theory, yielding broader applications and providing an understanding that organizations, cities, and transportation networks had structural similarities to living systems.

The resolution of commonalities among subsystems provided a theoretical basis for the interrelated and complex nature of processes inherent in living systems (Miller 1978). According to Clayton and Radcliffe (1996:12), "the size and complexity of the earth system indicates that there could be, at any one time, a very large number of development paths and possible outcomes, a smaller subset of which would be relatively sustainable for the human species." There was a growing awareness that man's activities might unbalance natural ecosystems and create dysfunction within and among them. Some believed that this was leading to gargantuan problems that were unpredictable, entropic and potentially pathological. Sustainable development advocates were increasingly concerned with managing current events and their possible outcomes in a manner that increased the probability of more favorable outcomes, ultimately offering greater potential for mankind's sustained existence on Earth.

National governments began to respond to a ground swell of public opinion that actions to mitigate environmental problems needed to be undertaken as the linkages between the environment and sustainability came into focus. In 1969, the U.S. Congress passed the National Environmental Policy Act (NEPA), responding to the influences of "population growth, high-density urbanization, industrial expansion, resource exploitation" and declaring it a policy of the federal government "to create and maintain conditions under which man and nature can exist in productive harmony, and fulfill social, economic, and other requirements of present and future generations of Americans." The National Environmental Policy Act introduced the concepts of both environmental harmony and intergenerational equity. The law was enacted in 1970 and represented the expansion of environmental governance at the federal level in the U.S. Other provisions of the NEPA included enhancing the quality of renewable resources, recycling depletable resources, and maintaining an environment that supports diversity. The influences to which the NEPA responded are of even greater concern today.

The theory of sustainable development ultimately evolved to provide an even broader vision, addressing an even larger range of concerns. Yet its origins remain uncertain. Stephen Wheeler (Legates and Stout 2000:436) observes that:

> *"It is far from clear who was the first to use the term 'sustainable development' in its current sense. Rather, it seems one of those inevitable expressions—that so neatly express what many people are thinking—that once the words are mentioned they quickly become ubiquitous. The birth of the sustainability concept in the 1970s can be seen as the logical outgrowth of a new consciousness about global problems related to environment and development..."*

Wheeler believes that catalysts for the change in "consciousness" included events such as the rise of ecological problems and "the 1973 oil embargo during which millions of people suddenly realized that their fossil fuel use could not continue to expand forever" (Legates and Stout 2000:436).

As the stage was being set to address sustainable development during the 1992 United Nations Conference on Environment and Development, Gro Harlem Brundtland, then Prime Minister of Norway, asserted that "we should not be surprised that developing nations are approaching the Rio Summit with open economic demands... for them, it is essentially a conference about development and justice" (Panjabi 1997:282). On the other hand, there was a fear among the developed nations that they might be called upon to bear the primary financial burden of protecting the earth's biodiversity (Panjabi 1997:282).

THE UNITED NATIONS "EARTH SUMMIT"

The United Nations 1992 Conference on Environment and Development is now referred to as the Rio (or "Earth") Summit in reference to the host city of Rio de Janeiro. A product of the conference was the Rio Declaration on Environment and Development, a manifesto later referred to as Agenda 21. The central rationale noted in the preamble of Agenda 21 is the desire to work "towards international agreements which respect the interests of all and protect the integrity of the global environmental and developmental system" (United Nations 1999:1). While 179 heads of state and national governments have agreed to Agenda 21, the U.S. is notably absent.

Sustainability was considered to be "the arrangement of technological, scientific, environmental and social systems in such a way that the resulting heterogeneous system can be maintained in a state of temporal

2-1. Rio de Janeiro, Brazil

and spatial equilibrium" (Hens 1996). The idea of "sustainable development" combines the desire for environmental protection with the need for continued material and economic prosperity (Moffatt 1996). Hempel believed that sustainability was achievable by policy implementation at the supra-national level. He identified three conceptual approaches to global environmental governance: 1) a limited world federalist system; 2) reform of the United Nations and its agencies into a confederated model; and 3) a mixed form of "nationalism and nascent supra-nationalism" (Low *et al.* 2000:32; Hempel 1999:159-78). Administrators of some central governments perceived these suggestions to potentially threaten sovereignty by limiting a state's political alternatives.

The Agenda 21 Charter provided Principles that established goals that needed to be achieved in the world quest for sustainability. Agenda 21 deals specifically with development policies in Principles 4 and 8. Principle 8 suggests that "States should reduce and eliminate unsustainable patterns of production and consumption and promote appropriate demographic policies" (United Nations 1999:2). Environmental protection policies are addressed in Principles 2, 4, 7, 11, 15, 16, 24 and 25 of Agenda 21. Chief among these is Principle 4 that suggests "to achieve sustainable development, environmental protection shall constitute an integral part of the development process" (United Nations 1999:2).

The idea of polluters paying for the costs of their pollution is introduced in Principle 16. This concept was disturbing to many in a few developed countries, such as the U.S., who happened to be significant contributors to global air and water pollution, and also to others in developing (such as China and India) and third world countries who at that time lacked resources to cure their environmental problems. The Agenda 21 charter also addresses concepts such as individual rights and responsibilities (Principles 1, 10 and 23), equity issues including intergenerational equity (Principles 3, 5, and 6), and the roles of women and youth (Principles 20 and 21). The charter advocates international cooperation among the states (Principles 7, 14 and 27). A significant balance of the document concerned the mechanics of how to implement and promote sustainable development.

While energy usage is a primary contributor to global pollution and resource depletion (Roosa 2002a:315-329), it is remarkable that the term "energy" is not specifically mentioned in the body of Agenda 21. However, the term "resource" is broadly applied and energy resources are a relevant concern of Agenda 21 (United Nations 1992:56-58). Regardless, Principle 7 encourages "a spirit of global partnership to conserve, protect and restore the health and integrity of the Earth's ecosystem" (United Nations 1999:2). This is in loose agreement with Harken, who defined sustainability in terms of the "carrying capacity of the ecosystem, and described with input-output models of energy and resource consumption. Sustainability is an economic state where the demands placed upon the environment by people and commerce can be met without reducing the capacity of the environment to provide for future generations" (Harken 1993:139).

In the U.S., the President's Council on Sustainable Development (1996:49)[1] broadly defined parameters concerning how sustainable development policies might be achieved. The council noted that "effective investments in energy efficiency... lead to economic, environmental and equity benefits by reducing energy costs and environmental effects." However, policies in the U.S. have not been broadly reoriented to this path. In fact, the recommendations of the Council stopped short of suggesting changes that would reduce use of nonrenewable fuels and instead suggested unspecified shifts in "tax policy, reforming subsidies, and making greater use of market incentives" (President's Council on Sustainable Development 1996:49). It would require another 10 years before substantive changes in tax policy of this nature would be implemented in the form of the Energy Policy Act of 2005.

The President's Council suggested gauges of progress toward sustainability such as: 1) increased share of renewable energy use; 2) increased efficiency of electrical generation; and 3) a reduction in greenhouse gas emissions (President's Council on Sustainable Development 1996:49). Lacking legislative enactment, results based on these recommendations in the U.S. have been mixed. Since 1996, there has been: 1) a small increase in the use of renewable energy; 2) no significant change in electrical system efficiencies while a number of widespread blackouts over the subsequent years suggests decreased transmission system reliability; and 3) continued increases in greenhouse gas emissions.

In Europe, England's Prime Minister Tony Blair once called the movement the "sustainability revolution" citing the ultimate goal of achieving economic and social goals in tandem with continued economic growth. Couch notes that Agenda 21 is the primary plan for sustainable development for the 21st Century. He suggested that "much of the plan requires action at the local level and all local governments are therefore expected to produce a Local Agenda 21 (LA21)" plan (Couch and Dennemann 2000:141). The hope is that LA21 plans will address the principles of Agenda 21, develop community interaction, and subsequently achieve a greater degree of equity. Confounding such objectives are various interpretations of how sustainable development is operationally defined.

Regardless, the idea of sustainable development serves to value nature as an ethical issue. It is a conceptual launching pad from which to evolve global solutions to problems involving sustainability, development approaches, population growth, adverse ecological impacts, and patterns of energy consumption. As this evolutionary process has continued into the 21st century, sustainability is now considered to be "a universally established urban development goal" (Burges and Jenks 2000:91).

SUSTAINABILITY AND LOCAL GOVERNMENTS: ACTING LOCALLY

While sustainability might be a global ideal, *thinking globally* is irrational. It is possible to study things of global significance and to consider global solutions. On the other hand, *acting locally* is a proven means of effecting change. Perhaps for this reason, sustainable development gained momentum and filtered to local governments. Is it possible that the concept of "sustainable development" provides an overarching set of guidelines that can be effective for local governments? Berke (2002:23) concluded that sustainability may be the next paradigm or "framework to dramatically shift the practice of local participation from dominance by narrow special interests toward a more holistic and inclusive view." This seems important as many of the solutions proposed by Agenda 21 required regional and local involvement and legislation.

Urban sustainability deals with the application of sustainability at the urban scale. It refers to a somewhat idealized model of urban development while attempting to address the wider set of concerns about urban growth, patterns of urban development and environmental issues that arise as urban development occurs. Portney views sustainability as involving the "Ecological-environmental-natural resource issues; the performance of the local economy; a variety of quality of life issues; and long term governance issues" (Portney 2003:53). If true then what guidance is there that further defines what processes are involved? According to Beatley (2000:17; EC 1996), the four Principles of urban sustainability in the European Community's (EC) Sustainable Cities Agenda included:

- *The principle of urban management*—requires planning, and impacts the governing of urban areas. The process of sustainable urban management needs tools that address environmental, social and economic concerns.

- *The principle of policy integration*—implements means to stimulate the synergetic effects of social, environmental and economic dimensions of sustainability.

- *The principle of ecosystems thinking*—emphasizes the city as a complex system incorporating aspects such as energy, natural resources and waste production.

- *The principle of cooperation and partnership*—considers the crucial importance of interactions among various levels of government, organizations and interest groups.

Beatley (2000:16) cites examples in European Union (EU) documents which advocate an integrated approach and argue for an ecosystems view of cities. He notes that cities affect their local environments (e.g., with their regional hydrological systems) while existing simultaneously as habitats for plants and animals. The EU suggests that cities must be viewed as complex "interconnected and dynamic systems. The challenge of urban sustainability is to solve both the problems experienced within the cities themselves… and the problems caused by cities" (European Commission 1996:6-7). To this end, "Energy efficiency and renewable energy are increasingly being considered in connection with EU policies on climate change, as well as on the security of supply, employment and industrial competitiveness" (Klevas and Minkstimas 2004:309). As a policy, energy efficiency programs in the EU are formalized by the Energy Charter Treaty, the Accession Partnership, the Europe Agreement, and various other protocols and directives (Klevas and Minkstimas 2004:309).

While the U.S. had extensive environmental laws in place at the time, the country lacked a comparable national sustainability agenda and an operable national energy policy to springboard local efforts. Efforts by the U.S. legislative branch to pass a national energy policy during 2003 and early 2004 were initially unsuccessful. However, in August 2005, the Energy Policy Act was signed into law. The Act provided tax incentives and loan guarantees for both carbon-based and alternative energy production, plus incentives for certain types of energy-related projects.

Drakakis-Smith (2000:8) suggested that "sustainability also emphasizes the interlinked nature of the individual components of rapid urbanization" and broadly defines the components of sustainability as:

1)   equity, social justice and human rights;

2)   basic human needs, such as shelter and health care;

3)   social and ethnic self-determination;

4)   environmental awareness and integrity;

5)   awareness of linkages across both space and time; and

6)   not seeking gain at the expense of someone elsewhere in the world or of the generations to come.

Urban sustainability involves a broad range of issues. The nature of the "concept of sustainability which integrates environmental, economic and social forces into one (not balancing them off against each other) must therefore be integrative."[2] Cities are considered to be open systems impacting areas much larger than their geographic areas. The role of cities and local governments in developing models for sustainability cannot be understated.

DEFINITION OF SUSTAINABLE DEVELOPMENT

For the purposes of this book, sustainable development is defined as: The ability of physical development and environmental impacts to sustain long term habitation on the planet Earth by human and other indigenous species while providing:

1)   An opportunity for environmentally safe, ecologically appropriate physical development;

2)   Efficient use of natural resources;

3)   A framework which allows improvement of the human condition and equal opportunity for current and future generations; and

4)   Manageable urban growth.

Non-sustainable development is the antithesis of sustainable development. It implies persistently unmanageable growth that is environmentally unsafe, that consumes resources such as energy ineffectively while degrading the human condition. Extreme cases of non-sustainable development are characterized by ecosystem disruption, paralyzed communications, dysfunctional transportation systems, persistent lack of resources and materials, pervasive environmental mismanagement, prolonged poverty and lack of medical care, perpetual destruction of infrastructure, and recurring military conflict. Any and all of these conditions, either individually or in combination, can be disruptive to environments, yielding non-sustainable results.

This new definition incorporates social concerns, such as those held by Perloff (1980:182), who noted that, "A price must be paid for material progress in the destruction of irreplaceable natural resources, in air, water and noise pollution; and in millions of disadvantaged individuals and families who are left behind; and in cities that are also left behind with large sections in decay and poverty." Al-Homound (2000:21-38) believes "Usable resources are made available to mankind to be utilized for their benefit and well-being. Every individual bears the responsibility of not wasting or misusing usable resources. All moral codes are against such actions. Therefore, every individual should be educated and trained to become part of the management of resources for the benefit of generations to come." Having a moral belief that resource efficiency is beneficial certainly has global social implications. In addition to moral beliefs, there are other direct linkages between sustainability and resource management.

LINKING SUSTAINABILITY AND ENERGY

Next, this book explores the idea that energy is a key component of sustainable development. What does energy have to do with sustainable development policies? Can sustainability be implemented in the face of continually increasing energy consumption? The linkages among energy, sustainability and their relationships are an important component of achieving sustainability goals. According to Saha (2203:1), "The main challenge to energy policy makers in the 21st century is how to develop and manage adequate, affordable and reliable energy services in a sustainable manner to fuel social and economic development."

Political opposition to sustainable development policies at the national level was recently brought to bear by the U.S. government on the world stage, despite having adopted components of its definition in its 1969 NEPA. Part of the action plan of the World Summit on Sustainable Development included a proposal that the use of renewable energy technologies be increased to 15% of world-wide energy production by 2010 (Verrengia 2002a:A4). This relatively modest increase would have generated employment in the U.S. at a time of employment losses due to a recession and fostered a reduction in oil imports during a period of increasingly negative trade imbalances due, in part, to increasing oil costs. Delegates from the United States (the world's largest oil importer), Saudi Arabia (the world's largest oil exporter) and other states "were lobbying to eliminate the provision and set no specific goals" (Verrengia 2002a:A4). It is doubtful that the nature of the alliance in opposition to the proposal was lost on the delegates.

The U.S. solution took the form of a counter proposal. It included a series of hastily organized industry and foundation partnerships with an ambiguous agenda, involving (at most) only $600 million per year in expenditures with a commitment of not more than four years. This figure dwarfs the hundreds of billions of dollars expended annually by the U.S. on foreign oil alone. It is regrettable that a more viable alternative model was not offered. Given the continuing debate, it is important to delve into the links between sustainability and energy use.

Sustainability, in terms of energy use, can be considered to be: 1) the ability to provide energy that is environmentally neutral; 2) not use energy in an overly consumptive manner; 3) provide options for renewable energy; 4) reduce the need for energy from non-renewable energy sources; and 5) distribute energy resources equitably. To explore more fully the relationship of energy to sustainability, it is important to consider how this relationship developed.

Energy can be defined as the "capacity of doing work and overcoming resistance" and as "strength or power *efficiently* exerted" (Guralnik 1972, italics added). Energy can also be defined as "the work that a physical system is capable of doing in changing from its actual state to a specified reference state" (Morris 1969). The term *conservation* is the "act or practice of conserving; protection from loss, waste, etc.; preservation; official care and protection of natural resources" (Guralnik 1972). In apolitical terms, consuming natural resources in a strategically appropriate manner without adverse environmental impact provides economic benefits. The phrase *conservation of energy* in a physical sense refers to the "principle that energy is never consumed but only changes form and that total energy in a physical system, such as a universe, cannot be increased or diminished" (Guralnik 1972).

The Second Law of Thermodynamics (a.k.a. The Law of Increased Entropy) validates that the fundamental processes in nature are essentially irreversible. While the quantity of energy remains constant, the quality deteriorates. Debating the idea of running out of energy is a fruitless and counter-productive endeavor. There will always be energy available.

Suggesting that there may not be adequate and useful forms of energy or the means of transforming the available energy into usable product is far more rational. This is not about energy resources but the services available from their use. For example, burning oil for heating is not the end product or service. The end product might be usable heated space, heated water or process heat. These products are actually services that can be provided by multiple means—they do not necessarily need to be provided by burning oil.

Consumer behavior creates patterns of *consumption* ("the using up of goods or services either by consumers or in the production of other goods," Guralnik 1972) that can economically provide the means to satisfy socially beneficial needs. *Consumptive* behavior ("consuming or tending to consume; destructive; wasteful," Guralnik 1972) can increase product costs, cause shortages, equity imbalances, or contribute to economic externalities.

Conceptual thinking regards decision-making processes and management options as steering functions, while development and energy are considered production forces (Choucri 1993:449). As a result, the pervasive concerns regarding energy use cause urban energy consumption to have broader global implications. Interestingly, energy usage is often highly decentralized while energy generation and production tends to be comparatively centralized.

Yet centralized, capital intensive options may not always be the most viable alternatives. Such options often require complicated distribution networks that are subject to systemic disruption. Externalities include those not only associated with water and air pollution but also economic availability and equity issues. Energy usage in the built environment has been increasing due to a number of causes. Transportation systems, rapid growth in population and increases in the number and scale of buildings augment demand for usable energy. Energy is required for primary services such as the need for conditioned space that complies with upgraded standards for human comfort (e.g., in the U.S, ASHRAE Standards 55 and 62), especially in the workplace.

CITIES—THE LARGEST CONSUMERS OF ENERGY

Urban development is creating ever-greater demands for all forms energy. Cities are now responsible for consuming more than three-quarters of the world's energy (Rogers 1997:2-27). As a result, many successful strategies to reconfigure energy use tend to focus on urban areas. Perhac (1989) noted that the availability of inexpensive energy supplies has not only impacted the urban experience but has also been responsible for facilitating economic development. Despite changing developments and views, the importance of energy is too often overlooked in urban literature.

David Rusk in his book *Cities Without Suburbs* provides a valuable theory of urban elasticity. Rusk offers detailed descriptions of 24 lessons that "can be drawn from a broad look at what has been happening in urban America over the last 40 years" (Rusk 1993:5-49). Despite a history of oil price shocks, power outages, a lack of utilities in the homes of the needy, and the skyrocketing costs for energy-associated environmental remediation, none of the 24 lessons discussed the potential impact of energy on urban areas. The fact that growing, elastic cities require significant and often increasing amounts of energy plus extensive energy related infrastructure was entirely missed by Rusk as a "characteristic of metropolitan areas" (Rusk 1993:51-54).

Kenneth Jackson, in his thought-provoking and seemingly comprehensive book *Crabgrass Frontier*, details the suburbanization of the U.S. and discusses at length the impact of the automobile. Look for "energy" in the Index and the only reference you find is typical: "see oil." Peter Hall's *Cities of Tomorrow* omits "energy" and instead offers only "energy crisis" as a listing in its Index. Despite the birth of the "Hydrocarbon Age" populated by the "Hydrocarbon Man" (Yergin 1991:389-541) and the oil supply disruptions of the 1970s, energy is obviously not only about oil and can be reasonably labeled a "crisis" when infrastructure (market or equipment) failures occur. The cause of the "energy crisis" in California in 2001-2002, a notorious example, was due to unstable and imperfect markets caused by a mismatch between supply and demand in the newly re-regulated electrical market (Brown and Kooney 2003:860; Borenstein 2001:191-212).

The relationship between energy and sustainability focuses on what types of fuels are used, how they are extracted, how they are transported, how and where they are stored, how they are transformed, how they are consumed and the efficiencies of their use. It is about identifying resources and selecting methodologies. Sustainable energy use concerns not only alternative selection of fuels and the means of using them, but also their economic, environmental and social impacts. In a broader context, the sustainability of cities is complex and must consider energy flows.

It is arguable that the availability of low cost energy has made areas of the U.S. habitable. Houston was constructed on swampland, Las Vegas in a desert basin and Orlando is in the humid and hot center of the Florida peninsula. Water must be pumped (using electrical pumps) up and out of New Orleans to prevent localized flooding in the city and when the pumps and levees fail, the city will flood. The availability of energy-intensive technologies has provided the means for the rapid urbanization of cities. Railways, automobiles and new roads paved the way for rapid urban development.

Inexpensive, portable, low voltage window unit air conditioners (nicknamed "window shakers") became available in 1951 (Banham 1969:185-187), allowing interior spaces to be habitable during hot and humid ambient conditions. The expanding use of residential air conditioning in the U.S. has been explosive. In 1960, only 18.2% of homes in the South had air conditioners, growing to 50.1% by 1970 and to 73.2% by 1980 (Mohl 1990:186). By 1980, the U.S., a country with only 5% of the world's population, was managing to produce and consume as much cooled air as the rest of the world combined (Jackson 1985:241). By 2005, hardly a new home or business space in the southeastern regions of the U.S., areas with hot summers and high relative humidity, is constructed without a central air conditioning system. The number of air conditioned homes in the United States has increased to 83% of all housing units of which 54% now have central systems (Air Conditioning & Refrigeration Institute 2002).

The essence of the issue is this: *without energy, sustainability is unachievable.* According to Diner and Rosen (2004:2), "attaining sustainable development requires that sustainable energy resources be used, and is assisted if resources are used efficiently." This is not just about energy *per se*, but about the *forms* of energy and how they are deployed. This concept is often overlooked. For example, measured air and water quality are often citied as amenities worthy of study using hedonic models. Yet access to energy is typically not considered an amenity (Bartik and Smith 1987).

In the developed world, access to energy resources is assumed. Modern cities cannot grow, develop or even function without power. The lights go out, computer systems are unusable, communication is disrupted, food spoils, water isn't pumped, elevators halt, most modern housing and office buildings become nearly uninhabitable, transportation systems fail, sewage is unprocessed, social stress occurs and markets collapse. Major electrical outages are rare, but they are not uncommon. Examples include the events that occurred in the summer of 2003 when a wide-spread electrical outage occurred in the UK, another in Switzerland, Italy and France, and yet another in the U.S. and Canada that impacted over 50 million people. One of the more devastating outages, caused by a single lightning bolt, occurred in New York City in July 1977. The blackout lasted 25 hours creating "a night of terror."[3] The outage led to arson, the looting of 1,700 businesses, over 3,000 arrests and more than $150 million in property damage.[4]

Power outages cost their regional economies billions of dollars, euros and yen. In extreme cases for prolonged periods, even anarchy is possible. Most of the theories of sustainability assume that basic human necessities will be available or can be made available to large portions of urban populations. While many perceive energy to be plentiful, ever-larger populations today live without access to modern energy services and for others, access is severely limited. If energy is required for a sustainable lifestyle— then where is the research that explores the connections between energy and sustainability?

Until recently, academic research assessing energy and its relationship to sustainability has been sparse. According to a recent United Nations report (2003), "There is a need to analyze and understand energy systems within the dimensions of:

- Sustainability—how much and at what rate energy is consumed, and its effect on long term sustainability; the quality and quantity of available alternative/renewable forms of energy; and the effect of existing energy use on the global environment as a whole.

- Efficiency—the technology, planning and management of energy systems that will facilitate efficient use of energy for human activity, particularly transportation.

- Equity—the appropriate financial mechanism for research, development and use of finite and alternative energy forms and their equitable distribution for all mankind."

Implementing targeted projects that achieve the goals of such policies is clearly feasible. Arguments that suggest such initiatives are economically infeasible have been countered by suggestions that the cost of inaction can be greater (Roosa 2002a:320).

It is challenging just trying to imagine how sustainability goals could be achieved without access to relatively affordable energy. According to Dincer and Rosen (2004:10), "sustainable development within a society requires a supply of energy resources that, in the long term, is readily... available at a reasonable cost and that can be utilized for all required tasks without causing negative societal impacts." In the Greenhouse Gas Policy Statement issued by Puget Sound Energy (2007:1), the following is found, "Energy drives the economy. Sustainable energy is an essential component of sustainable development. Global and national problems ultimately require global and national solutions."

By any measure energy usage in the U.S. is increasing, taking an ever-greater share of economic resources. In 1998, the U.S. energy bill was $527.0 billion, an average of $1,911 per person (Energy Information Administration 2000) or 6.0% of the U.S. Gross Domestic Product (GDP). By the year 2000, the U.S. energy bill had grown to $703.2 billion, an average of $2,499 per person (Energy Information Administration 2000) or 7.2% of the U.S. Gross Domestic Product (GDP). This amount equals the GDP of Brazil and is greater than the combined GDPs of Russia, Turkey and Austria. While exact data are unavailable, it can be estimated that due to increases in the prices of electricity, oil, and other fuels in the U.S., total energy costs in 2006 exceeded $1 trillion. In the same year, the average household expended $2,227 for gasoline (3.6% of pre-tax income) alone, a 78% increase from 2001.[5] Costs of gasoline are literally "hitting home." According to U.S. Representative Bart Stupak from Michigan, "Gas prices are causing Americans significant financial hardship."[6]

Despite financial hardship, the U.S. continues to import more and more oil, mostly from Canada, Mexico, Nigeria, Saudi Arabia and Venezuela. From 1980 to 2000, the importation of crude oil by the U.S. has increased from 6.9 to 11.4 million barrels daily (Energy Information Administration 2001). From 1990 to 2000 the net U.S. merchandise trade deficit from the purchase of oil alone increased from $54.6 billion to $109.1 billion (Energy Information Administration 2004b), growing to an estimated $122.3 billion by 2003. Table 2.1 lists the total crude oil imports annually from all countries.

By 2005, total imports of crude oil had increased to an average of 13.7 million barrels of crude daily. Using the 2005 figure for imports and the

Table 2-1. U.S. Total Crude Oil Imported Annually (Thousand barrels)

Decade	Year-0	Year-1	Year-2	Year-3	Year-4	Year-5	Year-6	Year-7	Year-8	Year-9
1980s		2,188,420	1,866,358	1,843,744	1,989,935	1,849,508	2,271,582	2,437,359	2,709,140	2,942,099
1990s	2,926,395	2,783,763	2,886,897	3,146,454	3,283,621	3,224,753	3,469,128	3,708,970	3,908,446	3,961,074
2000s	4,194,086	4,333,038	4,208,538	4,476,501	4,811,104	5,005,541				

Table from the U.S. Energy Information Administration, http:/tonto.eia.doe.gov/dnav/pet/hist/mttimus1a.htm, accessed 19 May 2007.

June 2007 average cost of $63/barrel, the 2005 U.S. tab for imported oil can be projected to be roughly $315.3 billion in 2007... but only if total oil imports were held to 2005 levels. Importing oil is certainly big business. Are expenditures of such magnitude a wise investment in America's future? Think of the economic impact if these dollars could be invested in domestic energy services.

While regaining U.S. energy independence may have once been thought possible, it may no longer be achievable without economic redirection and major infrastructure restructuring. Today the "idea of 'energy independence' is a myth, and attempts to insulate U.S. consumers from world crude oil markets and could actually drive up domestic energy prices" argues Linda Stuntz, a former U.S. Department of Energy official.[7] Yet policies designed to increase U.S. dependence on foreign oil have failed to keep domestic energy prices from increasing as well.

The benefits of energy independence for the U.S. would be economically and politically important, providing the opportunity to offset trade imbalances and improve the national economy. According to Flynt Leverett, a director at the New America Foundation, "Energy security could be the Achilles heel of U.S. global competitiveness."[8] He continues, "The most profound challenges to U.S. pre-eminence over the next 25 years flow from the strategic and political consequences of on-going structural shifts in global energy markets."[9] Indeed, it is difficult to imagine sustainability on a national scale as being coupled with over-dependence on fickle foreign governments for strategic resources such as oil. An international food fight over oil is not a future event—it is happening already. It is likely to continue to strain U.S. relations with its allies and trading partners.

While a goal of energy independence might appear elusive and unachievable, alternatives to oil are certainly worthy of serious consideration. As is often the case, it is local governments that are the testing ground for novel solutions. The city of Denton, Texas partnered with BTE Biomass Energy to become the first in the world to use landfill gas from their city dump to power a biodiesel production facility. The biodiesel plant has a capacity of three million gallons annually which is used to fuel 386 city buses, garbage trucks and utility vehicles. By the end of 2004, approximately 380 landfill gas energy projects were operational in the U.S., most of which were generating electricity (Gordon and Ode 2005:4).

ALTERNATIVE ENERGY, ENERGY CONSERVATION AND ENERGY EFFICIENCY

More efficient use of energy in the built environment can have a significant impact in meeting certain urban objectives. These objectives include enabling more effective development, modifying urban density, reducing demands on utility infrastructure, providing more appropriate housing solutions, improving transportation systems, reducing negative environmental impacts and lowering the costs of governance. Reducing energy usage has been identified as an indicator of improved sustainability (Bell and Morse 1999:63). Over the past century, technological improvements have provided opportunities to dramatically increase efficiency. The oil industry has proved exemplary in this regard, finding ways to use almost every drop of oil in the process of converting oil to usable products, and improving the efficiency of extracting additional oil from existing wells. The industry has also used new technologies to improve its environmental performance.

This is not true for all industries. Increased energy use by some industries has led to varying degrees of "hyper-inefficiency." This is a counter-intuitive process in which industries in countries with the technological capabilities and resources to implement improvements in strategic resource efficiencies simply fail to so do and routinely choose to employ less efficient technologies that are actually more energy intensive. One example is the U.S. electrical production industry. Due to heat recovery techniques used in the early 1900s, the electrical industry marked its highest production efficiency at 65% in 1910, declining to 33% by 1959 (Casten 2003:63-64). Despite the availability of a variety of technological innovations, the efficiencies of the electrical power production industry have hardly improved since.[10] The resulting record of productivity decline in the electrical industry means that the U.S. now produces electricity at a rate of efficiency lower than in 1880. Rather than focusing on system efficiencies, the industry has instead chosen to focus on keeping its resource costs low and has failed even at that. Thankfully, this failure is becoming widely acknowledged and the electrical utility industry is expanding its focus.

Today many electric utilities operate in protected markets with little incentive to provide real gains in efficiency improvements. Power factor correction in regions of the U.S. is at a lower reactive efficiency than can be found in regions of developing countries such as India. Author N. Low sums the issue succinctly, "To overcome the ecological crisis of the 21st century, cities must consume much less of the natural environment than they do today" (Low 2002:44). The focus must ultimately turn from *energy production* to *energy resource management*.

The theory of energy conservation "did not develop historically without paradigm destruction" (Kuhn 1996:97-98). It developed "from a crisis in which an essential ingredient was the incompatibility between Newtonian dynamics and some recently formulated consequences of the caloric theory of heat. Only after the caloric theory of heat had been rejected could energy conservation become part of science" (Kuhn 1996:97-98; Thompson 1910:266-281). An outgrowth of systems theory, "energy conservation provides...links between dynamics, chemistry, electricity, thermal theory, and so on," evolving into a "theory of a higher type, one not in conflict with its predecessors" (Kuhn 1996:95-98).

More efficient use of energy in the built environment can have a significant impact on reducing direct economic costs. These include the ability to provide for urban expansion without necessarily constructing additional power generating facilities. Further, environmental impacts can be more directly managed. A vocabulary of technologies and methodologies began to develop in the 1970s and 1980s that responded to such concerns. Driven by ever increasing energy costs, energy engineers began to develop innovative solutions, such as use of alternative energy, more efficient lighting systems, and improved electrical motors. Controls engineers developed highly sophisticated digital control systems for heating, ventilating and air conditioning systems. There is mounting evidence that energy can be obtained less expensively by investing in efficiency and conservation than in new generating capacity, meaning that "'negawatts' are more valuable than megawatts"(Choucri 1993:199). Pollution can be avoided. Successfully achieving such solutions requires that the utility industry upgrade their facilities, broaden their services and retool their offerings and processes.

Technologies are available that provide more efficient use of energy. In the upside-down world of energy economics, cost-effective, energy-efficient technologies are often not implemented even when the financial return on their investments are greater than other investments involving similar risks. Why is this? An extensive literature has evolved concerning barriers to implementation (Brown 2001:1197-1207; DeCanio 1993:906, 1998:441; Fisher and Rothkoph 1989:397; Jaffee and Stavins 1994:43; NPPC 1989; Roosa 2002a: 320-322). While no clear consensus has emerged, market failure, bureaucratic inertia, lack of consistent policy agendas, failure of strategic decision-making processes, and political influence and inaction have all been suggested as likely barriers to broader implementation of energy efficient technologies. There are also practical limitations to increased energy efficiency (Rosen 2004:36-37).

According to N. Desai, "A shift in energy systems towards environmentally sound and efficient energy technologies would contribute substantially to sustainable development" (Inoquchi *et al.* 1999:245). This involves developing alternatives to carbon-based energy sources. "Energy resources such as fossil fuels are finite and thus lack the characteristics needed for sustainability, while others such as renewable energy sources are sustainable over the relatively long term" (Dincer and Rosen 2004:4). Often, alternative approaches require collaborative solutions to the problem of meeting energy needs by focusing on decentralized energy strategies rather than a centralized production approach. While electrical fuel generation is *regionally centralized* (e.g., coal-fired electrical power plants), alternative energy can be *locally decentralized*. There are literally billions of individual decisions made on a daily basis as to what energy to use and when to use it. As a result, alternative energy, energy conservation and energy efficiency require local and regional urban involvement and planning to avoid relying on centralized energy production solutions.

Energy conservation approaches sustainability with the premise that incremental improvements can be achieved with focused efforts to reduce or eliminate the consumption of targeted fuels used for specific purposes at strategically selected times. Resource conservation provides the opportunity to accommodate urban expansion while simultaneously mitigating environmental impact. Simple examples include turning off equipment when it is not needed for production or individual use.

Energy conservation might also involve facility redesign, facility modification, local planning policy initiatives or governmental decisions. Examples include initiatives such as installing extra building insulation or energy efficient windows in existing structures. Today, window technologies are available that are 35-40% more energy efficient than typical double glazed windows (DeSimone and Popoff 1997:181). Local governments might choose development of locally accessible rapid transit rather than opting for new highway construction.

Alternative energy offers a range of answers to the questions of sustainability. Fuel substitution includes solutions such as replacing coal or oil-fired powered generation equipment with natural gas. While this can be classified literally as an "alternative energy" approach, the category of energy sources which are typically referred to as "alternative fuels" or "renewable energy" includes:

1) solar power which is used directly for space heating, with heat exchangers for water heating, or for direct electrical energy generation using photovoltaic cells;

2) wind power which is used for water pumping and direct conversion of electricity;

3) biomass (such as wood or landfill gases) fuels for heat production;

4) hydrogen fuel cells for various uses;

5) hydropower for electrical generation; and

6) geothermal energy which makes use of heat from the earth.

Notice that the distinction between conventional energy sources and alternative energy resources is often viewed as the distinction between fossil and non-fossil fuels. This distinction can also be viewed as the difference in carbon intensity. Accounting for the contribution of alternative energy can be problematic. The contribution of wood heat, for example, to total energy used by an economy is often not calculated. One might be able to determine the total sales and rated capacity of solar collectors but not the actual output of the installed capacity in operation at any given time. As a result, the actual contribution of alternative energy is often questioned or minimized.

Energy efficiency refers to the concept of replacing less efficient equipment with equipment of greater efficiency, improving the efficiency of existing equipment, or satisfying the energy requirement in an alternative manner. Energy efficiency reduces costs and can free funds for capital improvements (Hansen 1998:5). Due to the costs embedded in existing infrastructure, energy efficiency improvements often occur after equipment obsolescence or failure. One example is the improvement in efficiency of electric motors. Electric alternating motors are significant energy users in the United States. Remarkably, electric motors account for 64% of U.S. industrial sector electrical energy use (Malcolm 2006:1). Since newer motors may offer efficiency improvements on the order of 2-6% when compared to older models, reductions in aggregate electrical energy use are achievable when they are replaced. Due to the large amounts of electrical energy that motors require and the large numbers of electrical motors in service, replacing older motors is a more cost effective means of meeting electrical demand requirements than building power generation facilities. It also does not require additional land for power production facilities and does not generate additional pollution.

Vehicles with internal combustion engines provide yet another example. Incremental improvements in fuel efficiency results in reductions in gasoline use and reduces atmospheric pollution. This is due to the millions of vehicles in use and their rapid rates of obsolescence. Improving fuel efficiency has potential. During the 1990s, automobiles purchased in the U.S. used roughly 33% more fuel/km than those purchased in Europe (White 1992:39)

Alternative means of providing energy (or lowering the need for it) often result in fewer demands for urban land, reduced environmental pollution and less dependency on certain strategic fuels. Employment opportunities are known to result. As an example, the field of Energy Engineering did not formally exist 30 years ago and now has over 9,000 professionals worldwide (Association of Energy Engineers 2004).

ENERGY'S CRITICAL ROLE IN SUSTAINABILITY

Energy is related to sustainability in a number of ways. Consumption of hydrocarbon-based fuels has environmental impacts such as polluting the atmosphere, polluting natural water systems, and contributing to climate change. In addition, since there is a finite quantity of source fuels available to markets at any given time, use of the resources will ultimately reduce availability and expedite depletion. Deceases in supply have caused increases in the prices of certain fuels, making them too expensive for people in third world countries.

Countries that import oil allocate significant portions of their national income to their purchases, creating a need to reallocate resources, often with disruptive results. For example, in the U.S., hundreds of billions of dollars are expended annually for imported oil, and for national defense to protect oil supplies. Historically, energy policies in the U.S. have largely focused on these principles:

1)   Secure oil supplies are required, meaning that they must be under U.S. economic or military control;

2 Inexpensive energy must be provided for economic progress, meaning that government policies need to assure abundant supplies at low costs; and

3) Attention is directed toward ways to manage resulting surpluses, meaning that concern for the problems of managing scarce resources becomes secondary (Andrews 1999:295).

The theory being that when energy costs are kept relatively low, the costs of goods and services that are dependent on energy will also be kept low. Regrettably, this type of thinking has hindered our resourcefulness in allocating income by diverting attention away from improving efficiencies. In fact, the policies have failed to achieve their goal of minimizing energy costs. These policies have instead created a system of incentives and subsidies that have resulted in our dependence on infrastructure that is costly to operate and maintain. Costs of oil, natural gas and electricity have all continued to increase unabated.

During the 1970s and 1980s comments were made that "energy independence" was the "moral equivalent of war" (President Jimmy Carter, in a televised a 1977 speech). Yet since then, the net impact of U.S. policies have led to higher energy usage, higher energy costs, greater dependence on foreign oil, externalization of environmental impacts to unregulated global economies, and a corresponding decrease in sustainability. In countries that pursue a policy of non-intervention, such as the United States, "Sustainable development has gone largely unnoticed and non-supported" and has had "virtually no significant impact on the operations" of the central government which can be effectively categorized as "disinterested" (Lafferty and Meadowcroft 2000:415).

Elsewhere in the world, sustainability policies are being instituted. Sweden and Norway support an agenda based on best sustainability practices, offering policies and incentives to achieve objectives such as reducing carbon emissions (Lafferty and Meadowcroft 2000:397:416). The Netherlands has developed an innovative and workable national environmental policy agenda. Germany has implemented legislation promoting vertically integrated product stewardship (Andrews 1999:285) that includes requirements for component recycling. To understand how *central governments can have such divergent attitudes toward sustainability*, the views of those who question sustainable development will be considered.

CRITIQUING THE IDEA OF SUSTAINABILITY

There are a number of arguments in opposition to the idea of sustainable development. The arguments can be summarized as follows:

1)   The concept of sustainable development is new and remains both untested and unproven;

2)   The concept's primary belief structure is unfounded, undemocratic and questionable;

3)   Economic analysis criteria must be applied and implementation of the concept of sustainability will result in far-reaching economic disruption;

4)   There are only limited ways to implement sustainable development; and

5)   It is a myth to suggest that sustainability can be achieved (Kirby 2000:3).

The idea of sustainable development in its current form is relatively new, having gained wide-spread popularity only in the 1990s. As a result, physical manifestations of its principles (such as buildings and developments) are rare and those that exist have not endured extensive evaluation over an extended period of time. Architects and planning professionals on occasion default to routine design solutions with which they have become most accustomed, regardless of their appropriateness. Alternatively, they may resort to idiosyncratic shock tactics intent on calling attention to the individuality of their masterpieces. This process typically involves a limited site-based focus toward problem solving without considering the wider implications of planning and design decisions. Neither design approach addresses the sustainability of buildings. As a result, most design professionals have simply not adopted the recent concepts upon which sustainable development is based. This is changing.

Local codes and ordinances, either adopted wholesale from those used by other cities or newly written by lawyers and developers who may lack a knowledge of the languages of conceptual or physical design, often unwittingly preclude or indiscriminately penalize the use of sustainable design approaches and technologies. In both design and legal approaches, the promise of sustainable solutions is often lost somewhere in the software. With these and other complications, are the conceptual goals of sustainability valid from a practical perspective? Where are the successful empirical examples? Finding few, there are those who believe that sustainable development may not be a goal worth pursuing and may be inherently unachievable.

Holland summarizes the views of those that believe sustainable development is undemocratic by stating that the concept is "incapable of uniting the hopes of those who fight for justice, those with concerns for the future and those who want to defend nature, but also that it has the potential both to frustrate and to marginalize these very causes" (Holland *et al.* 2000:2). Interestingly, Holland does not propose pursuit of policies that are truly antithetical to the idea of sustainable development, such as a "use it once, throw-it-away" solely profit-motivated and resource intensive strategy.

Ko (2000:2), in an international study to investigate the feasibility of sustainable development that included studying energy use and ecological footprints in Costa Rica, South Korea, Mexico, the Netherlands and the U.S., found that the efficiency of turning energy use into agricultural and other forms of gross domestic production had declined in all countries except the U.S. What then is the relationship between energy and sustainability?

In the Americas and elsewhere, economic efficiency increases may in part be due to increased international trade and globalization. To maintain competitiveness, companies may choose to innovate, cooperate, emigrate from the host country, evaporate from the scene or opt for a strategic combination. Corporations can morph their operations, move their operations to undeveloped countries, and launch ectogenous companies. More pointedly, globalization creates the opportunity for industrialized countries to shift energy-intensive, pollution-creating production to developing or third world countries. This may be interpreted as "dilution is the solution to pollution" on a world scale; further that trade efficiency may be diminishing real value on a global scale. The reality is that the economic feasibility of *sustainable development* is diminished.

Others have expressed stronger opinions. Michael Kraft believes that "a transition from the regulatory regime initiated in the 1970s to one based on environmental sustainability and new policy approaches will be ill-defined, at any given time, and its effect unpredictable" (Desai 2002:29). Well... the regulatory regimes in place prior to the application of sustainability solutions yielded their own set of unintended and unpredictable results (e.g., higher than anticipated costs, misinterpretation of regulations, legal quagmires, confrontational disputes and heavy reliance on carbon-based energy sources).

In a disheartening review and less hopeful assessment, Berke believed that "increasing per capita consumption and population increases along with increasing environmental impacts have made the idea of sustainable development unfeasible" (Ko 2000:2). In this view, the forces and observed trends driving the ecosystem problems are essentially uncontrollable and already potentially not resolvable. There may not be solutions to the challenges that we face.

Yet another argument against implementing sustainable or alternative technological solutions is that such investments fail the "simple payback test" or some other imposed economic assessment standard. This refers to a need for investors to obtain a given rate of return in order to accept the risk of implementing new technologies. Such tests are more applicable to short-term commercial investments. The benefits from many investments are not immediate and are spread out over a period of time, often years. What this argument fails to consider is that most of the conventional technologies being used today fail these criteria as well and that the costs of their related externalities remain unbalanced and unqualified. The benefits of the U.S. Space Program were not clearly known upon inception and it is unlikely that a benefit-cost analysis would have justified moving forward with the program. Yet the benefits have been remarkable. When the costs of externalities are factored into the equation (which seldom occurs), less sustainable technologies fail miserably. The status quo is maintained primarily due to inertia and to the fact that the infrastructure supporting it is already in place and being amortized.

Richard Levine (1990:24) believes that "sustainable development does not alone lead to sustainability" and "to the extent that sustainable development agents move from crisis to crisis, using technological fixes to patch up larger structural problems, they tend to strengthen the systematic relations supporting unsustainability" and that "a change from the old and new can only come from a catalyst—specifically, design and institution of sustainable cities." Though the assertion is true that sustainable development alone may not lead to sustainability and while instituting new sustainable cities may be an admirable endeavor, the path of such logic is structurally flawed. As will be shown, there are actually a number of paths that cities, institutions and governments can choose that will enhance their potential for sustainability by using incremental approaches.

Generalized, exclusionary, anti-incremental, "tear down the old, install the new" infrastructure approaches, such as Levine's, tend to represent a sophomoric philosophy of implementing examples of sustainable development only on a grand scale, rather than an assault on the concept of sustainability itself. They disregard the idea that existing cities can evolve to become ectomorphic in their use of energy and in other ways. The theory of policy-making by developing and incorporating mutual adjustments has an influential history in planning (Lindblom 1958, 1965). Regardless, if a system (be it a building, community or city) is identified as having certain unsustainable characteristics, then why not incorporate technological innovations rather than simply denounce the entire system as inherently malignant? After all, the infrastructure of an existing system is the embodiment of the energy and resources consumed during its development. It often requires greater resources and costs to dispose of this infrastructure and then begin anew, than to rehabilitate it. If not carefully implemented, there is no actual guarantee that the new replacement will not impose a new set of sustainability problems.

With their longer-term view of incorporating strategic solutions, few sustainable development advocates move from crises to crises. The more conservative proponents discount suggestions that crises actually exist (see as an example *The President's Council on Sustainable Development* 1996). Most advocates of sustainability view themselves as realists, looking for ways to incorporate appropriate technologies that cure the malignancy while providing solutions that simultaneously address multiple economic and social issues. In fact, there are examples of projects or programs that successfully employ focused technologies to improve urban environmental sustainability (Roosa 2002a:324-325), that offer empirical evidence that such actions can provide environmental benefits (Roosa 2002b:57-73), that spawn environmental recovery (Burgess, Carmona and Kolstee 1997:208), and that demonstrate how local focus can trigger larger structural improvements (Register 2002:125-129).

A DIFFERENT SET OF ARGUMENTS
AGAINST SUSTAINABILITY

Gordon and Richardson (1997:95-107) level a set of arguments not directly against sustainable development but against certain of its central beliefs. They contend that:

1)   There is a global energy glut and that markets "are the real sources of energy shortages";

2)   Sustainable development "does not support the case for compact cities";

3)   That the movement to the suburbs "has been the dominant and successful mechanism for reducing congestion"; and

4)   Per capita energy use "has actually declined in the U.S. since 1973."

Their arguments discount the idea that urban environmental impacts and other economic externalities are worthy of attention.

Gordon and Richardson imply that a market-driven energy shortage is required before increased efficiency would be considered viable. In the U.S., shortages can result from embargoes and events that include disruption to oil production or pipeline capacity due to a natural disaster (such as Hurricane Katrina in August 2005) or a simple blockage of a critical harbor or shipping lane (such as the shipwreck of the *Lee III* in February 2004 that blocked the Southwest Pass of the Mississippi River in Louisiana). On the other hand, when a "glut" occurs and prices fall, is it inherently wise to continue to consume energy inefficiently and indiscriminately?

Regardless, gluts and shortages of energy resources are not the primary concern. The fact remains that such events are typically short-lived, often have localized impact, and result in markets eventually achieving balance. Compact cities may be desirable and appropriate solutions for some, but not all, urban configurations, as there are many exogenous factors contributing to their morphologies. The point is that not all options work for all cases. In the same paper, Gordon and Richardson (1997:97) admit that "the absence of congestion pricing and emissions fees is a widely acknowledged problem; it constitutes an implicit subsidy to auto users." If the movement to the suburbs has been such a successful mechanism for reducing congestion, then why is the lack of congestion pricing and emissions fees considered problematic? This unresolved contradiction seems beyond their ken.

While it is certainly true that per capita energy consumption declined from 1973 to 1997, Gordon and Richardson fail to ask how this was possible during a long period of generally sustained economic growth (yes, recessions occurred along the way). The answer lies in that "invisible energy resource" which has continued to outstrip the "contributions of natural gas, coal, nuclear, or hydroelectricity" –*that "Fifth Fuel" called energy efficiency* (Nemtzow 2003:22).

By 2005 energy efficiency improvements, one of the heavy lifters in the toolbox of both neo-liberals and sustainable development advocates, contributed approximately 22% of U.S. energy consumption or a total of 24.8 quadrillion kilojoules (27 quadrillion Btus) when compared to 1999 data (Nemtzow 2003:22). The future potential for continued efficiency improvements cannot be understated. For example, U.S. legislation that improved air conditioning and heat pump efficiency requirements from a Seasonal Energy Efficiency Rating (SEER) of 10 to 13 by 2006 will eliminate the need to design, develop and construct approximately 150 electrical power plants by the year 2020 (Hebert 2004). *Are there other incremental improvements that offer similar potential?*

Given the difficulties of implementing alternatives, perhaps doing nothing at all needs to be considered. However, the concerns which sustainable development attempts to address suggest that doing nothing may *not* be the wisest approach. Historical trends indicate that our populations will continue to grow. Cities will require multiple and dependable supplies of energy, both renewable and non-renewable, to ensure that development is sustained. While the cost of certain energy resources had moderated in the 1990s (Holland, Lee and McNeill 2000:117-126), this changed dramatically by 2007. The prices of oil, natural gas and electricity are now more expensive.

Taking no action effectively defaults to the status quo and a continuation of "do nothing" policies. This tends to exacerbate current problems and sets into motion a series of events that not only support the status quo but also cause the implementation of new technologies to be stymied. Doing nothing can result in wasted resources, and increased life-cycle costs. In addition, employment reductions can occur in cities and towns that fail to provide renewable energy production, energy efficiency services, and alternative designs and planning solutions.

Finally, there are those who believe that the idea of sustainability and perhaps sustainable cities is unachievable. History has shown this to be true in the long term. After all, cities have a natural life of their own and few survive more than a couple of thousand years before being abandoned or lost to destruction or natural disaster. Another paradox is that cities require a high degree of effort and resources to construct but can be destroyed or made uninhabitable by a number of means, disruption of their power systems being one. Unlike the cities of old, cities today cannot exist without dependable sources of high-quality energy. The potential obsolescence of today's cities might be a function of the types of energy they use. Sustainability and energy are inherently interlinked.

POLITICAL VIEWS ON SUSTAINABLE DEVELOPMENT POLICIES

Interpretations of policy are often influenced by political views. As Kelbaugh (2002:9) so succinctly states, "Hidden political values and social agendas are inevitably embedded in calls for sustainability." Sustainability is linked to energy by both practicalities and politics. A problem central to "urban research today is how to combine economic and political logic into a coherent theory of urban politics" (Swanstrom 1988:87). Political theory often manifests itself in the form of regimes that evolve to promote their respective agendas. In the U.S., three primary camps hold different political perspectives about sustainable development:

1) **The conservatives**, justified by a narrow interpretation of market based, supply-side economics, who believe little or no action in regard to sustainable development is necessary;

2) **The U.S. neo-liberals** who advocate marginal intervention, mitigation and measured response;

3) **The advocates of sustainability**, who tend to favor a more energetic, broad-based effort which requires concerted action on an international scale.

How Conservatives View Sustainability

The view of today's conservatives in the U.S. starkly contrasts with the beliefs held by conservatives in the 1960s and earlier. Then, conservatives tended toward non-intervention, held a belief in self-reliance, felt that U.S. self-sufficiency was an important goal, and believed that foreign control of strategic resources was most likely a part of a communist plot. Today, this stance has shifted more to the beliefs of neo-liberal interventionists—with the one exception being that rather than communists, the plotters are the oil-exporting countries.

Today's neo-conservative view is firmly entrenched. The conservatives believe that there are few real problems with urban areas that require intervention and there is need for neither consternation nor further debate. After all, the main problem with life in suburbia is traffic. Cities solve problems and don't really create them. Social inequities may exist but always will. In their view, there is little real cause for alarm and no need to redirect constrained resources, as any inequities will only be redistributed. The conservatives believe that there will always be inadequate resources to resolve inequities. Regardless, urban problems, like environmental problems, tend to be self-correcting. Clean air and water are "free goods." The conservative theory is that "any tilt toward equality harms economic efficiency and leaves society with decreased productivity and fewer benefits to be distributed" (Stone and Sanders 1987:275). Conservatives argue that their opponents, whom they would characterize as "no-growth elitists" would "freeze the underdeveloped countries out of the benefits of economic growth" (Porter and Brown 1996:25). Some believe that there exists an "energy—gross national product" link, and that without increasing energy usage, a county's GNP would stabilize or decline.

According to conservatives, the primary problem to worry about concerning energy is whether or not a shortage exists, and if none is apparent, things are best left for market conditions to sort out. Surprisingly, the views of conservatives are mixed in regard to the security of energy supplies. For some, "the amount of oil that a country imports is not an indicator of its energy security" (Morris 2002:367). Others insist, "For too long we have been dependent on foreign oil" and we have a vital need "to diversity our energy supply."[11] Regardless, conservatives tend to believe that there are political, diplomatic and military means to control any critical strategic resources that must be obtained from abroad.

Conservatives believe there is little empirical evidence to prove conclusively that environmental issues are causing difficulties (e.g., global warming). They assert that "many consequences of climate change are not expected to be harmful" and that "there are likely to be many ecological and social benefits of a moderately warmer and wetter world, not to mention... a fertilization effect of higher CO_2 concentrations on plants and agriculture." (Morris 2002:165). Many consider the world's environment to be basically self-adjusting regardless of the stresses applied by human activities. The aphorism "the solution to pollution is dilution" best illustrates of the view of the Un-Earth Day crowd. It continues to be man's "manifest destiny" to dominate nature and modify the environment, as he feels appropriate (Andrews 1999:79). In addition, epigenous population growth should be allowed to continue unabated. There are some conservatives who feel environmentalists are misguided alarmists and that some are eco-terrorists.

In the conservative perspective, future generations can fend for themselves with new and better technology. Developmental problems in cities are of an academic nature and are limited to concerns such as whether or not a new sports arena is affordable while maintaining minimal urban amenities to attract new businesses. Agenda 21 is viewed as little more than a program for supra-national government using a communitarian legal basis. *Sustainable development* to many conservatives is an aphorism that really means *no development*.

According to Harken (1993:7), "The conservative view of free-market capitalism asserts that nothing should be allowed to hinder commerce." The idea of "market failure" is an oxymoron. This view exemplifies laissez-fare capitalism (Andrews 1999:155) and consumer-based normative economic theory. "Real energy problems stem from acts of government, not acts of the market" (Morris 2002:165). Some even believe that, "sustainable development is a dangerous notion that can bring the prosperity of the U.S., Western Europe, and Japan crashing into a prolonged economic dark age."[12] The holders of this view often take an interventionist posture leaning toward the control of energy resources. The posture in the short term benefits those who directly market energy resources such as those companies who sell oil, coal or natural gas. The conservatives often recommend additional research as the solution, using "study the problem" as a stalling tactic rather than a means of planning for responsive action. Remarkably absent from their argument is the assertion that increased use of imported carbon-based energy will improve equity, balance national trade, improve national security, reduce atmospheric pollution, cause less water pollution, and alleviate global climate change problems. One of the logical faults of the conservative view is the failure to understand that lack of a policy can actually be a policy—yielding its own set of implications. In the view of those who advocate a *laissez-faire* approach, planning theories and systems theories are of no consequence and are themselves problematic.

Advocates who supported the conservative stance flourished in the 1970s and 1980s. They were beginning to flounder at the turn of the 21st century as the weight of objective evidence concerning the effects of prolonged atmospheric pollution, greenhouse gases, ozone depletion and resource inequities bore down upon them. They have resurfaced today by redefining sustainability—not in its broader context—but in more focused terms such as setting aside national parkland, land conservation programs, and developmental water quality improvement plans such as infrastructure for sewage plants (all of these efforts preceded the current idea of sustainability).

The conservative solution to global energy problems is for developed nations to gain permanent economic or military control of the world's oil supplies and perhaps build more environmentally "clean" atomic power plants—just not in their back yards. Conservatives feel comfortable with this since they believe that "we are moving away from the renewable energy era" (Morris 2002:167). Therefore, the authority of central governments should be maintained—as long as there are no tax increases.

How Neo-liberals View Sustainability

The interventionist approach lies between the conservatives and the advocates of sustainable development and is often labeled *neo-liberal*. The neo-liberal view "assigns a central role to the market system along three interconnected dimensions—economic, social and political" (Self 2000:159). Neo-liberals believe that there is no positive relationship between energy usage and GNP. In fact, "the relationship between energy use and GNP does not have any economic and statistical meaning, because their statistical scopes are different" (Sun 2001:1). In practical political and economic applications, neo-liberals "endorse the value of democracy" yet believe that "substantial market regulation is necessary" (Self 2000:22). According to Gore (2007:193), "Anyone who believes that the international market for oil is a 'free market' is seriously deluded… it is subject to periodic manipulation by the group of countries controlling the largest recoverable resources… sometimes in concert the small group of companies that dominate the global production, refining and distribution network."

The interventionist position is neo-liberal in the sense that it tends to minimize the "contribution of urban primacy to urban problems," arguing that "no causal relationship has been established between the incidence of urban problems and city size" (Burgess *et al.* 1997:9). However, neo-liberal theory appears to emphasize the issue of city size without focusing on other urban issues. Increasing urban density, even on the mega-city scale, can be beneficial (Soleri 1971). In the interventionist view, the benefits of technological and engineering solutions have not yet been given the opportunity to have long-term impact. Expanding the application of technological solutions must be seriously considered. Thus, many urban problems can be primarily resolved by increases in efficiency and productivity.

The neo-liberal (tree-huggers—and proud of it) view on the environment is that we have "liquidated our environmental trust fund in the currency of pleasure, convenience, profit or environmental indifference" (Kelbaugh 2002:33). Anyone who believes that cities are not environmentally problematic hasn't been reading the newspapers lately. However, many environmental problems can be mitigated by policy measures that directly impact efficiency... so the best policy is to improve urban infrastructure, implement efficiency improvements and institute urban environmental management (Burgess *et al.* 1997:9).

The neo-liberal view is that an urban environment is unique and "like no other environment on earth" (Register 2002:86). The world's wealthy cities have "transferred the environmental costs that their concentrated production and consumption represents from their region to other regions in the global systems" (Habitat 1996:8). Today, energy and potable water resource shortages have resulted in challenging urban health problems. In extreme cases, pervasive and persistent resource depletion is also of concern. These conditions are deleterious to the ideals of sustainability and require mitigation.

Equity issues can be resolved by appropriately redesigning social programs. Cities are responsible for creating their problems and can redirect resources to resolve them independently, by concerted effort or with state and federal intervention. The interventionist solution to global problems is to improve efficiency and implement new technologies more quickly and on a broader, local scale, increasing taxes if necessary. Due to political ineffectiveness, mismanagement of funds and unremarkable results, "the idea that the central government—one huge mainframe—is the most important part of governance is obsolete" (Naisbitt 1994:47). These interventionists agree with the conservatives on one issue: both regard technology as our most important means of ultimate salvation.

The Pro-sustainable Development View

Proponents of sustainable development tend to agree with the concepts of improving efficiency, and with certain policy approaches of the neo-liberals, while severely discounting the conservative view. While markets can efficiently allocate resources, they do so within a time frame that devalues "scarcities or overuse of resources" (Self 2000:184). Externalities such as environmental pollution remain unaccounted. That is why per unit of volume, a gallon of gasoline in the U.S. costs less than water when purchased by the pint.

Advocates of sustainability believe that while technological innovations are available and can be carefully implemented, they can be costly to deploy and may have unintended and unwanted environmental consequences. Technological solutions can backfire. The use of "appropriate technology" is a key (Roosa 2002a:315-327) since it appears that "some of the important functions of the natural world cannot be replaced within any realistic time frame, if ever, by human technology, however sophisticated" (Ayres 1998:8). Examples include water, energy resources, topsoil, rain forests, etc. Allowing a "live and let die" view of species extinction is simply unacceptable.

According to "sustainable development theory, there is a specific environmental rationale that has to be considered in the choice and construction of urban, regional and national" policy frameworks (Burgess *et al.* 1997:81). Sustainability advocates believe that "considerable concession can be made to equality without harming efficiency" of production (Stone and Sanders 1987:275). There is no logical reason not to take action now. As David Clarke noted, it is actually difficult to believe that "urban populations can double over the next 25 years without some form of economic or ecological breakdown" (Legates and Stout 1996:582).

The holders of the pro-sustainability view propose a posture that leans toward the management of demand. Waste management and energy conservation are two valid examples. Like the neo-liberals, this posture also benefits companies who market products such as computer control systems, wind turbine generators, solar collectors, alternatively fueled vehicles or services such as energy saving performance contracts. These products and services are more likely to be domestically provided and manufactured.

In accord with the ideal of sustainability, targeted policy measures to reduce inequities, improve waste management, and address environmental problems are appropriate issues of concern. However, they may

be insufficient to resolve problems on a global scale. Not only are policies widely variable, sustainability theorists argue "that even with existing population size, consumption levels and inequities, many mega-cities
are already exhausting their environmental support capacity, with water
consumption exceeding the replacement capacity of primary sources; the
destabilization of ecosystems; and air pollution levels that are highly inju-
rious to human health and safety" (Burgess *et al.* 1997:80).

Their argument suggests that urban problems not only exist but
may not be resolved quickly under any planning or intervention scenario.
Those favoring sustainable development consider the economic view that
"the free market will take care of the problem" without intervention as
"false and unwarranted" because there is simply inadequate evidence "to
establish a plausible case that technical solutions" exist that can be rapidly and successfully deployed without unintended complications (Ayres
1998:47). According to advocates, plans need to be immediately initiated,
significant resources allocated, and mitigation policies implemented with
concerted effort *on a global scale*. Negative externalities will continue to
cause markets to be less productive, discounting if not ignoring, their environmental impact. Furthermore, it is impossible to take city size "out of
the equation for dealing with urban environmental issues" (Burgess *et al.*
1997:80).

Sustainable development advocates suggest revamping policies at
all levels of government in order to successfully mitigate damage to already impacted ecosystems and prevent further ecosystem disruption.
Regimes need to be "built to help pressure and document the transition
from a carbon-intensive economy to an economy fueled by cleaner energy sources" (Choucri 1993:469). Theories of how sustainable development can be technically implemented are available (Devuyst 2001:54-61)
and many suggest changing energy usage patterns. Recent developments
include techniques to gauge indicators of urban sustainability (McClaren
1996:182-202).

What conservatives refuse to admit and what neo-liberals begrudg-
ingly accept is that that sustainable development advocates decentralization of authority, plus customization of policies and localization of agendas. Thus, the notion of sustainability appears in step with traditional concepts of American pluralism. "The theory and practice of American pluralism tend(s) to assume… that the existence of multiple centers of power,
none of whom is sovereign, will help to tame power, to secure the consent
of all and to settle conflicts peaceably" (Dahl 1967:24). In addition, "ques-

tions of policy... are also effectively beyond the reach of national majorities... In fact, whenever uniform policies are likely to be costly, difficult or troublesome in pluralistic democracies, the tendency is to find ways by which these policies can be made by small groups of like-minded people who enjoy a high degree of independence" (Dahl 1967:23). Pluralists, according to Crenson (1971:19), tend to "argue for the independence of local political systems." The ideals of sustainable development policies are best suited to local interpretation and action.

Despite the various elusive operational definitions, mixed political views, and divergent agendas, the concept of sustainability seems neither ephemeral nor implausible. On the contrary, the idea has proved to be both penetrating and appealing. According to Peter Newman, "sustainability is now the most significant global political movement to challenge the dominance of neo-liberalism."[13] He notes several products and services that are considered no longer sustainable. These include production of greenhouse gases, car-dependent cities, and physical construction that is energy-inefficient. Regardless, while many view sustainable development as a compelling ideal, it is challenging to pursue it in the real world. Inherent in the very concept is an endless struggle to achieve a balance of "competing legitimate claims for economic growth, environmental health and social justice" (Eckstein and Throgmorton 2003:4). *Can local governments, each married to the overarching goals of sustainability, develope their own local solutions?*

IMPLEMENTING SUSTAINABLE DEVELOPMENT POLICIES

Approaches to implementing sustainable development are varied. The process involves selecting the problems to be resolved, developing a strategy for resolution, implementing interventions and resolving these problems. The problems being considered have global implications, social or environmental characteristics, and are easily politicized. For example, in September 2002, the World Summit on Sustainable Development was held in Johannesburg, South Africa. At the conference, U.S. Undersecretary of State Paula Dobriansky stated that, "The United States is the world's leader in sustainable development. No other nation has made a greater and more concrete commitment" (Verrengia 2002b:A17). She was seemingly unaware that there is no national policy concerning sustainable development.

There exists in the U.S. a policy tradition that frames sustainability as a need to trade economic development for environmental solutions as if both were always mutually exclusive. This sort of illogical "either-or" thinking has stymied development of creative efforts to achieve sustainability since the 1970s. Consider industrial air pollution. Crenson (1971:165) studied 18 variables prominent to air pollution problems and found that "The correlation coefficient... shows that where business and industrial development is a topic of concern, the dirty air problem tends to be ignored." In fact, when objective measures such as energy resource consumption and the emission output of critical pollutants are considered, it seems the U.S. fails to score highly as a sustainable economy, especially when compared to other industrialized nations.

Using a common system input measure such as size of energy footprint (often a more conservative measure than gross energy used or per capita energy consumption), *the U.S. is the third least-efficient country in the world*, behind the United Arab Emirates and Kuwait, both major producers and net exporters of oil. Based on this measure, Canada and most every other industrialized country in the world is more efficient in using energy than the U.S. Consider Japan, a country that imports 100% of its oil and taxes it to a far greater extent, and yet consumes only half as much energy per unit of economic output as the U.S. (Holland, Lee and McNeill 2000:125). In fact, when emission output measures are considered the U.S. also ranks poorly. Table 2.1 (compiled from Desai 2002) ranks selected industrialized countries using 1998 (kg) output of CO_2 per Gross Domestic Product (GDP) in base 1990 U.S. dollars and gross tons of emissions per capita.

Table 2-1. Select Industrialized Countries Ranked by CO_2 Emissions (1998)

	Carbon Dioxide Emissions kg/1990 US$ GDP	Carbon Dioxide Emissions Tons/capita	Nitrogen Oxide Emissions kg/capita	Sulfur Oxide Emissions kg/capita
United States	0.78	20.30	80.0	69.0
Canada	0.72	15.80	68.0	90.0
Germany	0.47	10.70	1.2	18.0
United Kingdom	0.50	9.60	35.0	34.0
Italy	0.41	7.40	31.0	23.1
Japan	0.45	9.30	11.0	7.0

Economic policies in the U.S. divert revenues from social issues such as housing, education, health and welfare to supporting energy producing and marketing companies (that, not coincidentally, wield far greater political influence).

It has long been suggested that even a small level of taxation can significantly reduce demand for imported oil if the proceeds are used to promote energy efficiency and energy conservation (Roosa 1988:41-42). However, the opposite often occurs in the U.S. For example, federal gasoline taxes and funding policies have long subsidized highway construction to a far greater extent than mass transit, enabling less efficient transportation systems to be supported. As a result, countries such as the U.K., The Netherlands, Germany, France, Spain, Italy and Japan are among the international leaders in the design, production and export of high-speed regional rail services, magnetic levitation transport, submersible tunneling systems, merchant shipping and cruise ship construction. These advantages provide their economies with expanded manufacturing employment and technological capabilities while the U.S. has excluded itself from these lucrative, high technology markets.

Contrary to assertions by U.S administrations, the U.S. Department of Energy freely admits on its Energy Efficiency and Renewable Energy Network[14] (which it co-sponsors) that, "The complex problems shared by cities throughout the U.S. are evidence of the impacts of urban sprawl—increasing traffic congestion and commute times, air pollution, inefficient energy consumption and greater reliance on foreign oil, loss of open space and habitat, inequitable distribution of economic resources, and the loss of a sense of community." Never underestimate the ability of central governments to be self-contradictory.

The President's Council on Sustainable Development provided a report in 1996 entitled *Sustainable America—A New Consensus* which included ten sustainable development policies and recommendations for changes at all governmental levels (Holland, Lee and McNeill 2000:83). It appeared that the council's recommendations were simply importing to the U.S., albeit belatedly, policies widely understood and being seriously implemented in other developed countries (DeSimone and Popoff 1997:162).

Despite this effort, achieving sustainable development is not yet an adopted goal of the U.S. federal government and there are no specific targets for implementation (Holland, Lee and McNeill 2000:93). To date, none of the ten recommendations have been effectively implemented by

the central government, either internally or externally. The efforts of local governments aside, this evidence is compelling—assertions that the
U.S. leads the world in sustainable development are not supportable.

Interestingly, concrete examples of planning for sustainable development are more easily found elsewhere in the world. The development plans of cities in the European Union serve as one example. In
China, Kisho Kurokawa, a Japanese Architect, has been a proponent of
the planning concept of an "ecological city." His "eco-media city" plans
are a more "developed version of the eco-city" (Kurokawa 1998:2). He
proposed a plan for Futian in China based on a central eco-media city
park concept to symbolize urban sustainability (Cartier 2002:1521). In
a suburb of Linz, Austria called Pichling, a new solar powered district
housing development for 25,000 residents, has been constructed (Beatley
2000:275). In the Netherlands, Ecolonia—a demonstration town for ecological development—is yet another example.

It is clear that many policies being implemented both locally and
internationally with a goal of improving local sustainability.

OBSERVATIONS ABOUT SUSTAINABILITY

Sustainability can be so broadly defined that potential solutions for
its core goals remain unfulfilled, leaving a disconnect between world
needs and the means of fulfilling them. One example might be low-income housing, a problem which urban literature has long considered
(Gilderbloom and Capek 1992; Healy 1974). If the housing occupants are
unable to afford the costs of utilities for their home, then they may lose
hot water, water supplies, toilet facilities, space heating, etc., thus negat-
ing the benefits of housing. In some U.S. cities, winter utility costs now
exceed apartment rents. Sustainable housing cannot be provided under
such conditions. As a result, the ideal of housing as both shelter and as a
package of housing services is diminished in value.

The relationships between sustainability and energy are so inter-
laced that it is unrealistic to consider sustainability without dealing with
the issues and impacts of energy. Urban literature is rich in the discussion of inequities, special aspects of poverty, central city decline, urban
housing, social needs, de-industrialization, suburbanization, globalization, urban regimes and economic impacts of development in cities (Benfield, Terris and Vorsanger 2001; Bluestone and Harrison 1982; DiGaeta-

no and Klemanski 1999; Duany, Plater-Zyberk and. Speck 2000; Gilderbloom and Capek 1992; Hall 1988; Healy 1974; Jackson 1985; Koven and Lyons 2003; Savitch and Kantor 2002). The idea that energy is central to urbanization seems to be frequently devalued, minimized, if not omitted entirely from urban literature. We find that urban theorists tend to ignore or minimize the impact of energy on development. However, energy is directly linked to sustainability. In order to evolve to a broader understanding of urban development, energy use, energy conservation, and energy efficiency must be key components of any worthwhile sustainability solution.

CONCLUSIONS

Sustainability is now a core concept in major policy-making processes, in the environmental aspects of population growth, and in the redevelopment of existing facilities and infrastructure. Sustainable development can be considered to be a higher form of environmental policy. Due to the broad implications of resource management, sustainability is functionally dependent on energy use, on the ways energy is transformed, and the types of energy selected for a given task.

As populations expand and energy use increases, policies diverge and sustainable development becomes increasingly difficult to achieve. Indicators of sustainability can be theorized and measured, and sustainable solutions can be implemented—but sustainable development can also unleash economic and political interests that threaten it. Energy-saving consumer products can be readily marketed to the public, but the benefits from alternative energy sources are often difficult to measure and promote.

It is not the concept of energy *per se* that is critical to sustainable development. Rather it is the selection of the type of energy involved and how it is converted to use that impacts sustainable development policies. Energy policies will ultimately prove to be the most important component of sustainable development. An understanding of how energy affects cities is critical to comprehending urban sustainability. Sustainable development is a newly evolving agenda that has entered the world debate since the 1980s. Achieving sustainability requires consensus and concerted action. Both remain elusive. Sustainability has its critics and it has yet to mature into mainstream acceptance in energy-intensive countries such as the U.S. Regardless, it is clearly both alluring and gaining popularity. Sustainable development certainly provides a fundamental vision and framework for future planning and development.

In the U.S., policies have been directed toward minimizing energy costs. The policies have been based on the assumption that the cost of goods and services that are dependent on energy will also be minimized. Regrettably, this type of thinking has hindered our resourcefulness in allocating income by diverting attention away from improving efficiencies. It has also reduced our international competitiveness. In fact, the policy has failed at achieving its goal of maintaining low energy costs and has back-fired. It has contributed to a system of incentives and subsidies that have effectively maintained high-cost, energy-intensive infrastructure.

The time has come for an alternative agenda. We can see that sustainable development provides a set of overarching principles that can be used to broadly assault certain pervasive problems, such as imprudent development policies, wasteful energy practices and environmental degradation. Implementing sustainable solutions that involve energy conservation, energy efficiency, and alternative energy production can reduce energy use, improve urban sustainability, and begin to address environmental problems. Lowering energy consumption, especially of carbon-based fossil fuels, has beneficial results that include reducing greenhouse gas emissions, improving the quality of life and improving health and the quality of life in cities. Policy efforts toward such ends are beginning to occur at the local and urban levels in the United States, Europe and elsewhere.

SUMMARY

The Rio Agenda 21 provides a set of principles regarding sustainable development. It is concerned with unsustainable growth patterns, environmental protection, intergenerational equity, and resource conservation. These are keys to sustainability. The concept, while representing a new paradigm that evolved from a convergence of ideals, is fraught with ambiguity and is subject to misinterpretation. The concept of sustainability can be polarizing, divisive, and politically confrontational. Three political views were identified to illustrate this divisiveness—conservative, neo-liberal and advocates of sustainable development.

A long-term view is necessary. Sustainable development "can be seen as based on a new recognition of the complex web of interconnections between different issues, fields, disciplines and actors," requiring a "holistic and interdisciplinary perspective" (Legates and Strout 2000:438). Regardless, achieving sustainability by modifying energy generation and consumption patterns may be among the most viable approaches.

With the increased use of carbon-based energy, earlier policies that failed to support sustainable development are now unworkable. The idea of doing nothing and maintaining the status quo is increasingly unacceptable. Such arguments were found to be weak and often categorically unsupportable. More importantly, advocates of the "do nothing" alternative fail to offer substantive evidence of effectiveness, nor do they suggest adjustments to existing policies that could resolve their shortcomings.

The U.S. is the world's largest user of energy. It is important to clarify the relationship between energy and sustainability. Another problem is how can catalysts for change at the local level be identified? What are the solutions? One might be to use governance and policy approaches to reduce hydrocarbon-based energy use. If energy policies are a means of achieving local sustainability goals, solutions should include implementing new technologies and embracing energy conservation and efficiency improvements.

Sustainability, energy, and environmental impacts are linked. The next chapter, *The Environment and Sustainable Development*, elaborates on this vital relationship.

Endnotes

1. Among the members of the "blue ribbon" panel on sustainability were Kenneth Lay, former Chairman of Enron Corporation, Kenneth Derr, Chairman of Chevron, Richard Clark, retired Chairman of Pacific Gas and Electric, and Samuel Johnson, Board Member of Mobil Corporation.
2. Newman, P. *Sustainability and planning: a whole of government approach.* ahttp://www.sustainability.dpc.wa.gov.au/docs/SustainabilityandPlanning.pdf, accessed 7 July 2007.
3. McShane, L. (2007, 15 July). Memories of 1977 still sharp. *The Courier-Journal.* p. A9.
4. Ibid.
5. Hahenbaugh B. (2007, 17 May). Gas gulps more of family pay. *USA Today.* p. 1A. Data from the Federation of America and Consumers Union.
6. Ibid.
7. Baltimore, C. (2007, 10 January). Oil keeps U.S. vulnerable, lawmakers said. *The Washington Post.*
8. Ibid.

9. Ibid.

10. See Dincer and Rosen (2004:16-17) for an example of an energy and exergy analysis of the coal fired electrical generation plant at Nanticoke, operational since 1981. Though it is cooled by Lake Erie and uses pulverized coal, the overall energy efficiency was found to be 37% and the corresponding exergy efficiency was 36%.

11. U.S. President George W. Bush in his *State of the Union* address. 23 January 2007.

12. Davis, T. *What is sustainable development?* www.menominee.edu/sdi/whatis.htm., accessed 1 May 2007.

13. Newman, P. Sustainability and planning: a whole of government approach. ahttp://www.sustainability.dpc.wa.gov.au/docs/SustainabilityandPlanning.pdf, accessed 7 July 2007.

14. See U.S. Department of Energy (2002, 22 May). http://www.sustainable.doe.gov/landuse/luintro.shtml.

Chapter 3

The Environment and Sustainable Development

By Stephen Roosa and Matt Hanka

"...another enormously powerful threat to the flattening of the world is on the horizon. It is not a human resources constraint or a disease, but a natural resources constraint. If millions of people from India, China, Latin America, and the former Soviet Empire, who for years have been living largely outside the flat world, all start to walk onto the new flat world platform—with his or her own version of the American dream of owning a car, a house, a refrigerator, a microwave, and a toaster—we are, at best, going to experience a serious energy shortage. At worst, we are going to set off a global struggle for natural resources and junk up, heat up, garbage up, smoke up, and devour up our little planet faster than at any time in the history of the world. Be afraid. I certainly am."

FRIEDMAN (2005:494)

"Human beings are carrying out a large scale geophysical experiment of a kind that could not have happened in the past nor be produced in the future. Within a few centuries, we are returning to the atmosphere and the oceans the concentrated organic carbon stored in sedimentary rocks over hundreds of millions of years."

REVELLE, R., AND SUESS, H. (1957). CARBON DIOXIDE EXCHANGE BETWEEN AT-
MOSPHERE AND OCEAN AND THE QUESTION OF AN INCREASE IN ATMOSPHERIC
CO_2 DURING THE PAST DECADES. TELLUS, 9, PGS. 18-27.

There is a powerful relationship between the environments that we inhabit and the idea of sustainability. The natural environment and the endowment it provides is crucial to human existence. The concept of sustainability has been linked to land development practices, population growth, fossil fuel usage, forest management, aqua-culture, pollution, global warming, limited water supplies, species diversity and extinction, and the types of resources being consumed (Kocha 2004; Clayton and Radcliffe 1996; Layzer 2006). All of these impact our environment and the sustainability of ecosystems. Our stewardship of the Planet's environment and its resources is the most important legacy we will leave to our descendants. Causing pollution that permanently damages the natural environment is the antithesis of sustainability.

According to Rosenbaum (2005:362-363), "environmental science, embodied in the technical underpinnings of current understandings of climate warming, ozone depletion, and intergenerational equity, is compelling policy makers to think in terms of problems and impacts, of the consequences of present decisions and future undertakings, and on a time scale almost unthinkable a few decades ago." There has been a shift in viewing environmental issues as a set of local problems to viewing them as international in scope.

Natural resources that require preservation and stewardship "can be seen as the sum total of the ecological systems that support life, different from human-made capital in that natural capital cannot be produced by human activity" (Harken *et al.* 1999:151). Natural capital depletion occurs when resources in the natural environment are consumed or made unusable. Natural capital depletion is transnational in scope and has international importance. Within the last few decades, the idea that natural capital resources are limited and can become depleted or exhausted has become plausible. In addition to energy resources such as oil, strategic mineral resources are being depleted rapidly. Supplies of Indium, which is used for liquid crystal displays (LCDs) may run out within 10 years.[1] The development of the next generation solar panels, which are twice as efficient as current collector systems, may not be possible due to a lack of gallium and indium.[2] At current rates of use, supplies of hafnium, used in computer chips, will be exhausted by 2017; zinc by 2037 and copper by 2100.[3]

This is a radical change in perspective. The environmental, scientific, governmental and academic communities, especially in developed countries, are now directing resources to investigate and understand environmental processes and their impact on sustainability.

For the U.S., the linkage between sustainability and the environment predated Agenda 21 and was embodied in the U.S. National Environmental Protection Act (NEPA). Public opinion regarding environmental protection has been strengthened by a seemingly endless stream of manmade environmental catastrophes including:

1)   The fires on the Cuyahoga River in Cleveland, Ohio that were directly caused by toxic pollution (1969);

2)   Toxic waste contamination at Love Canal in Niagara Falls, NY (late 1970s);

3) Union Carbide's (a U.S. company that is now a division of Dow Chemical) incident in Bhopal, India, that killed thousands (1984);

4) The *Exxon Valdez* accident which spilled 257,000 barrels of oil in Alaska's Prince William Sound (1989);[4] and

5) The massive "dead zone" in the Gulf of Mexico caused by the runoff of agricultural chemicals carried down the Mississippi.

Serious environmental disasters have occurred that have increased this awareness in other countries as well.

1) Oil well fires ignited by Iraqi troops retreating from Kuwait (1991) burned over 5 million barrels, causing regional air and water pollution.[5]

2) The Chernobyl nuclear power plant incident near Kiev, Ukraine that spread radioactive material over most of Turkey and Europe (1986). The accident required the immediate evacuation of 135,000 people, left an area of fertile farmland half the size of Italy uninhabitable and directly caused the deaths of at least 7,000 people.

3) The Chinese oil giant PetroChina's safety record—a natural gas explosion that killed 243 (2003) and a toxic benzene spill into a river that left millions in China without water (2005) (Engardio 2007:60).

4) An magnitude 6.8 earthquake off the coast of Japan triggered an electrical transformer fire at the Kashiwazaki Kaiwa nuclear power plant (the largest in the world) causing a leak of radioactively contaminated water that was flushed into the sea (2007).[6]

Widely publicized environmental catastrophes such as these heighten public awareness of the importance of environmental stewardship. These events, and others like them, are related to the environmental sustainability of eco-systems.

While the federal government establishes environmental regulations that states must follow, many mandates are unfunded. Most of the funding for environmental protection in the U.S. comes from state governments. Many states have the primary responsibility for issuing permits, monitoring environmental conditions and establishing enforcement procedures and actions. State funding for these initiatives is widely variable and unevenly distributed. By the early 1990s, state governments were re-

sponsible for about 75% of the U.S. environmental programs and responsible for about 94% of the environmental quality data submitted to the
federal Environmental Protection Agency (Desai 2002:48). By 1996, states
were expending far more than the federal government on environmental
and natural resource policies (Desai 2002:48). This continues to this day.

Clearly, "Environmental concerns are significantly linked to sustainable development... activities which continually degrade the environment
are not sustainable" (Dincer and Rosen 2004:10). According to the United
Nations Commission on Sustainable Development, "air and water pollution in urban areas are associated with excess morbidity and mortality...
Environmental pollution as a result of energy production, transportation,
industry or lifestyle choices adversely affects health. This would include
such factors as ambient and indoor air pollution, water pollution, inadequate waste management, noise, pesticides and radiation" (United Nations 2001:38). The environmental impacts due to reliance on hydrocarbons include eutrophication of fresh waterways caused by nitrogen fertil-
izers and land degradation (Morris 2002:159).

It can be theorized that as economies evolve, a corresponding evolution of interest in the environment emerges. Hunting and gathering societies, dependent on the land for survival, often managed their local environments to maintain their resources. These societies migrated to ensure
survival, providing an opportunity for vacated areas to renew themselves.
Industrial societies are oriented toward production by converting resources into marketable products and adapting when key resources become
unavailable. The processes of industrialization often results in increased
levels of pollution. Industries such as mining, fishing and farming are examples. Often, the most direct approach to converting resources into a
product is preferable, regardless of environmental consequences. Environmental degradation becomes an economic externality with avoidable
associated costs. Open pit mines are abandoned, marine life is harvested
to extinction and farmers can deplete their soil. A *tragedy of the commons*
occurs when natural resources become exhausted by human activity and
mitigation and environmental protection efforts are belated.

As countries evolve to "become wealthier and develop a more ser-
vice-based economy, the environment tends to become more of a priority"
(Clayton and Radcliffe 1996:99). Environmental damage becomes more
noticeable when it threatens the basic requirements of human existence
(e.g., food and shelter). As populations become more educated, interest
in environmental policies and regulations also increases. People become

more willing to allocate resources to improve environmental conditions when they feel their lives are personally affected. When such conditions are present, there is also a corresponding propensity for concerted action to avoid or mitigate adverse impacts. Sustainable development can effectively deal with the questions of how environmental resources are to be managed and what sorts of actions are appropriate.

WASTE MANAGEMENT

Where garbage goes and how it is handled has become a major environmental issue with no easy solution. Paradoxically, efforts to improve environmental conditions by changing waste management strategies may backfire creating new sets of problems. With a goal of improving sanitation, waste management can itself contribute to environmental degradation. How cities handle garbage disposal provides one example of how closely waste management is related to sustainability. Eliminating municipal waste incinerators and avoiding the cost of installing pollution control equipment has helped improve air quality. While incineration is usually more costly than disposing of wastes in landfills, eliminating incinerators increases the demand for landfill space.

"Sanitary" landfills are one common means of handling municipal waste in the U.S. A total of 245.6 million tons of municipal solids wastes were generated in 2005. Of this total, 54.3% was deposited in landfills.[7] Today, many landfills are being designed to enable future recovery of landfill gas as an energy resource. Regardless, managing municipal waste has become a major environmental problem.

Municipal wastes must be transported greater distances and landfill space is becoming costly. For example, New York City generates about 12,000 tons of garbage a day, sending 600 fully loaded tractor-trailers to distant landfills in New Jersey, Pennsylvania, Virginia and as far away as Ohio (Brown 2006:4). While this may be preferable to dumping urban wastes in the ocean or turning Staten Island into a mountain of trash, hauling garbage such distances requires manpower and consumes energy. Tipping fees at sanitary landfills are adequate justification for some people to dump wastes in valleys and streams, causing unsightly and toxic conditions.

Landfill capacity is becoming scarce and costs for disposal are increasing. The U.S. has roughly 20 years of available landfill disposal capacity, with the Western States having the most available capacity and the

Northeast having the least.[8] Alaska, Connecticut, Delaware, North Carolina, New Hampshire and Rhode Island have less than five years of capacity remaining.[9] Landfill tipping fees in 2004 averaged $34.29 per ton of waste (an increase of roughly 425% since 1985) with a low of $24.06 per ton in the South Central Regions of the U.S. to a high of $70.53 per ton in the Northeast (Repa 2005:2). In the upside-down world of environmental economics, waste disposal can be more costly than purchasing coal. In June 2006, the average cost of coal was $50.50 per ton, which is less than waste disposal costs in the Northeast.[10]

THE ASSAULT ON FRESH WATER RESOURCES

Water is our most precious resource. Like energy, the management of water resources is directly related to sustainability. Clean, freshwater sources are increasingly important for growing populations. Freshwater is being consumed and supplies have become constricted. Freshwater resources are under assault from growth in demand, contamination and environmental calamities. There are notorious examples. The diversion of feed-waters to the Aral Sea in Uzbekistan and Tajikstan, once the world's fourth largest freshwater lake, reduced its water volume to only 20% of its original size in just 40 years.[11]

Estimates indicate that 1.1 billion people worldwide do not have access to clean water.[12] Clean water is defined as "water from a pipe, public tap, borehole, protected dug well, protected spring or rainwater collector."[13] According to a joint report from UNICEF and the World Health Organization (WHO), "the situation is becoming particularly acute in urban areas, where rapid population growth is putting great pressure on the provision of services and the health of poor people."[14] The extent of pesticide contamination of groundwater and produce throughout Asia is unknown.[15] Using local water sources for processing, Coca-Cola and Pepsi have both been alleged to use unacceptable levels of pesticide-tainted water for products sold in India.[16]

In the U.S., a country blessed with abundant water resources, drought conditions have plagued the west for the last decade, and now have extended to one-third of the country. From June 2006 to May 2007, California and Nevada recorded their driest year since 1924.[17] In Florida, a record low water level was recorded for Lake Okeechobee, the state's largest freshwater source.[18] In Alabama, commercial catfish farms are under stress, and wheat and corn crops are failing due to lack of rain.[19] The water tables of aquifers in many areas of the U.S. are dropping because water is being consumed faster than they can be recharged. The Ogallala Aquifer, an important freshwater formation in Nebraska, Wyoming, Kansa, Colorado, Oklahoma, New Mexico, and Texas has been declining in the North Plains Groundwater Conservation District by .53 meters (1.74) ft annually.[20]

Contamination of water resources in the U.S. is commonplace. Streams in the Eastern U.S. have been contaminated with mercury from power plant emissions. During periods of heavy rainfall, storm water run-off from parking lots and roads goes untreated, contaminating water supplies with oils and solids. Erosion from construction sites also causes damage to fish and wildlife habitats and contaminates drinking water supplies. Water supplies at nine U.S. military sites has been contaminated for years, knowingly exposing service personnel to chemical toxins, some of which cause cancer.[21] At Camp Lejeune in North Carolina, an estimated 1,000,000 people over three decades "drank and bathed in water contaminated with as much as 40 times the amount of toxins allowed today."[22] Toxins found in the water included trichloroethylene, a degreasing solvent, and dry-cleaning agents.[23]

China, a country bent on economic development, now has water-related environmental problems that may never be resolved. Amazingly, 383 of China's 661 major cities lack sewage treatment facilities leaving 70% of its rivers "severely polluted" (Schell 2007:84). In the coal producing Shanxi province many rivers have dried up, and 80% of those that still flow are rated by Chinese officials as "unfit for human contact" and cannot be used for irrigation or industrial purposes (Schell 2007:84). The diversion of water for irrigation and other purposes has left the Zhi, Ming, Anyang, Sha, Zhang and Huai and "many other legendary rivers... almost bone dry" (Schell 2007:84). Regardless, more water is needed for commercial and industrial development. Since per capita water consumption in China is only 1/8 of U.S. consumption, water conservation programs similar to those being proposed in the U.S. would be ineffective. Desperation fuels creativity. Attempts to access new sources of water have caused groundwater tables to drop in some areas of the county. In addition to water diversion projects, such as the $62.5 billion North Water Transfer Project which will move water (now mostly polluted) from the once mighty Yangtze River to the "breadbasket of China," the Chinese government is investing in cloud seeding projects using aircraft, rockets

3-1. Air Pollution in the Los Angeles Basin

and artillery shells (Schell 2007:84).

Continuing to pollute fresh water endowments, rendering them un-drinkable, means that they must be purified to be potable. Otherwise, we must desalinate seawater. Desalination technologies have been proven to be workable but are expensive—costing over three times more than purifying fresh water. A plant proposed in Texas will cost $150 million to build and another proposed in South Africa will cost $220 million.[24] The process of removing salt from sea water and purifying it is energy intensive. Today, only .1% of all drinking water has been made from sea water.[25]

AIR QUALITY

Air quality is inextricably linked to sustainability, issues include ozone depletion, increases in greenhouse gases, pollution induced by human activities, and pollution caused by natural events. Air pollution from coal-fired power plants causes acid rain (e.g., sulfuric and carbonic acids), mercury contamination, and dispersion of particulates.

At one time air quality standards were not nearly as stringent as they are today. Chicago's "Little Hell" in the early 1900s, the soot-laden air over the English cities of London and Birmingham in the 1920s, and the highly leaded toxic soup over Los Angeles in the 1950s, provide notorious exam-

3-2. Air Pollution and Haze over Santiago, Chile

ples. The cities of Santiago, Chile and Mexico City are renowned for their poor air quality.

The U.S. Clean Air Act of 1990 established national air quality standards. States that did not comply risked loosing federal highway funds. The efforts of the Environmental Protection Agency have been successful in improving outdoor air quality. Today air quality standards are more stringent, regulations are more widely enforced, and air quality has improved. Allowable tailpipe emissions of nitrogen oxides have been reduced from 1.0 grams per mile in 1981 to .07 grams per mile in 2004.[26] In the U.S. 70% of the 238 million vehicles on the roadways were built in 1996 or earlier and they are responsible for a quarter of the total miles driven—yet they account for two-thirds of automobile-generated air pollution.[27] We have also become more knowledgeable about the health effects of poor air quality in cities and their regions. Regardless, air quality continues to be a problem in metro areas throughout the U.S. and in other countries.

CLIMATE CHANGE

Regardless of the label you call it, "greenhouse effect," "climate change," "global warming," "planetary emergency," or "atmospheric pyrogenesis," the Earth's atmosphere is warming, and man's activities are partly responsible. Recognition of this phenomenon is not at all new. Climatic changes contributed to desertification and loss of topsoil in ancient Mesopotamia, causing agricultural production to decline to the point that the region was unable to sustain its population. In 1827, Jean-Baptiste Fourier observed that the atmosphere was trapping heat from the sun like a "glass vessel" (greenhouse effect).[28] In 1896, Svante Arrhenius quantified the extent of Earth's warming that was resulting from carbon dioxide emissions (global warming).[29]

Fast-forwarding about a century or so, there has recently been a shift from lack of acceptance to a greater understanding of the effects of global climate change, and the need to reduce emissions of greenhouse gases (Smith and Warr 1991). A scientific consensus has emerged regarding this once-contentious issue. The scientific community, once divided, has accumulated overwhelming evidence that the earth's atmosphere is warming. Politicians have taken notice. According to Gore (2007:205), "There is no longer any credible basis for doubting that the earth's atmosphere is heating up because of global warming. Global warming is real."

Given no major changes in the trend of emitting increasingly larger amounts of greenhouse gases, climate change will cause further increases in global temperatures. According to a report by the Intergovernmental Panel on Climate Change, human-caused warming will "continue for centuries... even if greenhouse gas concentrations were to be stabilized."[30] Climates are changing at a surprisingly rapid rate. The Earth's atmosphere is storing more heat and air temperatures and sea levels are rising. Natural disasters, such as more intense storms, drought, fires, and firestorms will become more widespread.

We may have already waited too long to completely stop global warming from further environmental damage. What options remain? According to John Holdren, co-chair of the National Commission on Energy Policy, "We have three choices: mitigation, adaptation or suffering... and we are already starting to do a little of each."[31]

Corporations are taking notice and businesses are changing their behaviors. Lehman Bothers published a report in February 2007 entitled "The Business of Climate Change." According to the report, the planet's well-being has evolved from a "fringe concern" of the scientific community to a "central topic" for corporate CEOs and their investors (Shell and Krantz 2007). Jim Rodgers, Chairman of Duke Energy, has been quoted as saying, "the science of global warming is clear... We know enough to act now. We must act now."[32] Edward Kerschner, a strategist for Citigroup, expresses it differently: "We are reaching a tipping point when it comes to climate change" (Shell and Krantz 2007). His research resulted in "Climatic Consequences," a report that focuses on 74 companies positioned to profit from changes in weather patterns.

So... why is there a sudden interest in climate change? Evidence shows that the size of the polar ice caps is decreasing and that glaciers are retreating in Alaska and Switzerland. A new waterway will likely open between the North Polar ice cap and the North American continent. The Antarctic ice sheet is shrinking with large pieces "breaking off and forming free-floating iceburgs."[33] According to Roland Psenner of the University of Innsbruck, glaciers in western Austria have decreased at the rate of 3% annually. If the trends continue, he concludes that "there won't be any more glaciers in the year 2050 except for high ones that lie above 4,000 meters" (approximately 13,000 feet) in elevation (Healy 2007). Climate change is raising ocean water levels, accelerating desertification, and having microclimatological impacts throughout the world. The results of man-made global warming can be measured to some extent and are likely to worsen.[34]

There are multiple causes for climate change, some of which are due to the activities of growing human populations. For a number of years, it has been widely accepted that current levels of atmospheric emissions of greenhouse gases are generating concern worldwide (Burgess and Jenks 2000:338; Chasek 2000:302-304; Roseland 1998:99-104). Some causes are difficult to control. Soot and dirt falling on snow and ice packs reduce solar reflectivity, absorb heat, and contribute to global warming.[35] The pollutants known as "greenhouse gases" that contribute to the "greenhouse effect" include CO_2, CH_4, CFCs, halons, N_2O, and peroxyacetyinitrate, all of which are produced by both industrial and manufacturing activities (Dincer 1999:848). The greenhouse effect refers to the trapping of heat in the Earth's atmosphere. The heat in the atmosphere is increased by the addition of these chemical compounds, which in turn further increases atmospheric temperatures.

Power plants and vehicles account for a significant portion of Sulfur Oxides, Carbon and Nitrogen emissions that are added to the atmosphere. Regarding carbon dioxide emissions this is indisputable:

- An estimated two-thirds of all carbon dioxide emissions have been generated by the U.S. and western European countries.[36]

- The amount of CO_2 held in the atmosphere has risen from 280 ppm prior to the industrial revolution to 380 ppm today and continues to increase (Begley 2007:65).

- Molecules of CO_2 currently in the atmosphere can remain suspended for up to 200 years.

- Rather than decreasing CO_2 emissions, the world's economies are substantially increasing them.

Poor countries will fare the worst from the impact of global warming. The 840 million people in Africa face problems from drought and loss of water sources due to changes in climate. Since 1900, their continent has accounted for less than 3% of global carbon emissions.[37] While the average Briton produces 48 times more carbon dioxide than the average Bangladeshi, residents of low-lying areas of Bangladesh face the greatest threats from climate change, including loss of land and increased deaths from diseases such as malaria.[38] Bangladesh is a country that "will simply disappear" if current "sea level predictions are true."[39] Emissions from transportation-related sources account for roughly one-fifth of total carbon emissions and are increasing rapidly in developing countries such as China and India (Kaihla 2007:70). In addition, CO_2 emissions and overall resource use are highly correlated (Holland, Lee and McNeill 2000:87). This means that increases in resource consumption are often directly related to increases in CO_2 emissions.

Avoiding further impacts of the greenhouse effect will require a 75-80% reduction in global CO_2 emissions (Daly *et al.* 1991:56)—a challenging task as 82% of anthropogenic greenhouse emissions come from the burning of fossil fuels.[40] U.S. emissions have risen from a total of 6 billion metric tons of CO_2 annually in 1992 to 7.9 billion metric tons annually in 2005 (Begley 2007:65).

According to the U.S. Environmental Protection Agency (USEPA 2000), the ten warmest years of the 20th century all took place in its last 15 years, with 1998 being the warmest on record. The three years with the warmest recorded surface temperatures all occurred during the period from 1998 through 2003, with 2003 being the third warmest.[41] More recently, 2006 was the sixth warmest year worldwide; it was also the warmest year recorded in the U.S.[42] Overall, the Earth's temperature has risen by 0.6° Celsius (1.1° F) since 1900, causing an estimated 20 cm (7.9 inch) rise in sea levels (Colonbo 1992:3-14). If no action is taken, carbon dioxide levels are projected to double and raise global temperatures by an additional 1° C by 2025 and 1.8° to 3.5° C (3° to 6.3° F) by 2100 (Chourci 1993:51; The Courier-Journal 2002:A4; Union of Concerned Scientists 1999:4).

It seems our intent is to increase greenhouse gas emissions rather than reduce them. World emissions of CO_2 totaled 26 billion metric tons in 2004, and according the International Atomic Energy Agency (IAEA) are expected to grow to 40 billion metric tons by 2030.[43] According to a recent report by the Intergovernmental Panel on Climate Change, full stabilization of atmospheric greenhouse gas emissions by the year 2030 would cost an extra $100 (U.S.) per ton.[44]

The potential impact of climate change is the new shot heard around the world. Global warming could cause a series of domino effects, forcing the release of greenhouse gases stored in the Arctic tundra and causing indigenous plant-life elsewhere on the planet to become stressed. Rising temperatures in metropolitan areas due to the urban heat island effect will increase air conditioning loads along with the demand for electrical energy. This will further raise CO_2 levels, and create a spiraling cycle of even greater energy use—and even more temperature increases. Shrinking glaciers and ice caps can reduce the amount of solar radiation reflected into the atmosphere, expanding areas of exposed land-mass, increasing Earth's heat absorption, and causing even higher temperatures.

There is growing international concern that changes in global temperatures may trigger a series of cataclysmic weather-related events such as longer periods of drought, more wildfires, flooding, and stronger storms. In the U.S., it is predicted that the American Southwest will get drier (it has experienced below average rainfall since 1999), causing Dust Bowl-like droughts in the plains and reducing the flow of the Mississippi River and its tributaries.[45] At some point, the warming of the ocean might trigger a slowing—but not a shutdown—of the "Atlantic conveyor" which cycles heat and water throughout the Atlantic Ocean. If this were to happen, the Gulf Stream could fail to bring warmer water to Iceland and Europe, causing those areas to experience colder conditions. One team of scientists monitoring this situation has found that Atlantic meridional overturning circulation patterns have weakened by 25-30% over the last decade at 25° North Latitude.[46] To study this phenomenon, the team has installed arrays of instrumentation to measure the North Atlantic's temperature, salinity, currents, and pressure and plan to collect data through 2008. As the flow of the North Atlantic Drift, the northward branch of the Gulf Stream, begins to taper off, Greenland's ice sheet (by melting up to 50% of its mass) will gradually release fresh water over several decades into the North Atlantic, slowing the ocean conveyor and temporarily offsetting the effects of global warming in Northern Europe.[47] According to Bogi Hansen, a scientist researching this, the process "would mitigate the global warming" rather than create conditions for "abrupt and cataclysmic cooling."[48] However, ocean levels will rise and marine life will likely be affected due to the redistribution of oxygen and nutrients in the deep ocean.

A quarter of China is now desert. The country already faces difficult challenges. Environmental problems include air and water pollution and further temperature increases could have fatal effects. "The most direct impact of climate change will be on China's grain production," says Luo Yung of the county's National Climate Center.[49] China is already coping with "severe water shortages... the results of pollution and over-use of surface water resources."[50] In addition, it is feared that "rising temperatures will aggravate... extreme weather events."[51]

The solution to global warming in the long term is easy enough to describe—simply reduce the amount of greenhouse gases released into the atmosphere. Reducing these emissions appears to be impossible in the short term for the U.S. and China, the two largest emitters, and also for many other countries. Burdensome tradeoffs would be required, such as reducing electrical generation from coal, reducing oil consumption, using more alternative energy, and implementing fuel substitution technologies. Coal emissions are difficult to address, since there are roughly 1,700 coal fired plants in the world that produce comparatively inexpensive electricity. More are scheduled for construction in the near future and few of these will use "clean coal" technologies or plan to sequester the carbon that they generate.

At the dawning of the twentieth-first century, climate change is the most important environmental problem we face. Climate change transcends the territorial borders of nations, and affects both developing and industrialized counties. For the most part, U.S. policymakers and politicians at the national and state levels seem ill-equipped and have failed to develop public policies regarding climate change. Existing political structures appear unable to handle these issues alone, and few existing foreign bureaucracies have been created to deal with problems of this sort (Cusimano 2000:7). This is changing, but the effects of policies to date have yielded only marginal improvements.

THE KYOTO PROTOCOL

The United Nations Climate Change Summit in Kyoto, Japan in 1997 was a follow-up to the Framework Convention on Climate Change at the Earth Summit in Rio in 1992. The Kyoto Summit brought 160 nations together in an effort to reduce carbon emissions and devise ways to mitigate global warming. In order to preserve their interests and protect their sovereignty, nations brought their own environmental agendas to these events. The countries involved offered varying perspectives on the issues, perspectives based on their potential costs, benefits, resources, and differences in scientific knowledge. At Kyoto, the success of the negotiations depended on the individual nations' willingness to comply with the agreement of the treaty. This proved to be a difficult task since each nation presented different methodologies to reduce emissions based on their asymmetries of costs, benefits, and resources.

The agreement reached is known as the Kyoto Protocol. The Protocol required commitments from the 39 industrialized countries to reduce emissions of greenhouse gases to a level 5% below their 1990 emission levels by 2012 (Kraft 2004:274). Canada agreed to reduce emissions by 6% below 1990 levels. Included in the Kyoto Protocol to reduce greenhouse gases is the Clean Development Mechanism (CDM). The CDM provides incentives to develop a global carbon trading market through which companies in the industrialized world "earn credit not for reducing their own emissions but for investing in energy efficiency improvements in the developing world."[52]

Projects using carbon "off sets" are becoming popular. Hundreds of projects, mostly in Brazil, China, and India, when fully implemented, will reduce CO_2 emissions by the equivalent of 115 million tons annually.[53] Rajesh K. Sethi, Secretary of India's CDM Authority, calls the program "one of the most successful ways we have found to reduce greenhouse gases."[54] The Kyoto Protocol, however, is not without its critics. Canadian Prime Minister Stephen Harper has as referred to Kyoto as a "socialist scheme designed to suck money out of rich countries."[55]

The U.S. emissions reduction target proposed by Kyoto for 2012 was a 7% reduction below 1990 levels (Kraft 2004:274). Meeting this goal would not have seemed impossible if energy consumption and carbon emissions had been stabilized at 1990 levels.

To meet this goal, the U.S. would need to increase automobile fuel economy standards by 10% and reduce coal use in power plants by roughly 10%. In 2006, U.S. manufactured vehicles had an average fuel efficiency of 10.8 km/l (25.4 mpg) while those manufactured by Toyota in North America averaged 14.7 km/l (34.7 mpg)—a 36% improvement in fuel economy.[56] Since most coal is used to generate electricity, a CO_2 reduction of the magnitude required could be achieved by improving energy efficiencies and with advances in residential and commercial building codes.

While the 15 members of the European Union (out of a total of 169 countries worldwide) have ratified the Kyoto Protocol,[57] the United States has not. President Bill Clinton was a signatory to the Kyoto Protocol, but President George W. Bush withdrew from it, claiming that it was fatally flawed because of its adverse economic impact and provisions that excluded developing countries from complying with its standards. G. W. Bush's campaign promise in 2000 assured the nation that "CO_2 would be regulated as a greenhouse gas" (Gore 2007:194). This changed quickly and this promise remains unfulfilled. Quin Shea of the Edison Electric Institute stated bluntly in a speech in 2001: "Let me put it to you in political terms...

the President needs a fig leaf. He is dismantling Kyoto" (Gore 2007:194).

In January of 2007, the President issued an Executive Order, mandating that federal agencies in the U.S. reduce energy use by 3% annually, or a total of 30% through 2015.[58] Fleet consumption of petroleum-based fuels was ordered to be reduced by 2% annually over the period and non-petroleum-based fuels are to be increased by 10% annually.[59] Section 3(a) of the Order requires agencies to implement "sustainable practices for energy efficiency, greenhouse gas emissions avoidance or reduction, and petroleum products use reduction."[60] While the Order was mandated, funding for implementation was neither identified nor provided.

While acknowledging the existence of climate change, the G. W. Bush administration continues to reject mandatory limits on greenhouse gases.[61] To date, the U.S. government has not yet offered a workable alternative to the Kyoto Protocol. CO_2 is not regulated as a greenhouse gas in the U.S. For its part, the G.W. Bush administration favored voluntary reductions in greenhouse gases and further research to study the issues of global warming and climate change (Kraft 2004:51). His administration claimed that adherence to the Protocol would cost the U.S. as much as $400 billion to implement. He estimated that 4.9 million jobs would be eliminated (The Courier-Journal 2002:A4). The perception today is that the economic payoffs for the U.S. as a free rider were simply greater than supporting the Kyoto Protocol and the coalition against climate change (Zhang 2004:443).

Estimates of economic and employment losses of the magnitude suggested by the G.W. Bush administration appear in retrospect to be patently unsupportable and politically motivated. The projected costs are more than half of the total U.S. expenditures for energy in the year 2000. Predicted job losses are a multiple of those estimated by the same U.S. administration to have been lost during the 2001 recession. Interestingly, the employment increases created by implementing sustainable development policies were overlooked in the analysis. Also omitted from the analysis were estimates of the employment that would be generated by developing renewable energy technologies, energy services, environmental mitigation efforts, and the exportation of equipment and engineering expertise that would at least partially offset any anticipated job losses. The net gains to the U.S. economy resulting from energy efficiency improvements at the macroeconomic scale were also overlooked. Finally, the analysis failed to consider the national security implications of energy supplies, the impact on trade deficits, the economic costs of currency devaluations, and the financial risks associated with escalating energy costs.

The experience of Britain provides a real-world indicator of the impact on employment. Rather than causing economic harm and significant employment losses, complying with the Kyoto Protocol has had the opposite effects. According to former Prime Minister Tony Blair, "there are great commercial opportunities to be had" (Breslau 2007:51). More than 500,000 "clean tech" jobs have been created since Britain began complying with the Kyoto Treaty (Breslau 2007:51). Many of these jobs are technology-oriented and have resulted in higher-than-average pay scales.

TXU, Texas' largest utility, provides an example of this distinctive shift in direction and the opportunities it can offer. As TXU faced concerns by environmentalists regarding its plans to construct 11 coal-fired, electrical-generating plants, a leveraged buy-out by Texas Pacific Group and Goldman Sachs was arranged—and it succeeded. After the deal was completed, plans for eight of the coal-fired plants were withdrawn. There were concerns by the utility's new owners that emission caps would eventually create substantial penalties for carbon-based generating facilities. As an alternative, TXU instead decided to invest $400 million on renewable energy projects and other conversion initiatives (Davison 2007).

Despite this dramatic change in direction for a single company, approximately 150 coal-fired power plants are planned for development in the U.S. However, while the U.S. Congress is seeking workable ways to reduce greenhouse gases, a program established in 1935 actually encourages rural electrical cooperatives to construct coal-fired plants. Many of the service territories of the co-ops, which were once rural areas, are now actually suburban areas of large cities. The program from the Agriculture Department's Rural Utilities Service provides subsidized, low-interest, federal loans to develop power plants. Rural electric co-ops plan to use the program to invest $35 billion in conventional coal plants over the next 10 years.[62] Rural co-op electrical production is mostly derived from coal, and electrical demand in rural areas is increasing at twice the national rate.[63] Since these coal-fired plants generate greenhouse gases, the Agriculture Department's program effectively provides financing to "offset all state and federal efforts to reduce U.S. greenhouse-gas emissions" over the next 10 years.[64] There are no comparable federal programs that subsidize long-term financing of electrical energy production from alternative energy sources.

Environmental problems are caused by the extraction of coal. Mining coal is a major industry and supports local economies worldwide.

There are many ways to mine coal such as deep mining and strip mining. The most controversial extraction technique is called "mountaintop removal," a method whereby the tops of hills are blasted away to expose the coal seams. Any debris from the removal is pushed down the hillsides, filling valleys and covering streams. It is used in states such as Kentucky and West Virginia—and natural ecosystems are destroyed in the process. However, the environmental destruction that accompanies coal mining generally devalues the surrounding area. Once the coal has been mined out (typically 10 to 15 years) the communities are left with unemployment and lower-quality ecosystems.

But no amount of environmental remediation will ever restore the quality of the ecosystems which existed prior to mining. According to the U.S. EPA, mountain-top removal accounted for less than 5% of U.S. coal production in 2001, yet hundreds of thousands of acres of forests have been destroyed and hundreds of mountains have been leveled.[65] Coal bed methane gas is also released into the atmosphere during this mining, contributing to greenhouse gas emissions.

Oil drilling also impacts the environment. Drilling for fossil fuels varies with location and drilling practices, but may include any of the following: disturbances to wildlife and vegetation, the release of drilling fluids, contamination of groundwater and surface water, seismic disruptions, and gaseous emissions.

Transporting these fuels creates opportunity for catastrophic accidents and terrorist acts. Oil spills have caused massive damage to the environment all over the world, whether accidental or intentional. Incidents include the 400,000 barrels of oil spilled from a major pipeline in the Amazon region of South America, the 38,000 barrels spilled in Nigeria, and the 28,500 barrels spilled in Columbia.[66]

The flaring of gas wells poses serious local environmental problems. Methane emissions from oil and gas operations, along with coal bed methane and agriculturally produced methane, make up an estimated 9% of total U.S. greenhouse gas emissions.[67]

There are alternatives such as nuclear power that have a greater role to play. Generating electricity from nuclear energy production generates no greenhouse gases except in fuel extraction and processing. According to Jerry Paul at the Howard Baker Center for Public Policy, "It is all about climate change and emissions. It is about economics and the recognition that nuclear power has the lowest operating cost of any form of (electrical) baseload generation."[68] TVA's Brown's Ferry Unit #1 nuclear plant came on line in 2007, after $1.8 billion in renovations (it was mothballed after a fire in 1985). It is the first "new" reactor since 1996 in the U.S. The U.S. Department of Energy (DOE) estimates that an additional 50 nuclear power plants will be required by 2030.[69]

While technologies are available, the goal of reducing greenhouse gases continues to be elusive. The bottom line is that the actual result of the Kyoto Protocol has been mixed—and in some cases ineffective. Most nations have yet to baseline their production of global warming gases. Data from a United Nations report indicated that from 2000 to 2004, 34 of the 41 countries it monitors (excluding most developing countries) had actually increased greenhouse gas emissions.[70]

Despite the past reluctance of the U.S. federal government to take broad-based initiatives on climate change, local and state governments are moving ahead. There are 16 states, plus the District of Columbia, Puerto Rico, and Guam that require greenhouse gases to be regulated like other atmospheric pollution.[71] Developers of large projects in Massachusetts are now required to "undergo state environmental reviews to assess how they contribute to global warming."[72] By June 2007, a total of 527 mayors of cities in the U.S. have signed the Mayor's Climate Change Agreement, in which they agree to meet *or exceed* the Kyoto Protocol's requirements for reducing global warming.[73] Actions being taken include instituting anti-sprawl land-use policies, developing urban forests, establishing public information campaigns, and encouraging state governments to become more involved.[74] In California, the 2005 Climate Action Initiative establishes ambitious greenhouse gas reduction targets for the state. It was followed in 2006 by the California Solar Initiative—the nation's most comprehensive solar energy policy, which allocates $3.2 billion for energy rebates over the next decade.

MONTREAL PROTOCOL

The ozone layer in the upper atmosphere protects life from the harmful effects of ultraviolet radiation. Processes that deplete the ozone layer magnify these effects. While Sweden became the first nation to ban aerosol sprays thought to damage the ozone layer, the Antarctic ozone hole was actually not discovered until 1985.[75] Human effects include severe sunburn and epidermal cancers.

One of the biggest obstacles in managing environmental problems such as the ozone layer is that many environmental issues are seen as international in scope rather than local. Existing institutions and foreign bureaucratic structures cannot handle and were not created to deal with these problems effectively (Cusimano 2000:7). Without the institutional capacity in place necessary to handle problems on an international scale, inertia and inaction are often the result. It was U.S. and British scientists who first determined that there was an opening in the ozone layer over the Antarctic. The effect of ozone depleting gases which contributed to the "ozone hole" in the atmosphere provides an example of a success story in which international cooperation has yielded positive results.

The Montreal Protocol on Substances that Deplete the Ozone Layer, known as the Montreal Protocol, was an important international environmental agreement. It was signed by 160 countries in order to control the production of ozone-depleting compounds. By establishing a timetable to phase out the production and use of ozone-depleting substances, it was hoped that further damage would be avoided, and that the ozone layer could be repaired.

Chlorine and bromine containing gases, such a chlorofluorocarbons (CFCs) and halon, are known to destroy ozone molecules in the troposphere. According to the terms of the Montreal Protocol, the manufacture of halogenated CFCs (known as Group I substances) would be reduced to just 50 percent of 1986 levels by 1999. In addition, manufacture of halon gases (known as Group II substances) were frozen at 1986 levels by 1992.

The Montreal Protocol incorporated a procedure for reviewing and refining the agreement if scientific information changed (Andrews 1999:328; Elliott 1998:56). Since the treaty was signed by the participating nations in 1987, and went into effect January 1, 1989, there have been five different revisions of the Protocol to update and strengthen the phaseout timetables of ozone-depleting gases, including a complete phase-out of CFCs, halons, and carbon tetrachloride by 2000 (Elliott 1998: 58).

The Montreal Protocol is widely viewed as one of the most successful environmental agreements ever adopted (Kraft 2004: 270). Regardless, the Protocol has flaws that have undermined its success. One clause effectively postponed implementation by including a 10-year grace period for compliance by developing counties, as long as their consumption of ozone-depleting substances did not exceed 0.3 kg per capita. Many of the developing countries chose not to participate in the initial negotiations of the protocol, although some eventually ratified the treaty (Elliott

2004: 56-57). While there are voluntary reporting provisions at national levels, there are no enforcement or verification procedures through international oversight (Elliott 1998:57). Since the initial protocol was approved, subsequent amendments adopted after the protocol's approval have yet to be ratified by many countries.

As a result of the Montreal Protocol, the ozone layer is improving. Measurements of ozone-depleting gases indicate that they peaked in 1994 and have since declined. By 2006, ozone levels had improved by 22% in the mid-latitudes. There was a 12% reduction in conditions that had resulted in the development of the ozone hole.[76] If present trends continue, the recovery of the mid-latitude ozone layer will occur about 2045-2055, while the full the recovery of the ozone layer is not expected until 2075-2080.[77]

While this is encouraging, the ozone layer recovery may be in jeopardy due to increasing levels of greenhouse gases in the atmosphere. We know that the problems associated with greenhouse gas emissions (e.g., global warming) may not be resolved during this century, regardless of any presently available mitigation actions that might be undertaken. This is unsettling to atmospheric scientists—recent evidence suggests that greenhouse gases also affect the ozone layer. By warming the atmosphere, greenhouse gases such as methane and CO_2, may cause further damage to atmospheric ozone molecules. According to a National Aeronautics and Space Administration (NASA) study, "Climate change from greenhouse gases can also affect ozone by heating the lower stratosphere where most of the ozone exists. When the lower stratosphere heats, chemical reactions speed up and ozone gets depleted."[78] While there is uncertainty, computer models indicate that an increase in greenhouse gas emissions may impede and delay the recovery of the ozone layer.

The U.S. Climate Action Partnership, which includes corporations such as BP (British Petroleum) America, Duke Energy, Caterpillar, Alcoa, PG&E, DuPont, and GE (General Electric), issued a statement that "we can and we must take prompt action to establish a coordinated, economy-wide, market-driven approach to climate protection."[79] Whether this approach alone is workable remains untested. To be truly market-driven, externalities must be fully valued and national markets must adjust to economic impacts. The efforts to eliminate ozone-depleting gases were not driven by market forces alone, but primarily by scientific research, indisputable evidence regarding the potential harm to human health, political consensus, and broad-based international cooperation. But such a combination of forces is often elusive and difficult to attain.

SPECIES DECLINE AND EXTINCTION

One goal of environmental stewardship is to maintain the planet's rich biodiversity. Many species can adjust to environmental changes by means of competition, adaptation, evolution or a combination of these—if they have time for adjustment and access to resources that they require to survive. Species unable to adapt to rapid environmental changes fall into decline and may become extinct. According to some biologists, we are experiencing the greatest mass extinction of species in 65 million years. This is questionable because we do not know how many species existed on our planet at that time or how many have become extinct. Another problem is that we have no count of how many currently exist, so we cannot accurately determine the present rates of extinction. According to Thomas Orrell, a biologist at the Smithsonian Museum of Natural History, "many are surprised that despite two centuries of work by biologists and the current world interest in biodiversity, there is presently no comprehensive catalog of all known species of organisms."[80] The estimated number of species on the planet varies widely—from 1.4 to 6 million. By March of 2007, a worldwide initiative to develop such a catalog of species has reached a count of 1,009,000 and the 3,000 or so biologists involved in the project expect the total to reach 1,750,000 by time it concludes in 2011.[81]

Despite the lack of a complete inventory of species, researchers continue to offer dire predictions. There is ample evidence to conclude that species extinction is a growing environmental problem. Estimates of species extinction rates vary from 18,000 to 55,000 species annually, with two-thirds of the Earth's remaining species likely to be extinct by the year 2100 (Rifkin 2006). There are a large number of animal species (e.g., tigers, panda bears, gorillas and others) whose populations are threatened or in decline.

The oceans absorb tremendous amounts of carbon dioxide. The world's oceans are now over 30% more "acidic than they were before the Industrial Revolution."[82] In addition to atmospheric consequences, increased acidification of the oceans due to higher CO_2 levels will likely "wreak havoc" on shellfish and coral colonies, causing disruption to food chains (Vergano and O'Driscoll 2007:D2). Acidification damages reef struc-

tures. Coral reefs, those rain forests of the seas, are particularly suscepti-
ble to damage from warming. Since the 1980s, an estimated one-fifth of
the world's coral reefs have already been eliminated—most likely caused
by a combination of factors including warming conditions.[83] Worldwide,
70% of the deep sea corals are likely to be damaged by conditions caused
by acidification and increased temperatures.[84] Coral bleaching can occur
when water temperatures increase by just 2°C for a period of two weeks.
On land, global warming could reduce the number of species in the Brazilian rain forests by as much as 30% (Watson 2007:A7). Recent studies indicate that warming conditions have directly contributed to the extinction
of at least 70 animal species, and have affected as much as 59% of Earth's
wild species (Vergano and O'Driscoll 2007:D2).

Man's activities have contributed to species decline in other ways.
Over-fishing of the world's oceans is one example. Bottom trawling, a
fishing technique that involves dragging mile-long nets along the sea floor
to capture sea life, disrupts food chains. In 2007, over 20 nations agreed to
discourage the practice in the South Pacific. It is paradoxical that despite
technological improvements, fish are becoming harder to find and catch.
The $158 billion commercial fishing industry has noted a steady decline
in the number of fish being harvested, with one recent study projecting a
collapse of harvested fish populations as early as 2048 (Durst 2007:78).

The impact is most difficult to assess in migratory fish species. The
migratory bluefin tuna, prized for its popularity in sushi and sashimi
dishes, is in decline. Once abundant, the high prices paid for this species caused devastating pressure on the bluefins. The Japanese consume
roughly 80% of the world's catch annually. The cost of a single tuna (they
can reach 4 meters in length and weigh 650 kg) is now higher than the cost
of most new cars (Pitzer 2007:b3).

There are indications that some species of fish are reaching their tipping points, the points where their populations may not recover. A number
of species of shark, relished for their fins, are also becoming endangered.
Coral reefs are being over-fished in many locations throughout the world,
causing the depletion of many species. A recent study that analyzed data
from 49 small island nations and territories found that 55% of their reef
fisheries now have production yields below sustainable levels.[85]

Populations of other species of animals are also declining. According to a report to be issued by the Intergovernmental Panel on Climate
Change, "Changes in climate are now affecting physical and biological
systems on every continent."[86] One result is changes in ranges and habi-

tats. Loss of habitat contributes to species decline and extinction. Climate change is occurring in the Arctic, causing polar bear populations to decline and populations to migrate as the ice caps recede. Extinction is possible by 2050 and the U.S. government wishes to add polar bears to the Endangered Species List.[87] "Hundreds of species have already changed their ranges, and ecosystems are being disrupted... it is clear that a number of species are going to be lost," according to University of Michigan scientist Rosina Bierbaum.[88]

The population of the African penguin (Spheniscus demersus) has declined from 1.5 million a century ago to only 120,000 today.[89] Living on the coasts of Namibia and South Africa, this species survives on anchovies and sardines. Seeking preferred colder water, the remaining sardine populations are migrating to Cape Agulhas, the southernmost point in Africa, which is reducing the food supply of the penguins.[90] Deserting hotspots in the oceans, the decline and migration of the sardines is most likely caused by commercial fishing and changes in the Benquela Current, an ocean flow of frigid water.[91] These changes in the current are likely due to a combination of natural causes and temperature changes in the current. Without the sardines as a food supply, the landlocked penguins are in peril.

Extinctions can be localized, meaning that as climates are modified, areas supporting certain species are no longer able to do so. In Hong Kong, a city whose temperature has increased 1°C (1.8°F) in the past 60 years, local warming could cause many of their unique species to become extinct by the end of the century. A colony of Palmation pelicans that wintered there for the last decade disappeared after their population declined precipitously.[92] According to Cheung Ho-Fai, Chairman of the Hong Kong Bird Watching Society, "The city may be getting too warm for these birds."[93] In the U.S., suburban sprawl, species invasion and climate change have been cited as causes for the decline of 20 common American songbirds including the meadowlark and the whippoorwill.[94] According to Greg Butcher of the American Audubon Society, "Most of these we don't expect to go extinct... we think they reflect other things that are happening in the environment that we should be worried about."[95] The decline of the bobwhite in the U.S. has been most precipitous—from a population 31 million in 1967 to 5.5 million today.[96]

There are other pressures on wildlife populations that may not be related to climate change. In Puerto Rico, the sounds of nocturnal frog populations in the rain forests are diminishing. In some areas of the Caribbean, certain species of turtles, prized for their shells and food value, are having difficulty maintaining their populations.

Hydroelectric dams have many benefits including retaining water to reduce water shortages, providing recreational opportunities, and generating electrical power. Despite these benefits, many hundreds of small dams in the U.S. have been removed or are planned for removal. Surprisingly, this is sometimes due to efforts to mitigate species decline. The decline of salmon in the Northwestern U.S., where very few new dams are now being constructed, serves as one example. Backed by both environmental and commercial interests, there is a movement to restore breeding grounds for salmon. The Elwha River basin in Washington State once supported over 1,000,000 salmon of 10 species, but now supports only 20,000 salmon (Ritter 2007:8B). On the Elwha River, the 64 meter (210 foot) tall hydroelectric Glines Canyon Dam is among the five dams scheduled to be demolished over the next few years in an effort to restore salmon to the area (Ritter 2007:8A). Within 20-30 years after the river is restored, there are hopes of growing the salmon population to over 400,000, creating new opportunities for recreation and sportsfishing. Other dam removal initiatives are being planned or considered in Southern California[97] and Oregon.

SPECIES INVASION

Ecosystems adjust to environmental changes by natural selection, an evolutionary process involving generational adaptation. Species in natural systems also adapt by occupying new niches when possible. Species invasion has become a world-wide phenomenon and includes wildlife, insects, parasites, and pathogens. Species are on the move, and migrating to more accommodating climates if they can. Local observations of these migrations can be unsettling.

In Hong Kong, "at least eight new songbird species native to Southeast Asia have appeared in the city" notably in the far south.[98] Dr. Cheung believes that "some of these bird movements are due to climate change... this seems to be part of a wider pattern in East Asia."[99] These observations have far-reaching implications. David Dudgeon from the ecology and biodiversity department of the University of Hong Kong noted that biodiversity in Hong Kong will be seriously affected "if global climate change continues."[100]

Tick populations are moving north in parts of Sweden. Infestations are spreading diseases previously uncommon in these areas, such as Lyme disease and encephalitis. It has not yet been determined if a number of species invasions, such as the proliferation of pine beetles in Canada and western regions of the U.S., is linked to global warming (Vergano and O'Driscoll 2007:D2).

In the U.S., exotic plants foreign to Michigan sand dunes and Florida swamps, are threatening or destroying habitats.[101] Species previously foreign to the states of Kentucky, Indiana, Ohio, Illinois and New Hampshire, such as Norway maples and Lombardy poplar trees, are invading landscapes.[102] Bamboo-like species such as kudzu, are proliferating and choking native species in Ohio and the Southeast U.S. Poison Ivy has become more prevalent in the South and more virulent due to increased atmospheric levels of carbon in the atmosphere.

The zebra mussel, transported in the ballasts of oceangoing transatlantic ships, invaded the Great Lakes in a period of only 10 years. It has since increased its range to the Mississippi, Tennessee, Ohio, and the Hudson River Basin. Mussels are consuming native species of algae, larval fish and invertebrates, damaging food chains and driving species into decline. Native American unionid clams in Lake St. Clair are nearly extinct due to the invasion of zebra mussels. In addition to being a nuisance, they have disrupted municipal water intakes and have damaged recreational areas. While chemicals are available to kill their populations, efforts are costly and only small-scale applications have been tried. Essentially, zebra mussels are "impossible to eradicate with the technology available today."[103]

CONCLUSION

There is a linkage between the environment and sustainability. Causing conditions that increase pollution and permanently damage the natural environment lowers the potential to achieve sustainability. Environmental resources such as air and water can no longer be considered "free goods." Without careful management of natural capital, it is possible to cause irreparable harm to the environment. The time available for mitigation efforts is diminishing.

Innumerable catastrophes have focused attention on the environment. Environmental problems tend to be difficult to resolve. Unresolved environmental problems include issues regarding waste management, air quality, water quality and species decline. One reason for failing to address them directly has been the economic costs associated with mitigation efforts. Mitigation solutions often have unintended consequences.

In the last few decades, new environmental issues have developed which are international, rather than local in scope. These include global warming, acidification of water, ozone depletion and species decline or extinction. Climate change due to global warming is of particular concern as it has international and regional implications. It is predicted to cause ocean levels to increase, glaciers to melt, storms to increase in intensity and life on the planet to be affected. Regional solutions are often ineffective. International treaties such as the Kyoto Protocol and the Montreal Protocol, have attempted to address greenhouse gas emissions and ozone depletion with mixed results. The Kyoto Protocol has been ineffective at reducing greenhouse gases. The Montreal Protocol has been somewhat effective at reducing emissions of ozone-depleting substances. The ozone layer seems to be improving as a result, but may be in peril due to increases in greenhouse gas emissions which raise atmospheric temperatures.

While federal programs in the U.S. subsidize the financing of coal-fired electrical power facilities, there are no comparable programs to subsidize electrical energy production from alternative energy sources. The Federal Energy Policy Act represents a positive step as it encourages energy conservation and the use of alternative energy. Incentives for alternative energy could be provided in the form of Production Tax Credits (PTC) and Renewable Portfolio Standards (RPS). These will provide subsidies to include alternative energy sources, such as wind energy, in utility portfolios. Utility customers can purchase or trade credits from providers marketing supplies of alternative energy sources (Pasqualetti 2002:35-36). Increasing subsidies for the development of alternative energy sources will reduce reliance on fossil fuels and help reduce dependence on carbon-intensive energy supplies.

Energy is linked to environmental impact. Drilling for oil and mining coal can cause irreparable environmental harm. Using energy from atomic power and alternative energy sources offer the best hope of lowering future atmospheric carbon emissions. The construction of hydro-electric dams impacts wildlife.

Maintaining biodiversity is an important goal of environmental sustainability. Species adjust to environmental changes by means of competition, adaptation, evolution or a combination of these. Rapid climate changes can cause species populations to decline and can lead to their extinction. Climate change also impacts the ranges of species and disrupts their food chains. Undoubtedly, climate change has contributed to the species decline and extinction.

Endnotes

1. From the New Scientist (2007, 25 May). What will we do when gallium runs out? *The Wall Street Journal*. p. B6.
2. Ibid.
3. Ibid.
4. According to *USA Today* (Iwata 2007), a U.S. Federal judge recently ruled that ExxonMobile mustpay damages of $2.5 million for this oil spill.
5. Trade and Environment Database, *Oil Production and Environmental Damage*, http://www.american.edu/TED/projects/tedcross/xoilpr15.htm#r3, American University.
6. Sasahara, K. (2007, 17 July). Japanese earthquake kills nine. *The Courier- Journal*. p. A5.
7. National Solid Waste Management Association (2006, 8 November). *MSW (Subtitle D) Landfills*. http://wastec.isproductions.net/webmodules/wearticles/anmviewer.asp?a=1127, accessed 5 May 2007.
8. Ibid.
9. Ibid.
10. Kentucky Office of Energy Policy (2006, 1 June). *Kentucky Energy Watch*. 7 (22).
11. The surface area of the Aral Sea was 68,000 m^2 in 1960 and was reduced to 17,160 m^2 by 2004. See http://en.wikipedia.org/wiki/Aral_Sea.
12. GAIA Movement Trust. *One billion people lack clean drinking water: U.N.* http://www.gaia-movement.org/Article.asp?TxtID=437&SubMenuItemID=118&MenulteMID=47, accessed 10 June 2007.
13. Ibid.
14. ABC News Online (2006, 6 September). *One billion people lack clean drinking water: UN*. http://www.abc.net.au/news/newsitems/200609/s1733920.htm, accessed 11 June 2007.
15. Mason, M. (2007, 17 June). Tainted foods a daily problem throughout Asia. *The Courier-Journal*. p. A19.
16. Ibid.
17. O'Driscoll, P. (2007, 6 June). Drought now covers at least a third of the U.S. *The Courier-Journal*. p. A3.
18. Ibid.
19. Ibid.
20. North Plains Groundwater Conservation District. *Ogallala Aquifer*. www.npwd.org/Ogallala.htm, accessed 20 June 2007.
21. Hofling, K. (2007, 131 June). Tainted water at Marine base studied. *The Courier-Journal*. p. A3.
22. Ibid.
23. Ibid.
24. Brezosky, L. (2007, 5 July). Desalination: the world's betting it's worth its salt. *The Courier- Journal*. p. A3.
25. Ibid.
26. Keen, J. (2007, 18 June). Older cars' emissions go unchecked. *USA Today*. p. A1.

27. Ibid.
28. Bain. M. (2007, 16 April). History of a hot topic. *Newsweek*. p. 54.
29. Ibid.
30. Borenstein, S. (2007, 2 February). Report: Global warming 'very likely' man-made, unstoppable for centuries. *The Courier-Journal*. p. A5.
31. Vergano, D. (2007, 3 May). Fixing climate carries big costs. *USA Today*. p. 9D.
32. Corporate leaders back action on global warming. *The Courier- Journal*. 23 January 2007.
33. Antarctic icebergs are "hot spots" for life. *The Courier- Journal*. 2 July 2007. p. E3.
34. Report: global warming will get worse. Review of a scientific report from the intergovernmental Panel on Climate Change. *USA Today*. 23 January 2007.
35. Warming the Artic? Blame the snow. *The Courier-Journal*. 18 June 2007. p. E3.
36. Revkin, A.C. (2007, 1 April). Poor nations bear brunt of warming. *New York Times*.
37. Ibid.
38. Yee, A. (2007, 2 February). Flood threat to Bangladesh a warning to the world. *Financial Times*. p. 3.
39. Ibid.
40. Energy Information Administration. From 2001 data. http://www.eia.doe.gov/oiaf/1605/ggccebro/chapter1.html, accessed 15 June 2007.
41. Records have been kept by the World Meteorological Organization since 1861. The normal earth surface temperature is 25.2 degrees Fahrenheit. The average earth surface temperature for 1998 was roughly one degree above normal (Fowler 2003: A12).
42. December makes '06 warmest year in U.S. *The Courier- Journal*. 10 January 2007. Data for the U.S. was accumulated from 1,200 temperature recoding stations across the country
43. Kuafman, M. (2007, 7 February). *Global Warming and Hot Air*. www.washingtonpost.com/wp-dyn/content/article/2007/AR200702060126.html, accessed 6 May 2007.
44. Kaufman, M. (2007, 5 May). Scientists rate costs of reducing global warming. *The Courier-Journal*. p. A7.
45. Kaufman. M. (2007, 6 April). Southwest getting drier, reports says. *The Courier-Journal*. p. A3.
46. Bryden et al (2005, 1 December). Slowing of the Atlantic overturning circulation at 25° N. *Nature* 438. p. 655-657.
47. Ritter, K. (2007, 14 June). Gulf Stream studies show European ice age unlikely. *The Courier-Journal*. p. A6.
48. Ibid.
49. Dyer, G. (2002, 2 February). Global warming debate begins to build in China. *Financial Times*. p. 3.
50. Ibid.
51. Ibid.
52. Robinson. S. (2007, 14 April). The global warming survival guide—38 Trade carbon for capital. *TIME*. www.time.com/specials/2007/environment/article/0,28804,1602354_163074_1603643,00.html, accessed 20 May 2007.
53. Ibid.
54. Ibid.
55. CBC News (2007, 30 January). *Harper's letter dismisses Kyoto as 'socialist scheme.'* http://www.cbc.ca/canada/story/2007/01/30/harper-kyoto.html, accessed 3 June 2007.
56. On 21 June 2007, the U.S. Senate approved legislation to increase fleet fuel econ-

omy standards to 35 miles per gallon by 2020 for cars, sport utility vehicles and some trucks.

57. The actual agreement of the Kyoto Protocol Conference of Parties (COP) in December 1997 was COP-3 which created a global target reduction in six primary greenhouse gases (including CO_2) of 5.2 percent from 2008 to 2012 (Chasek 2000:77-78).

58. Office of the Presidential Press Secretary. (2007, 24 January). *Executive Order: Strengthening federal environmental, energy, and transportation management.* http://www.whitehouse.gov/news/releases/2007/01/print/20070124-2.html, accessed 6 July 2007. Agency heads in intelligence, military, and law enforcement may exempt their departments.

59. Ibid., applies to agencies with 20 or more vehicles.

60. Ibid.

61. Borenstein, S. (2007, 2 February). Report: Global warming 'very likely' man-made, unstoppable for centuries. *The Courier-Journal.* p. A5.

62. Mufson. S. (2007, 14 May). Federal loans fuel new coal plants. *The Courier-Journal.* p. A3.

63. Ibid.

64. Ibid.

65. U.S. Environmental Protection Agency, *Mountaintop Mining/Valley Fills in Appalachia: Final Programmatic Environmental Impact Statement*, October 2005.

66. Trade and Environment Database, *Oil Production and Environmental Damage*, http://www.american.edu/TED/projects/tedcross/xoilpr15.htm#r3, American University.

67. Energy Information Administration, US Department of Energy, *Greenhouse Gases, Climate Change and Energy*, http://www.eia.doe.gov/oiaf/1605/ggccebro/chapter1.html.

68. Mansfield. D. (2007, 5 May). TVA sees future in nuclear power. *The Courier-Journal.* p. D2.

69. Ibid.

70. Samuelson. P. (2006, 10 November). *Greenhouse Guessing.* www.washingontonpost.com/wp-dyn/content/article/2006/11/09/AR2006110901768.html?nav=rss_opinion/columns, accessed 6 May 2007.

71. Ritter, J. (2007, 6 June). California sees sprawl as warming culprit. *USA Today.* p. 1.

72. Ibid.

73. U.S. Mayor's Climate Protection Agreement. www.seattle.gov/mayor/climate, accessed 11 June 2007.

74. Ibid.

75. *Ozone layer.* http://en.wikipedia.org/wiki/Ozone_layer, accessed 6 June 2007.

76. Data from the National Oceanic and Atmospheric Association, *NOAA/ERSL Releases Ozone Depleting Gas Index*, accessible at www.noaa.gov/hotitems/storyDetail_org.php?sid=3843, dated 15 January 2007.

77. Ibid.

78. Goddard Space Flight Center (2002, 4 June). Climate change may become major player in ozone loss. www.gsfc.nasa,gov/topstory/2002041greengas.html, accessed 6 May 2007.

79. Corporate leaders back action on global warming. *The Courier-Journal.* 23 January 2007.

80. One million organisms and still counting. *The Courier Journal.* 9 April 2007. p. A3.

81. Ibid.

82. Lovejoy, T. (2007, 16 April). The ocean's food chain is at risk. *Newsweek.* p. 80.

83. Philips, M. (2007, 9 July). The fading forests of the sea. *Newsweek.* p. 50.

84. UNESCO (2007, 10 April). *Climate change threatens UNESCO world heritage sites.* whe.unceso.org/pg_friendly_print.cfm?id=319&cid=82&, accessed 2 July 2007.
85. Overfished. *The Courier-Journal.* 2 April 2007. p. E3.
86. Associated Press, from a preliminary draft prepared by a network of 2,000 scientists and 100 governments. Species will be lost as world warns, UN says. *The Courier-Journal.* 1 April 2007, p. A11.
87. Ibid.
88. Ibid.
89. Wines, M. (2007, 11 June) Penguins in peril. *The Courier-Journal.* p. E3.
90. Ibid.
91. Ibid.
92. Heron. L. (2007, 4 February). Nature's warning: climate change is hitting HK. *Sunday Morning Post.* p. 1.
93. Ibid.
94. Borenstein, S. (2007, 15 June). 20 birds no longer so common. *The Courier-Journal.* p. A3.
95. Ibid.
96. Ibid.
97. The Matilija Dam in southern California is scheduled to be demolished in 2013.
98. Heron. L. (2007, 4 February). Nature's warning: climate change is hitting HK. *Sunday Morning Post.* p. 1.
99. Ibid.
100. Ibid.
101. Seewer, J. (2007, 3 June). Plant lovers, growers battle invading botanicals. *The Courier-Journal.* p. B6.
102. Ibid.
103. USGS (2007, 11 April). *Zebral Mussel.* www.glsc.usgs.gove/main.php, accessed 11 May 2007.

Chapter 4

Sustainable Buildings

"A design that is green is one that is aware of and respects nature and the natural order of things; it is a design that minimizes the negative human impacts on the natural surroundings, materials, resources, and processes that prevail in nature. It is not necessarily a concept that denies the need for any human impact, for human existence is part of nature too. Rather, it endorses the belief that humankind can exist, multiply, build and prosper in accord with nature and the earth's natural processes without inflicting irreversible damage to those processes and the long-term habitability of the planet."

GRUMMAN (2003:3)

Buildings and structures play an important role in meeting human needs. They provide shelter and can provide healthy environments. However, buildings can also provide unhealthy conditions and stress the environments in which they are built. Buildings require excavation for construction, can modify wildlife habitats, change rainwater runoff patterns, change landscapes, and require a cornucopia of materials. They also absorb and radiate heat, need paving for pedestrian and vehicles and, most importantly, consume resources—including energy. Energy consumption varies with building size, design and climate conditions. Chicago's Sears Tower alone consumes more electricity than Rockford, Illinois, a city with a population over 150,000 (Rifkin 2006).

In the U.S., buildings represent 39% of primary energy use (including production), 70% of electrical consumption, 12% of potable water use, and generate 136 million tons of construction waste annually (USGBC October 2006:1-3). If there were a way to reduce electrical demand in buildings by only 10%, most of the new electrical generating facilities scheduled for construction in the U.S. over the next 10 years would not be required. The carbon associated with the energy used in U.S. buildings "constitutes 8% of the current global emissions—equal to the total emissions of Japan and the United Kingdom combined."[1]

Buildings have an economic life span. Many are often financed for 30 years or more and have been designed to serve for 40-50 years. Over time,

113

4-1. Moving a Building – The Ultimate Recycling Project

buildings are subject to deterioration, exposure to the elements, changes in use and occupancy, changes in building standards, and changes in ownership. With this in mind the idea of a 'sustainable building' may seem to be a contradiction. However, buildings and building materials can often be recycled and reused. Older structures can be updated to serve new purposes, recycling their existing infrastructure. Perhaps the ultimate recycling project is moving a building to another location rather than demolishing it. In such cases, a substantial portion of the energy and material embedded in its original construction can be reused.

The idea of developing sustainable buildings stands in striking contrast to typical standard construction practices. *Sustainable* construction involves changes in the fundamental design processes, refocusing on-site development, rethinking materials and building components, considering environmental impacts, and determining how healthy interior environments can be configured. To further complicate the process of designing and developing green buildings, Huber (2005:216) concludes that "...building sustainability strategies have to consider basic societal conditions. Building regulations, financing models, rental legislation, environmental legislation, etc., significantly determine the construction design. The solution is approached by an eco-efficient optimization which considers the basic societal conditions."

Knowing that buildings require substantial inputs of resources to construct and maintain, what can be done to reduce their environmental impacts and energy requirements? The solutions include but are not limited to:

1) Changing land development practices;
2) Designing buildings with attention to improved construction standards;
3) Upgrading and reusing existing structures;
4) Providing more efficient buildings and higher quality equipment;
5) Changing the physical arrangement and configuration of buildings;
6) Carefully selecting construction materials; and
7) Harvesting on-site resources.

This chapter considers how buildings can employ green practices and how building technologies are incorporated into the design of buildings. "Green building programs" will also be discussed. These programs include efforts to establish energy and environmental standards for existing buildings and new construction.

LAND DEVELOPMENT PRACTICES

Land development practices have contributed to adverse environmental consequences and changes in infrastructure that have increased energy and water consumption. Large-scale developments such as manufacturing centers and residential subdivisions constitute a type of "terraforming," which cause permanent changes to their environs. The growing appetite of these structures for energy and environmental resources has contributed to ecosystem disruption and has forced us to rethink how buildings are sited and constructed.

While development is associated with localized environmental disruptions, scientific assessments of construction impacts have yielded mixed results. We know that construction is energy intensive and resources are consumed in the process of construction. Often, initial costs are minimized, and life-cycle cost analysis is not used in the selection of materials and mechanical systems. When initial costs are minimized, maintenance, energy, and water costs typically increase. Development that favors suburbanization and automobile biased transportation systems also increases resource costs. Green development practices coupled with design alternatives for structures with environmentally friendly and energy efficient attributes have now become a feasible alternative. When these are addressed in the site selection and design of buildings, life-cycle costs are typically reduced.

Sustainable development focuses on the built environment: buildings are seen as the primary building blocks of the urban infrastructure. Green construction practices provide alternatives. It is understood that "being green" implies a commitment to environmental protection and natural resource conservation."[2] If buildings can be constructed in a manner that is less environmentally damaging and more energy efficient, then they can be called "green" buildings. Being green means that construction needs to be "eco-efficient." Life-cycle cost assessments should be made in selecting building systems and construction components. "Eco-efficiency is achieved by the delivery of competitively priced goods and services that satisfy human needs and quality of life, while progressively reducing ecological impacts and resource intensity throughout the life-cycle to a level at least in line with the earth's carrying capacity" (DeSimone and Popoff 1997:47).

In regard to the built environment, architectural designers have recently renewed their emphasis on designing healthy buildings. Fundamental design issues such as site orientation, daylighting, shading, landscaping, more thermally cohesive building shells, and more energy efficient mechanical and electrical systems are getting renewed attention.

Drivers include experiences with "sick building syndrome" and illnesses linked to systemic malfunctions such as chemical exposures, legionnaire's disease, asthma and asbestosis. Medical conditions have been linked to mold exposures that result from indoor environmental conditions. These served to jolt architects and engineers into re-establishing the importance of indoor environmental conditions in general and indoor air quality in particular when designing their buildings. With concerns mounting as to product safety and liability issues involving the chemical composition of materials, manufacturers have began to mitigate the potential adverse impacts of construction building materials upon occupants. Concurrently, resource availability and waste reduction became issues influencing both building design and the design of construction components.

Decisions as to what types of buildings to construct, what construction standards apply, and what sorts of materials to use in the construction of buildings are typically made locally. Those who influence decisions regarding the physical form of a proposed structure include the owners, builders, developers, contractors, architects, engineers, planners and local zoning agencies among others. With so many parties involved in the decision making process, planning for green buildings can be a difficult, decentralized and often divisive process. In addition, all involved must abide by regulations that apply to the site and structure being planned. Planning ordinances and building codes often vary from one locale to another, further complicating the process. For these reasons, making green buildings requires a team of cooperative individuals. It is a comprehensive process that involves careful design and selection of building components and systems.

What is alarming is that past professional practice within the U.S. building industry has only rarely gauged the environmental or energy impact of a structure prior to its construction. Building types, especially for commercial construction, are often standardized, irregardless of differentiators such as local climate conditions, geography and site conditions. This lack of differentiation contributes to increased resource consumption. Green buildings are an attempt to improve the planning and design of new structures and major renovations.

THE CONCEPT OF GREEN BUILDINGS

There was a time in the U.S. when most construction material was obtained locally. Indigenous materials included accessible timber, fieldstone, locally quarried rock, adobe, thatch, slate, clapboard, and cedar shakes. Since construction materials were costly to manufacture (most were hand-made) and troublesome to transport, most components of demolished structures were reused in some manner.

Early in U.S. history, one- and two-room log houses were the norm. The central heating system was a drafty fireplace with a chimney constructed of local stone. When possible, design features, learned by trial and error, were added in an attempt to optimize thermal comfort. Examples included architectural features that would control lighting and temper the indoor environment with shading devices, take advantage of breezes, and carefully size and orient fenestration. Rainwater was often collected from

4-2. LEED Certified Building – Lincoln Hall, Berea College

roofs. Such were the humble beginnings of green construction practices.

Buildings today are infinitely more complex—and sustainable building practices can be even more so. Accepting the notion that sustainable, environmentally appropriate, and energy-efficient buildings can be labeled "green," the degree of "greenness" is subject to multiple interpretations. The process of determining which attributes of a structure can be considered "green" or "not green" can be viewed as inconclusive and subjective. A green design solution appropriate for one locale may be inappropriate for others, due to variations in climate or geography. Complicating the process, there are no clearly labeled "red" edifices with diametrically opposing attributes. While it is implied that a green building may be an improvement over current construction practice, comparison is often unclear and confusing. It can often be perplexing as to what sort of changes in construction practice, if imposed, would lead to greener, more sustainable buildings. At times, markets adjust and provide materials, components and products so that greener buildings can arise. Since standards are often formative and evolving, gauging the degree of "greenness" risks the need to quantify often subjective concepts. One attribute of green construction practices is the attempt to preserve and restore habitat that is vital for life, or to become "a net producer and exporter of resources, materials, energy and water rather than being a net consumer."[3] In Pennsylvania, the Governor's Green Government Council defines a green building as "one whose construction and lifetime of operation assures the healthiest possible environment while representing the most efficient and least disruptive use of land, water, energy and resources."[4]

There are qualities of structures, such as reduced environmental impact, improved indoor air quality and comparatively lower energy usage, which are widely accepted as evidence of green construction practices. For example, using recycled materials that originate from a previous use in the consumer market, or using post-industrial content that would otherwise be diverted to landfills, are both widely accepted green construction practices.

Grumman (2003:4) describes a green building by saying that "...a green building is one that achieves high performance over the full life cycle." Performance can be defined in various ways. "High performance" can be interpreted widely, and often in "highly" subjective ways. In an attempt to clarify, Grumman further identifies a number of attributes of green buildings:

- Minimal consumption—due to reduction of need and more efficient utilization—of nonrenewable natural resources, depletable energy resources, land, water, and other materials.

- Minimal atmospheric emissions having negative environmental impacts, especially those related to greenhouse gases, global warming, particulates, or acid rain.

- Minimal discharge of harmful liquid effluents and solid wastes, including those resulting from the ultimate demolition of the building itself at the end of its useful life.

- Minimal negative impacts on site ecosystems.

- Maximum quality of indoor environment, including air quality, thermal regime, illumination, acoustics, noise and visual aspects.

While goals and attributes of green construction practices are readily identified, developing construction standards to achieve such goals is another matter. High performance sounds like added value, for which a premium is likely to be paid. In fact, adding green building features is likely to increase construction costs by 1% to 5%, while lowering operating expenses and life-cycle costs. Energy-efficient electric motors may be slightly more expensive to install but the savings can be substantial. An estimated "97% of the life-cycle costs of a standard motor goes to energy costs and only 3% to procurement and installation."[5]

Buildings are further differentiated by the desires of the owners, the skills and creativity of their design teams, site locations, local planning and construction standards, and a host of other conditions. What may be a green solution for one building might be inappropriate if applied to another.

There are a multitude of shades of green in green building construction practices. Qualities of green construction practices have been variously identified.[6] Some qualities focus on exterior features and others on the types of green materials that are used. The 2006 International Energy Conservation Code (IECC), for example, requires that energy efficient design be used in construction and provides effective methodologies. Its focus is on the design of energy-efficient building exteriors, mechanical systems, lighting systems and internal power systems. The IECC is being included in building codes across the U.S. and in many other countries. ASHRAE Standard 90.1 (Energy Standard for Buildings Except Low-Rise Residential Buildings) deals with the energy-efficiency of buildings and HVAC components. Other standards concern improved air quality and better ventilation systems. Green construction opportunities are further described in the *ASHRAE Green Guide*. Green buildings may also be designed to focus carbon impact and some are classified as "zero-carbon," meaning that they have a net zero carbon emissions impact.

COMPARATIVE BUILDING ENERGY PERFORMANCE

Making buildings stingy in their use of energy is one aspect of green construction. Thankfully, there are many opportunities to improve energy performance in residential, industrial, and commercial buildings. Advances in construction practices have yielded striking results and provide the means to create more efficient buildings.

The author conducted a one-month (December 2006) heating and electrical system assessment of two occupied residences in the mid-western U.S. during the same time period. The homes are about 16 km (10 miles) apart. Only energy consumption was considered in this the assessment. Both residences are two story buildings, located in the same metropolitan area, served by the same utility companies, and on the same rate structures for electricity and natural gas. Neither used alternative energy sources. Both residences were heated with natural gas with thermostats set at approximately 22°C (72° F) for the period of comparison and neither used temperature setback controls during the assessment period. Both used high-efficiency (approximately 90-94%) natural gas fired, forced-air furnaces to heat interior spaces.

One building was a frame residence constructed in 1910 that was approximately 212 m² (2,280 ft²). This frame residence had limited insulation, single pane windows, electric water heating and used primarily incandescent lighting. The other residence (designed by the author) was a brick home in a suburban location constructed in 2001. Both residences used the conventional construction practices available at the times of their construction. The newer 427 m² (4,600 ft²) brick residence was designed with an expanded south-facing façade, reduced northern exposure, extra insulation in walls, an exterior infiltration wrap, insulated ceilings and foundation, high efficiency windows, natural gas water heating, extensive use of compact fluorescent lighting, and other features. To take advantage of topography and the thermal moderation available from the earth, it was set into the slope of a hill, minimizing the northwestern exposure and maximizing the solar gain from the southwest. The total energy bills for the older 212 m² frame structure for the period of study totaled $432, equating to $2.24/m² (19 cents/ft²). The total energy bills for the newer 427 m² brick residence for the period totaled $275, equating to $.58/m² (5.2 cents/ft²).

This comparison provides interesting findings. For the newer brick structure, total energy costs were 36% less than the older frame residence for the period of study despite the fact that it was twice as large. The energy costs for the newer residence (based on a unit area) were roughly 74% less for the period. The older frame residence used 556 kWh of electricity and 1,001.3 m³ (353.6 ccf) of natural gas while the newer residence used 1,182 kWh of electricity and 519.9 m³ (183.6 ccf) of natural gas. The newer residence consumed twice as much electricity and half as much natural gas during the study period. The increase electrical use was likely due to greater areas of space for lighting. The lower natural gas usage was likely the result of the improved thermal envelope.

While not a scientific study, the analysis provides empirical evidence that energy costs can be lower when energy efficiency technologies, using conventional construction practices, are incorporated in the design of residences.[7] It also demonstrates how the consumption of source fuels in buildings of similar function can vary significantly.

ENERGY STAR BUILDINGS

In an effort to provide information to improve the energy efficiency of buildings in the U.S., agencies of the central government co-sponsored the development of the Energy Star™ program. This program provides "technical information and tools that organizations and consumers need to choose energy-efficient solutions and best management practices" (USEPA 2003). The type of technical information available includes information about new building designs, green buildings, energy efficiency, networking opportunities, plus a tools and resources library. Energy Star offers opportunities for organizations and governments to become partners in the program. The program offers guidelines to assist organizations in improving energy and financial performance in an effort to distinguish their partners as environmental leaders. The multi-step process involves making a commitment, assessing performance, setting goals, creating an action plan, implementing the action plan, evaluating progress and recognizing achievements.[8]

Expanding on their success, Energy Star™ developed a building energy performance rating system which has been used for over 10,000 buildings. It does not claim to be a green building ranking system but rather a comparative assessment system that focuses on energy performance, an important component of green building technologies. The rating system is based not on energy costs but on source energy consumption. In the view of the Energy Star program, "the use of source energy is the most equitable way to compare building energy performance, and also correlates best with environmental impact and energy cost."[9] Using the Energy Star rating system, buildings are rated on their energy performance on a scale of 1 to 100 when compared to similar structures constructed in environments that experience similar weather conditions. Buildings that achieve a rating of 75 or greater qualify for the prized Energy Star label.[10]

GREEN CONSTRUCTION MATERIALS AND METHODS

Material and product recycling has a long history and green construction has its roots in byproduct recycling. During World War II, strategic materials, such as steel and aluminum, were recycled and reused to manufacture military equipment. After the end of the war, recycling programs fell into decline. Beginning anew in the 1970s, metals such as aluminum, copper and steel began to be recycled. By the 1980's, construction site wastes, such as steel frame windows and their glass panels, were being recycled rather than being sent to landfills. By the 1990's landfill space became more costly. In addition, once-flared natural gas from landfills began to be seen as a potential energy resource rather than a waste by-product.

Clayton (2003:486) defines a "sustainable" building philosophy as "to design, build and consume materials in a manner that minimizes the depletion of natural resources and optimizes the efficiency of consumption." Green construction practices have several categorical commonalities:

- Green buildings are designed to reduce energy usage while optimizing the quality of indoor air. These buildings achieve energy reductions by using more insulation and improved fenestration, by optimizing the energy usage of mechanical and electrical building subsystems, and by the use of alternative energy.

- There is an emphasis on reducing the costs of energy used to transport material to the construction sites. One means of achieving reductions in transportation costs is to use materials that have been locally manufactured. This provides the added benefit of supporting local employment and industries.

- There is a focus on using recycled construction materials (such as reusing lumber from demolished structures) or materials made from recycled products (such as decking materials that use recycled plastics). The idea is to reduce the amount of virgin material required in the construction.

- There is a preference for materials that are non-synthetic, meaning that they are produced from natural components such as stone or wood, etc. This often reduces the number of steps required for product manufacture and may also reduce the use of non-renewable resources employed: for example, reducing the use of oil required in the production of plastics. Extracted metals such as aluminum and copper may also be preferred as they can be more easily reused once the building's life cycle is completed.[11]

- There is a mandate of green construction to avoid the use of materials that either in their process, manufacture or application, are known to have environmentally deleterious effects or adversely impact heath. Examples include lead in paint or piping, mercury thermostats, and solvents or coatings that may outgas fumes and carcinogens.

- In an effort to reduce water from municipal sources and sewage treatment requirements, green buildings are often designed to harvest rainwater by using collection systems. Rainwater may be used for irrigation, toilets or other non-potable requirements. In addition, green buildings use technologies, such as flow restriction devices, to reduce the water requirements of the building's occupants.

4-3. Construction Waste

- The quantities of construction wastes are reduced. This occurs by strategically reducing wastes generated during construction and by reusing scrap materials whenever possible. The goal is to reduce the amount of material required for construction and to reduce the quantities of scrap material that must be trucked to a landfill.

New industries and entire product lines have emerged in an effort to provide construction materials that meet green building standards. There are numerous creative examples. Beaulieu Commercial has brought to market a carpet tile backing made from 85% post-consumer recycled content using recycled plastic bottles and glass, yielding carpet backing that is 50% stronger than conventional carpet backing. The Mohawk Group manufactures carpet cores using a recycled plastic material rather than wood. This change saves the equivalent of 68,000 trees annually. There are products on the market (e.g., ProAsh), made with fly ash—a once wasted by-product—which has been recycled into concrete. Armstrong now offers to pick up old acoustical tile from renovation projects and will deliver them to one of their manufacturing sites to be recycled into new ceiling tiles. PVC products became available to satisfy piping needs in domestic applications for drainage systems rather than copper, lowering the weight of products and their components. The installation times are reduced by eliminating the need to sweat pipes, which lowers labor costs. While PVC has its own associated environmental issues, it is now being used for exterior trim due to its durability. Many wood products are certified and labeled if they have used environmentally appropriate growing and harvesting techniques in their production. Pervious paving systems, that allow vegetation to grow and reduce the heat absorption from paving, are also available. The list of available green technologies and products seems endless. Triple pane, low "e" window glazing, solar voltaic roof shingles, waterless toilets and solar powered exterior lighting systems are among the green building products that are being used in building systems.

Green construction has provided opportunities to introduce new product lines and the movement to green construction practices offers ready markets for the products. Hydrotech manufactures a green roof system that provides a balance between water drainage and soil retention thus allowing roof gardens to flourish.

To learn which products are greener than others, computer software tools are available to assist in determining their environmental impacts. These programs also help ascertain product life-cycles.[12]

RATING SYSTEMS FOR GREEN BUILDINGS

While green building components are available, incorporating them into green buildings requires forethought, engineering, and creative design. Green building standards have been developed by both private and governmental organizations, all bent on finding ways to assess green construction practices. Comparing the degree of "greenness" from one building to the next is difficult. One solution is the use of categorical rating systems in an effort to reduce subjectivity. The development of green building attributes or standards by private organizations recognizes that decisions are to be based on stakeholder consensus. These stakeholders are often from widely diverse industries and geographic locations.

Developing a rating system for green buildings is both difficult and challenging. According to Boucher, "the value of a sustainable rating system is to condition the marketplace to balance environmental guiding principles and issues, provide a common basis to communicate performance, and to ask the right questions at the start of a project" (Boucher 2004). Rating systems for sustainable buildings began to emerge in the 1990s.

4-4. Pervious Paving System in Central Hungary

Perhaps the most publicized of these first appeared in the U.K., Canada, and the U.S. In the U.K., the Building Research Establishment Environmental Assessment Method (BREEAM) was initiated in 1990. BREEAM™ certificates are awarded to developers based on an assessment of performance in regard to climate change, use of resources, impacts on human beings, ecological impact and construction management. Credits are assigned based on these and other factors. Overall ratings are assessed according to grades that range from pass to excellent (URS Europe 2005).

The International Initiative for a Sustainable Built Environment (IISBE), based in Ottawa, Canada has a Green Building Challenge program which is now used by more than 15 participating countries. This collaborative venture provides an information exchange for sustainable building initiatives and has developed "environmental performance assessment systems for buildings" (iiSBE 2005). The IISBE has created one of the more widely used international assessment systems for green buildings.

The U.S. Green Building Council (USGBC), an independent non-profit organization, grew from just over 200 members in 1999 to 3,500 members by 2003 (Gonchar 2003), and to 7,200 members by 2006. The core purpose of the USGBC "is to transform the way buildings and communities are designed, built and operated, enabling an environmentally friendly, socially responsible, healthy and prosperous environment that improves the quality of life" (USGBC 2006:1).

Prior to the efforts of organizations like the USGBC (established in 1995), the concept of what constituted a "green building" in the U.S. lacked a credible set of standards. The USGBC's Green Building Rating System has a goal of applying standards and definitions which link the idea of high performance buildings to green construction practices. The program developed by the USGBC is called "Leadership in Energy and Environmental Design" (LEED™). Sustainable technologies are firmly established within the LEED project development process. LEED loosely defines green structures as those that are "healthier, more environmentally responsible and more profitable" (USGBC 2004).

The LEED Green Building Rating System is a consensus developed and reviewed standard, which allows voluntary participation by diverse groups of stakeholders interested in its application. The LEED system is actually a set of rating systems for various types of green construction projects, and ranks projects as part of the labeling process. LEED rating systems are available for new construction, existing buildings, commercial interiors, core and shell development projects, residences, and neighborhood development.[13]

Developed by the U. S. Green Building Council (USGBC), LEED rating systems evolved to help "fulfill the building industry's vision for its own transformation to green building" (USGBC 2004). The first dozen pilot projects using the rating system were certified in 2000. By 2006 there were 27,200 LEED Accredited Professionals, 3,539 LEED registered projects in 12 countries, and 484 projects that have completed the certification process (USGBC, October 2006). Additional LEED rating systems are being developed and separate rating programs for educational facilities, healthcare and retail new construction will soon be available. The LEED rating system is poised to become the new international standard for green buildings.

THE LEED-NC RATING SYSTEM

LEED for new construction and major renovations (LEED-NC) is perhaps the most widely adopted LEED program to date. LEED-NC provides an example of how rating systems for green buildings are structured. This section describes how LEED works, discusses the influences that shaped the development of LEED-NC for new construction and major renovations, and reviews how new projects are scored.

LEED-NC is the USGBC's standard for new construction and major renovations. It is used primarily for commercial projects such as office buildings, hotels, schools, institutions, etc. The rating system is based on an assessment of attributes and an evaluation of the use of applied standards. Projects earn points as attributes are achieved, and as evidence has been provided that the requirements of the standards have been followed. Depending on the total number of credits (points) a building achieves upon review, the building is rated as Certified (26-32 points), Silver (33-38 points), Gold (39-51 points), or Platinum (52 or more points) (USGBC 2003:6). Theoretically, there are a maximum of 69 achievable credits. However, in real world applications, gaining certain credits sometimes hinders the potential of successfully meeting the criteria of others. While achieving the rating of Certified is more easily accomplished, obtaining a Gold or Platinum rating is rare and requires both creativity and adherence to a broad range of prescriptive and conformance-based criteria.

The LEED process involves project registration, provision of documentation, interpretations of credits, application for certification, a techni-

4-5. LEED Certified Interpretive Center at Bernheim Forest, Shepherdsville, KY

cal review, rating designation, award and appeal. Based on variables such as project square footage and USGBC membership status, registration fees for the process can range up to $7,500 per building (USGBC 2004).

To apply for the LEED labeling process, there are prerequisite project requirements which earn no credits (points). There are also categories of initiatives and attributes that do earn credits. The prerequisite requirements are interesting. For example, in the Sustainable Sites category, prerequisite procedures must be followed to reduce erosion and sedimentation. In the category of Energy and Atmosphere, procedures are required for building systems commissioning, minimal energy performance standards (e.g., adherence to ANSI/ASHRAE/IESNA Standard 90.1-2004, Energy Standard for Buildings Except Low-Rise Residential Buildings, or the local energy code if more stringent) and there must be verification that CFC refrigerants will not be used or that their use will be phased out. In addition, there are prerequisite requirements outlining mandates for storage and collection of recyclable material, mandates for minimum indoor air quality performance (the requirements of ASHRAE Standard 62.1-2004, Ventilation for Acceptable Indoor Air Quality must be adhered to) and a requirement that occupants who do not smoke tobacco will not be exposed.

In addition to the prerequisite requirements, the LEED process assigns points upon achieving certain project criteria, or complying with certain standards. The total points are summed to achieve the determined appropriate rating. Projects can achieve points from initiatives within the following sets of categories: Sustainable Sites (14 points); Water Efficiency (5 points); Energy and Atmosphere (17 points); Materials and Resources (13 points); Indoor Environmental Quality (15 points).

Using a LEED Accredited Professional to assist with the project earns a single point (USGBC November 2002). Additional points are available though application for creativity in the Innovation and Design Process (maximum of 4 points).

Within each category, specific standards and criteria are designed to meet identified goals. In the category of Sustainable Sites, 20.2% of the total possible points are available. This category focuses on various aspects of site selection, site management, transportation, and site planning. The goals of this category involve reducing the environmental impacts of construction, protecting certain types of undeveloped lands and habitats, reducing pollution from development, conserving natural areas and resources, reducing heat island impacts, and minimizing external light pollution. Site selection criteria are designed to direct development away from prime farmland, flood plains, habitat for endangered species, and public parkland. To encourage higher development densities, a point is awarded for projects that are essentially multi-story. If the site has documented environmental contamination or has been designated by a governmental body as a brownfield, another point is available.

In regard to transportation, as many as four points are available for locating sites near publicly available transportation (e.g., bus lines or light

4-6. Recharging Station for Electric Vehicles

rail), providing bicycle storage and changing rooms, providing for alternatively fueled vehicles, and for carefully managing on-site parking.

Two points in this category can be obtained by limiting site disturbances and by exceeding "the local open space zoning requirement for the site by 25%" (USGBC 2002). Points are awarded for following certain storm water management procedures, increasing soil permeability and attempting to eliminate storm water contamination. Potential urban heat island effects are addressed by crediting design attributes such as shading, underground parking, reducing impervious surface areas, and using high albedo materials such as reflective roofing materials, or for using vegetated roofing systems. Finally, a point is available for eliminating light trespass from the site.

Water efficiency credits comprise 7.2% of the total possible points. With the goal of maximizing the efficient use of water use and reducing the burden on treated water systems, points are credited for reducing or eliminating potable water use for site irrigation, capturing and using rainwater for irrigation, and using drought-tolerant or indigenous landscaping. The LEED-NC standard also addresses a building's internal water consumption. Points are available for lowering aggregate water consumption, and also for reducing potable water use. Reducing wastewater quantities or providing on-site tertiary wastewater treatment also earns points.

The category Energy and Atmosphere offers the greatest number of points, 24.6% of the total possible. These include improving equipment calibration, reducing energy costs, supporting alternative energy, reducing the use of substances that cause atmospheric damage, and offering measurement and verification criteria. Optimizing the energy costs of regulated energy systems can achieve a maximum of ten points. To assess the result, project designs are modeled against a base-case solution which lacks certain energy-saving characteristics.

Interestingly, the unit of measure for evaluating energy consumption and performance in order to achieve credits is not a measure of energy (e.g., kilocalories or million Btus) but dollars. Points are awarded in increments as the percentage of calculated dollar savings increases. In addition to the ten points for energy cost optimization, a maximum of three additional points are available for buildings that use energy from a site-generated renewable energy source. Purchased green power is allocated a single point if 50% of the electrical energy (in kWh) comes from a minimum two-year green power purchasing arrangement. This category provides points for additional commissioning, and for eliminating the use of HCFCs and halon gases. Measurement and verification (M&V), a means of validating equipment performance and associated energy use, is allowed a point, but only if M&V options B, C and D, as outlined in the 2001 edition of the International Performance Measurement and Verification Protocol (IPMVP), are used.

The Materials and Resources category represents 18.8% of total possible points. This category provides credit for material management, adaptive reuse of structures, construction waste management, resource reuse, use of material with recycled content, plus the use of regionally manufactured materials (certain renewable materials, and certified wood products). A point is earned for providing space in the building for storage and collection of recyclable materials such as paper, cardboard, glass, plastics, and metals. A maximum of three points is available for the reuse of existing on-site structures and building stock. The tally increases with the extent to which existing walls, floor, roof structure, and external shell components are incorporated into the reconstruction.

LEED-NC 2.1 addresses concerns about construction waste by offering a point if 50% of construction wastes (by weight or volume) are diverted from landfills and another point if the total diversion of wastes is increased to 75%. A project composed of 10% recycled or refurbished building products, materials and furnishings gains an additional two points. Another two points is available in increments (one point for 5%, two points for 10%) if post-consumer or post-industrial recycled content (by dollar value) is used in the new construction. To reduce environmental impacts of transportation systems, a point is available if 20% of the materials are manufactured regionally (defined as being within 500 miles or roughly 800 km of the building site) and an added point is scored if 50% of the materials are extracted regionally. A point is available if rapidly renewable materials (e.g., plants with a 10 year harvest cycle) are incorporated into the project. Another point is earned if 50% of the wood products are certified by the Forest Stewardship Council.

The category of Indoor Environmental Quality allows 21.7% of the possible total points available. Goals include improving indoor air quality, improving occupant comfort, and providing views to the outside. With ASHRAE Standard 62-1999 as a prerequisite, an additional point is available for installing CO_2 monitoring devices in accordance with occupancies referenced in ASHRAE Standard 62-2001, Appendix C. A point is also available for implementing technologies that improve upon industry standards for air change effectiveness or that meet certain requirements for natural ventilation. Systems that provide airflow using both under-floor and ceiling plenums are suggested by LEED documentation as a potential ventilation solution. Points are available for developing and implementing indoor air quality (IAQ) management plans during construction and also prior to occupancy. The IAQ requirements include using a Minimum Efficiency Reporting Value (MERV) 13 filter media with 100% outside air flush-out prior to occupancy.

There are points available for use of materials that reduce the quantity of indoor air pollutants in construction caused by hazardous chemicals and by volatile organic compounds in adhesives, sealants, paints, coatings, composite wood products, and carpeting. A point is offered for provision of perimeter windows and another for proving individual controls for conditioned airflow and lighting in half of the non-perimeter spaces. Points are available for complying with ASHRAE Standard 55-1992 (Thermal Environmental Conditions for Human Occupancy), Addenda 1995, and installing permanent temperature and humidity control systems. Finally, points are gained for providing 75% of the spaces in the building with some form of daylighting and for providing direct line of sight vision for 90% of the regularly occupied spaces.

In the category of Innovation and Design Process, 7.2% of total possible points are available. Innovation credits offer the opportunity for projects to score points as a result of unusually creative design innovations, such as substantially exceeding goals of a given criteria or standard.

Despite the complexity of the scoring system and the challenges of constructing a LEED building, major corporations are lining up to use the LEED process for their buildings. Genzyme Corporation's headquarters building in Cambridge, Massachusetts features 18 indoor gardens, thermostatic controls in every office, and mirrors that reflect sunlight into an atrium. The company estimates that sick leave among employees is 5% lower than what is normally anticipated and that 58% of building occupants feel they are more productive in their new $140 million headquarters building (Palmeri 2006:96).

Assessing LEED-NC: Strengths

The LEED-NC process has numerous strengths. Perhaps the greatest is its ability to focus the owner and the design team on addressing site energy and environmental issues early in the design process. The LEED design process brings architects, planners, energy engineers, environmental engineers, and indoor air quality professionals into the program at the early stages of design development. The team often adopts a targeted LEED rating as a goal for the project. A strategy evolves based on selected criteria required and points required to achieve the rating. The team members become focused on fundamental green design practices that are often overlooked when traditional design development processes are pursued.

Furthermore, the LEED program identifies the intents of the environmental initiatives. The program requirements are stated in LEED documentation and acceptable strategies are suggested. Scoring categories directly address the criterion associated with the related energy costs and environmental concerns. When appropriate, the LEED-NC program defers to engineering and environmental standards developed outside of the USGBC. The components of the program accommodate local regulations.

The educational aspects of the program, which succinctly describe select environmental concerns, cannot be understated. Professionals must be accredited. Obtaining LEED accreditation can be challenging for professionals who have not been educated in green building requirements. LEED documentation and manuals concisely present information on LEED construction requirements. Case study examples, when available and pertinent, are described in the LEED literature. A web site (www.usgbc.org) provides updated information on the program and clarifies LEED procedures and practice. To expedite the process of documenting requirements, letter templates and calculation procedures are available to accredited professionals. Training workshops sponsored by the USGBC are instrumental in engaging professionals with a wide range of capabilities. These workshops also provide a forum to explain how the rating system works.

These strengths bring credibility to the LEED evaluation process. Advocates of the LEED rating system hope it will be the pre-eminent U.S. standard for rating new green construction. To its credit, LEED is becoming a highly regarded standard and it continues to gain prestige. Nick Stecky, a LEED Accredited Professional, firmly believes that the system offers a "measurable, quantifiable way of determining how green a building is" (Stecky 2004).

Requirements for buildings to meet LEED standards are taking root. The Washington D.C. City Council passed The Green Building Act of 2006 which is unique in that it applies to both public and non-public sector buildings. It calls for all new development in the city to conform to the U.S. Green Building Council's LEED standard beginning in 2008 for publicly financed buildings. Beginning in 2008, public buildings with more than 10,000 SF will be required to be certified LEED Silver. The law requires the city's mayor to review LEED for Schools, and also circulate rules that require public schools to fulfill or exceed LEED for Schools (or use a similar rating system that mandates full-building commissioning). Private buildings in Washington D.C. will have to meet or exceed the new standards beginning in 2012. In addition, all nonresidential buildings will have to be LEED-certified.[14]

Assessing LEED-NC: Weaknesses

Despite its strengths, the LEED-NC has observable weaknesses and is not without its critics. Auden Schendler of the Aspen Skiing Corporation comments that the LEED process is too costly, too bureaucratic, fails to reward the best environmental options, and requires too much documentation (Palmeri 2006:96). He believes that for final project approval, the fees for architectural consultants and computer modeling can add as much as $50,000 to the cost of a 930 m^2 (10,000 ft^2) building (Palmeri 2006:96).

The LEED-NC registration process can sometimes be burdensome, and has been perceived as slowing down the design process and creating added construction cost.[15] Isolated cases support these concerns.[16] However, there are few comparative studies to validate claims of significant cost impact. Seemingly minor changes in the design of structures or the selection of materials can have major effects on construction costs, while being unrelated to green construction. Alternatively, there are a few case studies that suggest that there is no construction cost impact as a result of the LEED certification process. Most indicate costs ranging from 1% to 5% more than traditional construction. However, savings resulting from the use of certain LEED standards (e.g., reduced energy use) can be validated using life-cycle costing procedures.

LEED-NC fails as a one-size-fits-all rating system for new construction. There are regions where the program has been successful and others where the program has not yet been validated. Many of the LEED Platinum and Gold buildings constructed to date are located in California, Washington state, Oregon, Pennsylvania and Michigan. Other areas of the country have no or few buildings that have obtained higher levels of certification. Given the differences in geography and climate in the U.S., the lack of empirical examples leaves open the question as to whether or not the program will prove to be universally applicable.

There are other valid concerns in regard to the use of LEED-NC. In an era when many standards are under constant review, standards referenced by LEED are at times out of date. For example, the ASHRAE Standard 90.1-1999 (without amendments) is referenced throughout the March 2003 revision of LEED-NC. However, ASHRAE 90.1 was revised, republished in 2001, and the newer version is not used as the referenced standard. Currently, LEED-NC 2.2 has been updated to use ASHRAE Standard 90.1-2004.

Since design energy costs are used to score Energy and Atmosphere points, and because energy-use comparisons are base-lined against similar fuels, cost savings from fuel switching is marginalized. In such cases, the environmental impact of energy use remains unmeasured since energy units are not the baseline criteria. Using only the dollar cost of energy as the criteria (nicknamed the "green-back" approach) is not necessarily related to green construction and therefore questionable. At some point in the future, source energy may actually be priced inversely to its carbon intensity. There is no standardized energy modeling software commercially available that is specifically designed for assessing LEED buildings. LEED allows most any energy modeling software to be applied and each has its own set of strengths and weaknesses when used for energy modeling purposes. It is possible for projects to comply with only one energy-usage prerequisite, applying a standard already widely adopted, and still become

LEED certified. This allows certification of buildings that are not energy efficient. In fact, it is not required that engineers have specialized training or certification to perform the energy models.

Though LEED aspires to international stature, its documentation currently lacks System International (SI) unit conversions, which limits its applicability and exportability to countries outside the U.S.

A number of the credits or points offered by the rating system seem questionable. For example, using extensive areas of glazing to improve daylighting may substantially increase energy use in some regions, off-setting the sustainability of the design. Increased daylighting may reduce electrical requirements for interior lighting. However, if a daylighting system is not carefully designed, cooling and heating loads can be increased. While indoor environmental quality is touted as a major LEED concern, indoor mold and fungal mitigation practices which are among the most pervasive indoor environmental issues, are not addressed by LEED and are not necessarily resolvable using the methodologies prescribed. It would seem that having a LEED accredited professional on the team would be a prerequisite rather than a valid credit.

LEED projects in locations with abundant rainfall or where site irrigation is unnecessary can earn a point by simply documenting a decision not to install irrigation systems, leaving the implication that scoring systems are difficult to regionalize. The ability of the point system to apply equally to projects across varied climate classifications and zones is also questionable and unproven.

Finally, the LEED process is not warranted and does not necessarily guarantee that in the end, the owner will have a "sustainable" building. While LEED standards are more regionalized in locations where local zoning and building laws apply, local regulations can also pre-empt green construction criteria. Of greater concern, is that it is possible for a LEED certified building to *devolve* into a building that would lack the qualities of a certifiable building. For example, the owners of a building may choose to remove bicycle racks, refrain from the purchase of green energy after a couple of years, disengage control systems, abandon their M&V program after the first year, fail to re-commission or maintain equipment and control systems, or remove recycling centers... yet retain the claim of owning a LEED certified building.

Despite the growing popularity of the LEED rating systems, only a small percentage of projects actually apply for the process in the U.S. The planned construction of a new $450 million dollar University of Kentucky hospital in Lexington, Kentucky presents an opportunity for a green building… yet the university is not seeking LEED certification. Typical of many building operators, Bob Wiseman, the University of Kentucky president for Facilities Management, states that "we are producing much greener buildings but haven't gone through the formal certification process yet… but (someday) we will."[17]

MEASUREMENT AND VERIFICATION IS VITAL

Increasingly, measurement and verification (M&V) is being used for green building projects. M&V refers to the process of identifying, measuring and quantifying utility consumption patterns over a period of time. Measurement and verification can be defined as the set of methodologies that are employed to validate and value proposed changes in energy and water consumption patterns that result from an identified intervention (e.g., set of energy conservation measures) over a specified period of time. This process involves the use of monitoring and measurement devices and applies to new construction and existing buildings and facilities.

Measurement and verification methodologies are used for LEED projects, performance based contracts, project commissioning, indoor air quality assessments and for certain project certifications. By establishing the standards and rules for assessment criteria, the concept of measurement and verification is a key component of energy savings performance contracts. In performance contracts, the performance criterion of a project is often linked to guaranteed cost saving that are associated with the facility improvements.

Technologies and methodologies are available to measure, verify and document changes in utility usage. Tools are available in the form of M&V guidelines and protocols that establish standards for primary measurement and verification options, test and measurement approaches, and reporting requirements. Using procedures identified in the guidelines and protocols, a measurement and verification plan is developed to validate savings and to serve as a guide as the process unfolds.

The process of measurement and verification typically involves five primary steps: 1) performing the pre-construction M&V assessment; 2) developing and implementing the M&V Plan; 3) identifying the M&V project baseline; 4) providing a post-implementation report; and 5) providing periodic site inspections and M&V reports.

The theoretical basis for measurement and verification in regard to assessments of resource usage over comparative periods of time can be explained by the following equation:

$$\text{Change in Resource Use}_{(adj)} = \Sigma \text{ Post-Installation Usages } +/-$$
$$\Sigma \text{ Adjustments} - \Sigma \text{ Baseline Usages}$$

Baseline usages represent estimates of "normal" utility usages prior to implementation of any cost savings improvements. Adjustments are changes in resource use that are not impacted by an intervention and are considered exceptional. The term intervention refers to the implementation of a project that disrupts "normal" or "projected" utility usage patterns. Examples of these interventions include electrical demand reduction measures and energy and water conservation measures. Post-installation usage refers to resource consumption after the intervention has been performed. Using this formula, negative changes in resource use represent declines in adjusted usage while positive changes represent increases in adjusted usage.

The International Performance Measurement & Verification Protocol (IPMVP, April 2003) is the most widely used M&V protocol.[18] Its standards will be used as an example. The measurement and verification options in the IPMVP provide alternative methodologies to meet the requirements for verifying savings. The four measurement and verification options described in the IPMVP are summarized as follows:

Option A: Partially measured retrofit isolation.

Using Option A, standardized engineering calculations are performed to predict savings using data from manufacturer's factory testing (based on product lab testing by the manufacturer) and a site investigation. Select site measurements are taken to quantify key energy related variables. Variables determined to be uncontrollable can be isolated and stipulated (e.g., stipulating hours of operation for lighting system improvements).

Option B: Retrofit isolation of end use,
measured capacity, measured consumption

Option B differs from Option A, as both consumption (usage) and capacity are measured (output). Engineering calculations are performed and retrofit savings are measured by using data from before and after site comparisons (e.g., infrared imaging for a window installation or sub-metering an existing chiller plant).

Option C: Whole meter or main meter approach

Option C involves the use of measurements that are collected by using the main meters. Using available metered utility data or sub-metering, the project building(s) are assessed and compared to base-lined energy usage.

Option D: Whole meter or main meter with calibrated simulation

Option D is in many ways similar to Option C. However, an assessment using calibrated simulation (a computer analysis of all relevant variables) of the resultant savings from the installation of the energy measures is performed. Option D is often used for new construction, additions and major renovations (e.g., LEED-NC certifications for new construction).

Depending on site conditions and the technologies being used, each approach has discrete advantages and disadvantages. For example, in cases where facilities have main meters in place, Option C may be preferred. In new construction, Option D is the favored alternative.

With recent advancements in monitoring and measurement technologies, it is possible for energy engineering professionals to log and record most every energy consumption aspect of the energy conservation measures they implement. Examples include the use of data loggers, infrared thermography, metering equipment, monitors to measure liquid and gaseous flows, heat transfer sensors, air balancing equipment, CO_2 measurement devices and temperature and humidity sensors. Remote monitoring capabilities using direct digital controls (DDC), fiber optic networks and wireless communication technologies are also available.

Measurement and verification costs vary as a function of the methodology, the complexity of the monitoring, the technologies employed, and the period of time that M&V needs to be performed. As applied monitoring technologies evolve and become accepted by the marketplace, costs for installed monitoring equipment will continue to decline as the capabilities of monitoring technologies continue to improve.

MEASUREMENT AND VERIFICATION FOR LEED PROJECTS

While an M&V credit is available for LEED projects, there is no requirement for a credentialed measurement and verification professional to be part of the M&V plan development or the review process. In fact,

LEED projects are not *required* to undergo the rigor of M&V—*it is optional*. To obtain a LEED credit, there is no requirement that M&V be performed to state-of-the-art standards—*only that it meets IPMVP 2001 requirements*. Regardless, measurement and verification is vital to the success of projects. Without a requirement that LEED projects perform M&V for an extended period of time, it is difficult to determine if predictive pre-construction energy modeling was accurate, or if predicted cost savings reductions were actually achieved. The lack of mandates to determine whether or not LEED buildings actually behave and perform as intended from an energy cost standpoint is a fundamental weakness of the LEED process. Prerequisite commissioning does not resolve this. Without M&V, any projected life-cycle cost savings resulting from the project cannot be accurately validated, risking illusionary energy cost savings.

There are a few problems with how the LEED certification process handles M&V. LEED excludes Option A as an acceptable M&V solution. Option A was updated in the 2003 version of the IPMVP and there was no rational argument to exclude Option A as an unacceptable M&V alternative. Another problem with LEED's use of the IPMVP is that no standardized requirements are provided for calibration when Option D is used. The calibration methodologies vary depending on the software, who is using it and how it is being used. Without a mandated M&V requirement based on actual performance data for a minimum period of two to three years, predictions of energy cost savings cannot be validated and verified.

GREEN CONSTRUCTION IN SCHOOLS

Schools are important. A "Green School" is a "high-performance" school. More and more, the professionals who design and construct new schools or major school additions are aware of their responsibilities. Between the ages of 5 and 18, students will spend roughly 14,000 hours of their lives in them. High performance schools require green construction practices. High performance schools provide comfort and a healthy environment for students and staff. They use energy and other resources efficiently and have lower maintenance costs. High performance schools involve a commissioning process, are environmentally responsive, are safe and secure, and feature stimulating architecture (Eley 2006:61)

To spearhead the "Green Schools" effort, standards are being developed and employed. A number of recently developed manuals are available that provide guidance on how to implement green construction practices. The Sustainable Buildings Industry Council (SBIC) has produced a *High Performance Schools Resource and Strategy Guide*[19] to show school building owners and operators how they can initiate a process that will result in better buildings—ones that provide students with better learning environments.

For new construction, Kindergarten -12th grade (K-12) school systems in New Jersey, California and elsewhere have adopted their own sustainable building standards. Kentucky is an example of one state that is developing new standards for "Green and Healthy Schools." The US-GBC is developing a LEED rating system for educational structures.

New Jersey

New Jersey has codified its construction practices for schools in a document entitled *21st Century Schools Design Manual*, developed by the New Jersey Schools Construction Corporation. Its performance objectives are structured to create schools that are healthy and productive, cost effective, educationally effective, sustainable, and community centered.[20] The manual establishes a set of 24 comprehensive design criteria for schools and mandates that design teams consider the following categories of issues in new construction:[21]

Engineering Aspects

Energy Analysis	Renewable Energy
Life Cycle Costing	Efficient HVAC
Efficient Building Shell	Efficient Electrical Lighting
Water Efficiency	Thermal Comfort
Commissioning	Indoor Air Quality

Architectural Design

Acoustic Comfort	Flexibility and Adaptability
Stimulating Architecture	Environmentally Responsive Site Planning
Daylighting	Learning Centered Design
Environmentally Preferable Material & Products	

Other Considerations

Information Technology	Visual Comfort
Accessibility	Safety and Security
Community Involvement	Catalyst for Economic Development
Community Use	

4-7. Solar Collector Array at Twenhofel Middle School, KY

Each topic covered in the manual provides a set of recommendations and identifies applicable standards. Interestingly, the manual's recommendations are similar to LEED-NC prerequisite requirements and elective credits. Schools meeting the standard can be considered to be high performance, LEED-like facilities—while avoiding the rigor and costs of LEED certification. The manual is a call for integrated design solutions to establish sustainable design as a cost effective means of achieving high performance schools in New Jersey.

California

The Collaborative for High Performance Schools (CHPS) is a non-profit organization established in 2000 to raise the standards for school facilities in California. Goals include improving the quality of education by facilitating the design of learning environments that are resource efficient, healthy, comfortable and well lit—amenities required for a quality educational experience (Heinen 2006:1). The CHPS program for high performance schools has been adopted by 14 school districts in California. The standard was recently updated and reissued as the *2006 CHPS Criteria* and applies to new construction, major renovation, and additions to existing school facilities (Heinen 2006:1). The CHPS offers a *Best Practices Manual* that details sustainable practices and resources available for the planning, design, criteria, maintenance and operations, and commissioning of high performance schools.

Kentucky

To implement sustainable technologies to help improve education in schools, Kentucky has developed its "Green and Healthy Schools Program," a voluntary program to encourage green standards for schools.

Twenhofel Middle School in northern Kentucky uses a number of technologies to reduce energy and water consumption. The building shell provides extensive use of insulation and high performance fenestration. The school uses a geothermal heat pump system for heating and cooling. Rainwater collected by a metal roof and drainage system is treated and used for non-potable needs. A central computer control system manages the energy used in the building. In addition, solar panels, installed on the roof, collect energy to generate a portion of the school's electrical requirements.

4-8. Display in Lobby of Twenhofel Middle School, KY

The building makes extensive use of daylighting in classroom areas. Sensors in the classrooms detect light levels to allow fixtures to adjust light output in response to the daylight available in classrooms. There is a touch screen monitor in the lobby of the school that allows students to monitor information concerning the water collection system, solar output and the geothermal loop. Science programs use the school as a learning laboratory. Students often provide guided tours to explain the sustainable technologies that were incorporated into the design of the facility.

CONCLUSIONS

Buildings are resource intensive in their construction and operation. Buildings are also complex systems. Today, buildings can be constructed with features that allow them to use less energy and consume fewer resources. The ideal of sustainable buildings is a response to the energy and environmental impacts which accompany every one of them. Developing a green building project is a balancing act and requires a series of trade-offs. It involves considering how buildings are designed and constructed—at each stage of the project delivery process.

Standards are constantly evolving. The 2006 International Energy Conservation Code (IECC), for example, requires that certain energy efficient design methodologies by used in construction. The Code "addresses the design of energy-efficient building envelopes and installation of energy-efficient mechanical, lighting and power systems through requirements emphasizing performance."[22] It is comprehensive and provides regional guidelines with specific requirements for each state in the U.S. New construction materials and products are available that offer new design solutions.

There are many opportunities to include green design features and components in buildings to make them more sustainable. Yet, there are differences in the standards for green construction. While energy assessment systems for buildings (the Energy Star program, for example) typically focus the analysis on source energy, the USGBC's LEED program considers the cost of energy as its primary rating criteria. A number of assessment systems for sustainable buildings are now being used throughout the developed world. LEED-NC is becoming a widely adopted standard for rating newly constructed "green" buildings and projects in the U.S. and elsewhere. LEED projects as rated as Certified, Silver, Gold or

Platinum using a credit-based system. LEED projects are credited for design attributes, costs of energy, environmental criteria and use of green building standards.

In the U.S., the LEED-NC program has proven successful by offering its credit-based rating system for green buildings. Its popularity is gaining momentum. A total of 49 localities and 17 state governments now encourage the use of green building practices, policies and incentives. Their number is growing. The USGBC estimates that 5%—almost $10 billion—of current nonresidential construction in the U.S. is seeking certification (Palmeri 2006:96). According to Richard Fedrizzi of the USGBC, "this movement has created a whole new system of economic development... *We are at a tipping point.*"

Perhaps LEED's greatest strength is its ability to focus the owner and her design team on energy and environmental considerations early in the design process. Its greatest weaknesses are its focus on energy costs rather than energy use and its lack of mandated requirements for measurement and verification of savings. Today, there are over 3,500 projects that have applied for LEED certification. Due to the program's success in highlighting the importance of energy and environmental concerns in the design of new structures, it is likely that the program will be further refined and updated in the future to more fully adopt regional design solutions, provide means of incorporating updated standards, and offer programs for maintaining certification criteria.

Establishing rating systems for sustainable structures, such as BREEM and LEED, is difficult due to the often subjective concepts involved, the evolving nature of standards, and the local variability of construction practices. Future research will hopefully respond to concerns about increased construction costs, and actual energy and environmental impacts. Measurement and verification has an important role to play, as it outlines procedures that can be followed to verify utility cost avoidance from energy and water saving projects.

In the states of New Jersey, California and Kentucky, standards are being used and developed for high performance schools. High performance schools have many qualities that include a fresh look at the architectural, engineering and educational aspects of school design.

Many green building technologies such as high efficiency windows, solar arrays and day-lighting applications are easy to find when walking by or through a building. On the other hand, it is discouraging to owners that many important engineered features of green buildings are hidden from view in the mechanical rooms and spaces not visible to the ordinary visitor. "Achievements in sustainable building design often go unnoticed by people who visit, work or study" in a green building (Ling 2006:1). Examples of these technologies include computer control systems to manage energy and water use, rainwater collection systems, lighting control systems, under-floor air flow systems, etc. Many green technologies used in structures require a trained eye to observe.

This chapter considered sustainable construction, the features of green buildings, and the importance of resource and energy use in green construction. There are a growing number of buildings, especially new ones which incorporate many aspects of green design. It is likely that the use of green construction techniques and technologies will continue to expand. In the future, more examples of green buildings will be available.

Endnotes

1. Koomey, J., Weber, C., Atkinson, C., and Nicholls, A. (2003:1209). Addressing energy-related challenges for the U.S. buildings sector: results from the clean energy futures study. *Energy Policy*. 29. Elsevier. This study provides an analysis of building sector energy use in 1997 and projects usage, carbon and dollar savings potential for the following 20 years based on two scenarios.
2. Bowman, A. (2003, August 28-31). From *Green politics in the city*. Presentation at the annual meeting of the American Political Science Association. Philadelphia.
3. Governor's Green Government Council. What is a green building? *Building Green in Pennsylvania*. http://www.gggc.state.pa.us/gggc/lib/gggc/documents/whatis041202.pdf, accessed 18 January 2007.
4. Ibid.
5. Siemens AG. (2007). *Combat climate change—less is more*. http://wap.siemens-mobile.com/en/journal/story2.html, accessed 3 July 2007.
6. The *ASHRAE green guide* is a primer for green construction practices.
7. The average outside air temperature during the study period was 46 degrees F. The assessment was based on billings provided by the local utility. It is likely that the substantially larger electrical usage experienced by the newer building was also due, in part, to two desktop computers which operated almost continuously during the study month, and to the larger area of interior lighting. The newer brick building also had 11 outside light fixtures while the older frame building only had 2.
8. USEPA. *Superior energy management creates environmental leaders*, Guidelines for energy management overview. 5www.energystar.gov/index.cfm?c=guidelines.guidelines_index, accessed 3 January 2007.
9. USEPA. *Be a leader—change our environment for the better*. Portfolio manager overview. www.energystar.gov/index.cfm?c=spp_res.pt_neprs_learn, accessed 3 January 2007.
10. USEPA. *Portfolio manager overview*. www.energystar.gov/index.cfm?c=evaluate_performance.bus_portfolimanager, accessed 3 January 2007.
11. Lincoln Hall, a LEED building at Berea College in Berea, Kentucky, actually uses copper downspouts.
12. International Design Center for the Environment. *eLCie—Building a sustainable world*. www.elcie.org, accessed 4 July 2007.

13. For a listing and explanation of the various rating systems developed for LEED, see the U.S. Green Building Council rating systems website. www.usgbc.org/Display/Page.aspx?CMSPageID=222, accessed 2 January 2007.

14. American Institute of Architects. (2007, 4 January). D.C. requires private buildings to go green. *The AIA Angle.*

15. Yudelson, J. (2006, 17 July). *Engineers and green buildings—Get green or be left behind.* http://www.igreenbuild.com/_coreModules/content/contentDisplay.aspx?contentID=2480, accessed 4 July 2007. Yudelson cites a survey by Turner Construction in 2005 involving 665 executives that indicated 68% believed "higher construction costs were the main factor discouraging construction of green buildings." In addition, 64% of the respondents cited a "lack of awareness of the benefits of green buildings" and 50% cited long payback periods associated with green construction practices.

16. Kentucky's first LEED-NC school, seeking a Silver rating, was initially estimated to cost over \$200/ft^2 (\$2,152/m^2) compared to the local standard costs of roughly \$140/ft^2 (\$1,505/m^2) for non-LEED construction.

17. State hopes to create more "green buildings." *The Courier-Journal*, 26 December 2006.

18. A new 2007 edition of the IPMVP has been developed by the Efficiency Valuation Organization (EVO). See http://www.evo-world.org/index.php?option=com_content&task=view&id=60&Itemid=79.

19. Available for purchase from the SBIC bookstore by accessing http://www.sbicouncil.org/store/index.php.

20. See *High performances schools New Jersey.* www.njhighperformanceschools.org/performance.html, accessed 7 January 2007.

21. See the *21st century schools design manual*, New Jersey Schools Construction Corporation, 30 September 2004, available at http://www.hpsnj.org/PDF/what/Design_Manual.pdf, accessed 8 January 2007.

22. International Code Council (2003). *International Energy Conservation Code.* p. iii.

Notes

Portions of this chapter (Sustainable Buildings) were previously published in the *Encyclopedia of Energy Engineering* in the chapter entitled "LEED-NC Leadership in Environmental Design for New Construction" by Stephen A. Roosa, edited by Barney Capehart. Permission granted by Taylor & Francis Group LLC, a Division of Informa, PLC, Copyright 2007.

Other portions of this chapter that outline measurement and verification methodologies are from Shirley J. Hansen's *Performance Contracting—Expanding Horizons*, Second Edition in the chapter entitled "An ESCO's Guide to Measurement and Verification" by Stephen A. Roosa. Permission granted by The Fairmont Press, Copyright 2006.

Chapter 5

Sustainable Energy Solutions

By Sieglinde Kinne

"The human appetite for energy appears to be insatiable. There are good reasons for the continued growth of energy consumption in the future: the survival needs of the under-privileged billions, increased adult life expectations stemming from the population boom... and the desirable goal of improved quality of life for everyone. But there are equally good reasons for carefully examining what the consequences of energy growth will be after pres-ent consumption rates have quadrupled. When do we melt the polar icepacks? How much land can we afford to set aside for energy plants? How much photosynthetic smog can we tolerate?"

ALFRED M. WORDEN,
PROCEEDINGS OF THE 1ST WORLD ENERGY ENGINEERING CONGRESS, 1978.

In the U.S., a country that has 5% of the world's population and con-sumes 24% of the world's energy, over 80% of energy requirements are met by fossil fuels. This shapes issues such as climate change, air pollution and exploration for oil in ecologically sensitive locations. While the U.S. lacks a central government mandate for generating energy from renew-able sources, at least 18 states have established minimum percentages of energy that must be derived from alternative energy sources.[1] In addition, several states have established regulatory requirements for greenhouse gas management.

Our fossil fuel supply is the lifeblood of civilization as we know it today. The relationship between energy consumption and technological advances is at the crux of the challenges we face and the solutions we can provide. Our use of more advanced technologies offers the opportunity to improve comfort and living standards. As fossil fuel resources are be-ing rapidly consumed, we are now on a collision course with the limits of available resources. Global reserve/production ratios of coal, natural gas and oil indicate that current reserves will last another 221, 61 and 40 years, respectively (Saha 2003:1053).

The resource we are most concerned about is our biosphere, the complex system of gases, living matter and bodies of water that sustains life on this planet. Carbon emissions and other green house gas emissions from man-made sources are now believed—with very high degree of certainty—to be disturbing the thermal equilibrium of our world. This has created a need to reassess how we consume carbon-based fuels. Humankind is awakening to the need to be more adaptive and resourceful.

The nature of different energy resources makes them more or less suitable for different uses. Coal, for example, is not a practical fuel for transportation uses. Because of the variable nature of different energy sources and the importance of transportation, liquid fuels with high fuel content have a greater intrinsic value. Battery powered ground vehicles may be practical but air transportation powered by batteries is unlikely to ever be practical.

Will technology be the hero or the villain in this scenario? This chapter outlines the ways that new technologies are part of the solution. Energy efficiency and the use of alternate energy sources play an important role. Two main challenges, depleting reserves of crude oil and environmental issues associated with fossil fuel, are first discussed.

PEAK OIL

The U.S. Department of Energy and scientists worldwide have been concerned for decades about the depletion of petroleum resources. Discoveries of new oil peaked in the 1960's, and since the mid-1980s oil production has exceeded new discoveries. The rising demand for oil, increasing by 1% to 2% per year, and the declining production of oil, declining by 3% per year, pose a challenge for the world's economies. Demand for petroleum resources from the rapidly growing populations and economies in China, India and other countries, is increasing.

The availability of inexpensive oil supplies has allowed 150 years of economic growth and supported a world-wide industrial revolution. It has also enabled a culture of consumerism and of suburban lifestyles in developed countries—contributing to dependence on automobiles. Increases in the price of traditional mobility fuels effects the ability of all but the wealthy to achieve higher living standards. Rising transportation costs affect the costs of all consumer products. The political, economic and social ramifications of large increases in petroleum prices are very serious.

For example, in Nicaragua increased bus fares caused by rising fuel costs affect a large portion of city population. In reaction, a student-led protest once shut down the city and busses were burned in the streets.

In 2006, at both the Association for the Study of Peak Oil Conference in Boston and the 7th International Oil Summit in Paris, scientists of the world came together to discuss the situation. Most energy analysts agree that oil production has either already peaked or will peak between now and 2040.

Oil peaking is not the same as "running out" of oil. The oil "peak" refers to the time when the production of oil begins to decline because oil deposits which are easily recoverable with today's technology have been depleted and the Energy Return on Energy Investment (EROEI) decreases. The EROEI for oil is already declining. The crude oil EROEI was once approximately 100:1, meaning that the energy equivalent of one barrel of oil was used in obtaining 100 barrels of crude from reserves. This EROEI has fallen over the years and now the average is closer to 20:1. This means that today, it requires five times more energy to extract oil than it did in the past.

"Unconventional" sources of petroleum are becoming more desirable, include oil shale, tar sand, and ultra-heavy oil deposits. Important oil deposits are contained in these geological formations, yet tapping them is costly. More energy is required to produce them. For example, the EROEI for oil shale is less than 3:1. Extraction and processing of crude oil from oil shale also requires large amounts of water and causes environmental damage as well.

ENERGY AND CARBON EMISSIONS

Concerns about the rising temperature of the biosphere are causing us to re-evaluate how we produce and consume energy. The last 150 years has been characterized by industrialization and the last 100 years by fossil-fuel consuming transportation technology (e.g., the personal automobile and the airplane) along with infrastructure to support its use. Fossil-fuel generated electric power, with its associated infrastructure, is another development of this era, providing unprecedented increases in productivity and in the quality of life.

During the past 20 years, about three-quarters of human-made carbon dioxide emissions were due to burning fossil fuels.[2] In the U.S., 82%

of carbon dioxide emissions result from fossil fuels combustion. Figure 5-1 shows the disproportionate impact of coal and petroleum consumption on these emissions.

"Clean coal" and carbon sequestration technologies are being developed in the U.S., Australia and elsewhere. Clean coal refers to technologies that pulverize and gasify coal to remove impurities such as sulfur and mercury prior to combustion. The particulate matter resulting from the combustion is then captured and either re-burned or stored. Carbon sequestration refers to the capture and storage of carbon dioxide (e.g., depositing it into underground storage) rather than releasing it into the atmosphere. Proponents of these technologies point to the potentially huge reduction in air emissions as compared to coal-fired generation of today. Critics point to its enormous processing costs and the many technical difficulties involved in scaling production to large-scale power plants. Despite the enormous costs and resources required, we

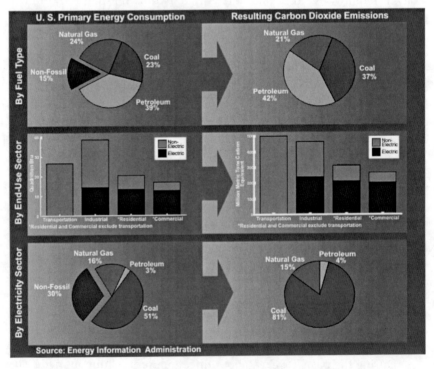

Figure 5-1. U.S. Primary Energy Consumption and Carbon Dioxide
Source: Energy Information Administration

can expect to see several commercial-scale facilities constructed within the next ten years. In the U.S., the FutureGen project, a $1 billion, 275 megawatt prototype plant will be the cleanest coal-burning plant in the world when completed.

While clean coal with carbon sequestration may resolve emissions problems from coal-fired power generation, these processes fail to mitigate environmental damage from coal mining. The hazardous materials in the slurry created during the washing of coal cause a new source of pollution. In summary, clean coal may develop into a viable short-term technology for addressing air emissions, but due to its high costs, the process cannot be considered to be sustainable. In fact, "clean coal" may be an oxymoron.

POTENTIAL FOR ENERGY EFFICIENCY

Energy efficiency in the U.S. and in other industrialized nations has been improving gradually since the 1970s. Many industries are now able to produce more products and services with less energy. Relative to gross domestic product, U.S. energy efficiency improved 49% between 1949 and 2000.[3] Nevertheless, as the U.S. population expanded from 149 million people in 1949 to 281 million in 2000 (an increase of 89%), total energy consumption grew 208%.[4]

Energy intensity can be defined in a number of ways. On a macroeconomic scale, it is described in terms of energy consumption per gross domestic product (e.g., Kcal or kBtu per $GDP). On a facility level, energy consumption intensity is quoted in units of energy cost per unit area. In a production facility, energy intensity is typically discussed as energy use per unit of production. The appropriate index depends on the context and scale being considered, yet all are intended to allow for energy use comparisons among varying time periods, processes and applications.

Figure 5-2 shows an overview of all energy consumed and illustrates losses in the systems. The electrical infrastructure, composed of resource delivery systems, power plants and the transmission "grid," is inherently wasteful. Of the total primary energy inputs, only one-third is delivered to the end users of the power. Two-thirds is lost due to conversion and transmission, with small amounts of input energy also consumed in lighting, heating, cooling and other operations at power plants.

The energy losses illustrated in Figure 5-2 predicate a need to address electrical energy efficiency, both from a building energy use per-spective and in transforming the infrastructure itself. Cogeneration and co-location are two approaches to reducing waste in power generation. Cogeneration is a way of generating electricity while simultaneously us-ing the heat normally lost during the energy conversion process for a use-ful purpose. Co-locating refers to finding synergies between multiple fa-cilities or processes so that wastes generated by one can be used in other nearby facilities.

Transportation energy consumption is also shown in Figure 5-2 as being grossly inefficient. There remains potential to improve transporta-tion efficiency by using vehicles that use less energy.

OPPORTUNITIES FOR ENERGY EFFICIENCY IMPROVEMENTS

A common perception is that pollution from cars and factories are the leading cause of global warming, but buildings account for 76% of electric consumption in the U.S. and nearly half (48%) of all

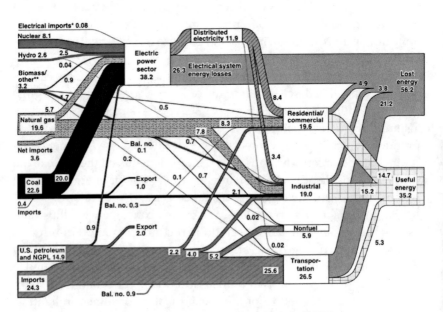

Figure 5-2. U.S. Energy Flow Trends (2002)
Net Primary Resource Consumption—97 Quads
Source: Lawrence Berkeley National Laboratory

greenhouse gas emissions—more than either transportation (27%) or industry (25%). It is possible that more than half of the projected future growth in electrical and natural gas demand by buildings can be met by improving the efficiency of existing and new systems.[5] Energy efficiency is an important energy *source*. Investments in energy efficiency, when compared to investments in renewable energy, are often a more cost-effective means of achieving economic and environmental benefits.

Opportunities for efficiency improvements are found in every architectural and mechanical facet of a home or facility. This section describes some of the most common and exciting opportunities for energy efficiency. These opportunities fall into six categories: 1) lighting; 2) HVAC (e.g., heating, ventilating and air conditioning); 3) building envelope (e.g., doors, windows, and insulation); 4) plug loads (e.g., appliances and electronics especially); 5) water heating; and 6) other opportunities. The category of "other" includes various specialized applications such hospital equipment and manufacturing equipment. Efficiency measures in each category generally involve equipment upgrades, replacement or improved scheduling and controls.

Lighting

Lighting technologies are being continuously improved. Enhancing the efficiency of lighting systems begins with good design. Natural lighting can be made available in every space of a home or building. In existing spaces, daylighting can be optimized by using window treatments, light shelves and daylight-sensing controls that adjust artificial light levels. Window treatments, such as tinted films, double-hung horizontal blinds (meaning they can be lowered from both the top and bottom), or the addition of sheer curtains, encourage occupants to take advantage of natural light. Light shelves are window features that are installed either on the inside or the outside of south-facing windows. Located slightly above eyelevel, light shelves reflect natural light toward the interior ceiling and into the other parts of the room. This creates a more pleasant light throughout the room, while reducing glare and preventing overheating of interior spaces near windows.

Daylight-sensing controls are often used with commercial and institutional lighting systems to ensure that when natural light is available, light fixtures near windows are turned off or have reduced the artificial light levels, thus lowering electrical energy requirements. These controls are becoming more widely specified for newer buildings, as the technology becomes less expensive and more reliable.

Appropriate fixtures and lamps are part of the lighting system efficiency solution. Lighting systems today are more energy efficient than lighting sold in the past. Further improvements in lighting system efficiency can be expected in the future. Commercial building codes that prescribe limits for the watts per unit area of floor space are setting aggressive standards for efficiency. It is important to investigate lighting options and install the best equipment available.

Fluorescent lighting systems are common for commercial spaces. New fluorescent lamps provide as much as 20% to 30% more light output per watt. Electronic ballasts that are available today consume less electricity than the magnetic ballasts that were the previous standard and tube efficacy (lumens per Watt) has also increased.

Compact fluorescent lamps (CFLs) are available now in many styles to replace nearly every type of lamp, including standard tungsten-filament incandescent lamps, flood lights, and decorative lamps. CFLs use 60 to 70% less power than incandescent lamps to provide equivalent light. This technology has improved in recent years and past concerns about outdoor use, buzzing, and compatibility with solid state controls have been resolved. The retail price of dimmable CFLs has declined to roughly 25% of the price just a few years ago. CFL lamps are increasing in popularity—to the point that some major manufacturers of incandescent lamps are considering phasing out production. Utilities are supporting the use of CFLs as a means of reducing electrical demand. In 2007, the Pacific Gas and Electric Company in southern California announced plans to distribute five million CFLs to its customers at no cost.

High intensity discharge (HID) lighting is commonly used outdoors and in spaces with high ceilings such as warehouses, manufacturing facilities and sports facilities. This type of lighting includes metal halides, mercury vapor and high and low-pressure sodium fixtures. HID lighting has undergone significant technological improvements, both in increased light output and in more efficient ballast technology. One breakthrough is the commercialization of pulse-start metal halide lamps. These lamps are 20% to 30% more efficient than previous designs. Metal halide fixtures are also making inroads into the areas of decorative and track lighting, using as much as 60% to 70% less energy when compared to incandescent or halogen lighting.

HVAC

Temperature conditioning and ventilation systems provide major opportunities for energy savings. The choice of HVAC systems, their associated controls, and the fuels they use are important for optimizing energy efficiency. It is important to select the best available HVAC options for the local climate and the needs of the interior spaces. In recent years, there have been a number of technology improvements that can improve efficiency and reduce the energy consumption of HVAC systems.

One technology involves the use of heat pumps. Heat pumps operate using vapor compression systems—not unlike standard air conditioning—but are designed to provide both heating and cooling. When these systems are designed to use a heat-rejection loop in bored out wells below ground, they are called ground-source heat pumps (GHP) and operate extremely efficiently. The winter coefficient of performance (defined as energy delivered in the form of heat divided by the energy consumed by the equipment) is from 3.0 and 4.0. These systems require an initial investment to drill the bore holes yet outperforms most other systems that are available for commercial and residential buildings. The use of GHP systems is becoming standard in areas of Colorado, Kentucky and elsewhere. In Kentucky, approximately 15% of new K-12 schools are using GHP systems. In moderate climates, summer cooling loads can be a challenge for these systems. One solution is to install a hybrid system that uses a cooling tower to supplement the cooling capacity when the weather is the warmest. This reduces the amount of bore holes required to meet the building peak cooling load. Since GHP systems are electrically driven, the cost to operate them needs to be carefully assessed before deciding to use them.

Natural gas, propane and fuel oil boilers and furnaces are now available that are 15% to 30% more efficient than the "standard" efficiency models. A breakthrough was achieved in the development of condensing systems. In the past, the peak efficiency was limited by the minimum temperature needed to safely exhaust combustion gases (approximately 120° to 132°C, or 250° to 270°F). This temperature ensured that natural convection would carry the gases safely to the outside and that the water vapor (a combustion by-product) would not condense inside the exhaust pipe and cause corrosion. This barrier was overcome with the design of special fans to facilitate the movement of the combustion gases in PVC exhaust pipes that contain the acidic water that condenses out of the cooler gases. The heat released as the water vapor condenses is now able to be captured and used, resulting in combustion efficiencies of up to 98%, an improvement from the limit of approximately 82% using conventional systems. For these systems, a small exhaust fan is required, resulting in typical system efficiencies of 92% to 95%.

Improvements in HVAC control technologies provide significant energy savings. Occupancy sensors, carbon dioxide sensors and better building automation systems can ensure that proper indoor temperature and air quality are achieved while minimizing HVAC energy use.

Other HVAC designs that offer both improved energy efficiency and comfort include radiant floor systems and in-floor delivery systems. For areas such as loading docks where high outdoor air infiltration rates are possible, energy savings can be achieved by using radiant tube heating instead of unit heaters.

Building Envelope

Creating a tightly sealed building with the appropriate amount of insulation for the roof and exterior wall areas is critical to building efficiency. Exceeding building code minimums by 15% or more is a recommended practice for an energy efficient building. Reflective roofs or green roofs can reduce the need for cooling by keeping the sun from heating the roof. The term green roof refers to a roofing system that incorporates a layer of soil or humus and live plants to provide excellent insulation.

New window technologies are being used in high-performance buildings. In the U.S., the energy lost through windows represents 30% of the heating and cooling energy consumed by buildings, or a total 4.3 kilojoules (4.1 quadrillion Btus) of primary energy.[6] In some buildings, heat losses and gains through windows can account for as much as half of the energy used. Using window technologies that have good thermal insulation properties reduces unwanted heat gain and loss. Since glass has poor insulation characteristics, past design practices have focused on adding storm windows or minimizing window areas. Improperly designed or installed windows create opportunities for air infiltration. The principle way to reduce these losses has been to seal frames with caulk and foam insulation. Installing double or triple glazed windows, some with a removable screen panel that allows another storm panel to be installed in extreme weather conditions, can substantially reduce energy losses. Since dead air spaces can help reduce heat losses, window manufacturers have increased the air space between glass panels.

Low conductivity (or "low-e," low-emissivity) glazing that uses transparent coatings of silver or tin oxide can be selected to maximize the infrared radiation reflected back into occupied spaces.[7] Using low conductivity gases in the dead air space between glazing panels, such as argon, carbon dioxide, krypton, or a mix of argon or krypton, further reduces the transfer of heat through windows.[8] The type of low-e glazing used varies depending on the climate. High solar transmitting glazing is used for climates where utility bills are highest in winter and low-solar transmitting glazing is designed for climates where utility bills are highest in summer.[9] In warmer climates, reflective tinted glazing, window films, and overhangs reduce undesirable solar heat from entering buildings. For cooler climates, there are now windows available that have positive energy ratings (ER), meaning that they actually contribute to heating buildings using passive solar heat gains.[10]

Entry vestibules and durable door seals reduce unwanted outside air infiltration. Providing for natural ventilation by using operable windows and fans reduces building energy consumption when conditions allow. The thermal mass of the building must be analyzed in order to understand how the building will react to weather conditions. Flooring materials in areas that receive direct sunlight absorb heat and can be intentionally designed to help control indoor temperatures. For example, a four-inch-thick slab of concrete covered with a dark-colored tile will reduce area temperature fluctuations. Properly sized overhangs on south-facing elevations can be used to provide summer shading for buildings in the northern hemisphere.

Plug Loads

Our homes and buildings are becoming increasingly electrified. Use of personal electronics and other personal appliances, such as mini-fridges and microwaves, is increasing. The use of computers has also increased cooling loads for most commercial and institutional buildings. These new electrical devices provide amenities, improving productivity while increasing energy consumption. Their energy impact can be minimized by choosing energy-efficient equipment and appliances. LCD monitors, computer equipment with power saving features, television equipment that draw less power when turned off, and Energy Star rated appliances can minimize energy use.

Water Heating

Water heating energy use can be reduced by installing efficient water heaters and water saving fixtures at faucets and shower heads. Instantaneous use water heaters and condensing water heaters are the best choices for efficiency. In some cases, the water temperature set point may be higher than necessary and can be lowered to reduce energy use. Water heating can be as much as 10% to 15% of total residential energy use. The costs of water heating can be reduced by installing timers to cycle water heaters off when not needed, adding insulating blankets to the outside of tanks, insulating pipes or using solar hot water heating.

Other Equipment

Over 20% of commercial energy consumption is by specialized equipment such as laundry equipment, kitchen equipment and medical equipment. Considering the energy use of the equipment prior to purchasing is important whenever choices are available. Avoiding equipment with unnecessarily high energy use and operating costs can yield substantial savings.

NEW BUILDINGS AND ENERGY EFFICIENCY

The U.S. Conference of Mayors and the American Institute of Architects (AIA) have launched the 2030 Initiative to reduce greenhouse gas emissions from buildings.[11] Resolution 50 calls for cities to set a goal to reduce the use of fossil fuels in buildings by 50% in 2010 and to reach full carbon neutrality by 2030. Carbon neutrality in buildings means that building operation will produce no additional carbon emissions. This is accomplished by attention to building efficiency and offsetting the carbon emissions of the fuel use with efforts such as planting new forests or investing in renewable energy resources.

Buildings have a life span of over 50 years, and it is estimated that by 2035 a surprising 75% of all buildings will either be rebuilt or undergo substantial renovations. Integrated design approaches, in which all of the trades and stakeholders are involved throughout the design process, can lead to innovations that optimize heating and cooling efficiency, utilize daylighting and renewable energy (solar or wind), incorporate control strategies to eliminate wasted energy consumption, and continually monitor and "tune" buildings. Building commissioning, a practice that ensures that the design intent of the new building systems is realized, is important in order to guarantee the efficiency of a new building.

ENERGY EFFICIENCY IN EXISTING BUILDINGS

Professional energy assessments (or audits) can identify opportunities to improve efficiency in existing facilities. Energy assessments provide a listing of opportunities and information on the scale and scope of the opportunities available to reduce utility and operating costs. Energy professionals can perform energy savings and financial calculations for investments in efficiency which can provide economic justification for facility upgrades.

Energy Savings Performance Contracting (ESPC) is a means of implementing and financing energy efficiency upgrades by partnering with an energy service company that guarantees savings that result from upgrading building systems. For example, by 2006 the U.S. federal Government had entered into over 400 performance contracts worth about $1.9 billion dollars using private sector investment ($3.5 billion including financing), which will have guaranteed energy savings of $5.2 billion through reductions in utility bills. The net benefit to the government is $1.7 billion dollars.[12]

Energy Service Companies (ESCOs) have been in existence for over a century. They are companies who specialize in identifying and implementing cost-effective energy efficiency upgrades to facilities. In an ESPC, the ESCO assumes some or all of the financial risk of the investment by guaranteeing that the host organization (e.g., a university, school district, hospital, etc.) will realize the anticipated savings. In some cases the ESCO will also assume the risk of financing the improvements.

The general structure of an ESPC is as follows:

1. An ESCO is selected.

2. An agreement is signed to proceed with the planning phase on the project.

3. An investment grade audit is performed.

4. The owner approves the measures to be implemented.

5. An ESPC is entered into by the owner and the ESCO and financing is secured for the upgrades.

6. A turnkey project is implemented with oversight by the ESCO.

7. Operations and maintenance procedures are established to manage the efficiency of the facilities.

8. A measurement and verification (M&V) plan is developed and procedures implemented in accordance with established protocols. Baseline measurements obtained during the energy assessment phase are used for savings comparisons.

9. Payments are made to the ESCO and the financier.

10. Annual reconciliations are performed to determine actual savings. If savings are not realized, the ESCO is responsible for paying the owner of the facility the difference between the guaranteed savings and the actual savings.[13]

ESPC provides an important mechanism for many cash-strapped organizations to upgrade building systems and equipment.

Cogeneration and Co-location

Cogeneration refers to systems that produce electricity and use heat that is normally exhausted, resulting in a much higher overall system efficiency. This type of system is also known as combined heat and power (CHP) or co-firing. Typical technologies involved use reciprocal engines, gas-fired combustion turbines, steam turbine systems with associated boilers, combined-cycle turbines, or fuel cells to produce power, coupled with heat recovery equipment.

Exciting possibilities for cogeneration exist where "opportunity" fuels may be burned to generate steam. A recent U.S. DOE study identified eight potential energy sources: anaerobic digester gas, biomass gas, coal-bed methane, landfill gas, tire-derived fuel, wellhead gas, harvested wood, and urban wood waste. The study found that over 105 gigawatts of electricity and 3.0 quadrillion kilojoules (2.8 quadrillion Btus) of potential exist for these fuels in the U.S.[14] The economics of many of these fuel sources will be favorably impacted by more stringent requirements for emission controls on new and existing coal-fired power plants and the prospect of limits on carbon emissions.

It is rare to find an application where the electricity needed matches the thermal needs at a single facility. More often, excess heat or power is purchased by neighboring facilities or communities. There will be increasing interest in "co-locating" facilities or communities to take advantage of opportunities to use cogeneration if energy prices continue to rise and as environmental concerns become more pressing.

The lack of existing infrastructure in many countries lends itself to opportunities for cost-effective cogeneration. As cogeneration and co-location technologies become more mature, they can often be successfully applied in countries that lack or have limited electrical infrastructure.

Transportation Energy Efficiency

Technologies exist to improve the efficiency of vehicles and aircraft. A study by the National Research Council and the National Academy of Sciences revealed a long list of developed and emerging technologies that could increase the fuel economy of ground vehicles by more than 30% by 2030, an average of 12 miles per gallon above the 2004 average of 26.7 miles per gallon.[15] A subsequent report for the Southern States Energy Board used those findings, as well as emerging technology improvements for other transportation modes, to project the potential for transportation efficiency improvements.[16] The report concluded that transportation efficiency improvements must be pursued as part of a comprehensive strategy for reducing oil dependency.

Renewable Energy Resources

Renewable energy is sometimes defined as energy from sources that are not significantly depleted by their use, either because they are regenerated or because of the vastness of the resource. The successful implementation of renewable energy technologies has been correlated to technological factors and geographic circumstances to a stronger degree than to financial or organizational factors (Nijkamp and Pepping 1998:1494).

One grassroots effort, the 25 x '25 Initiative, seeks to obtain commitments to generate 25% of U.S. energy consumption by 2025 from renewable energy sources. As of 2007, this Initiative has been endorsed by roughly one-third of federal elected senators and representatives and state governors. The focus of this initiative is to affect policy changes and to educate people in the U.S. about the potential for renewable energy.[17]

All energy available for use, including both renewable and non-renewable fuels, is derived from the sun with most of the heat stored inside the earth's crust. In addition to the direct use of solar energy technology, wind power, hydro power, wave power and bio-energy are all indirectly derived from the sun. Wind is created by the uneven heating of the earth by the sun. Hydro-power, in the sense of harnessing moving water in rivers and streams, is also caused by the sun through the formation and movement of clouds. Wave power utilizes the oscillating motion which results from wind on the surface of the water. Bio-fuels are essentially formed by photosynthesis in plants.

Tidal power, which is driven primarily by gravity and the relative motion of the moon and the earth, is also affected by the sun's energy to the extent that ocean currents are caused by temperature differences. Geothermal energy is derived from the heat contained within the earth's crust. While geothermal energy resources can be locally depleted if systems are not carefully designed, due to their vast availability, they are considered to be renewable.

Technologies Using Solar Energy

Solar energy is the most plentiful energy resource. The amount of sunlight that reaches the ground is 8,000 times greater than the total world energy consumption. The use of solar energy has several drawbacks, namely the intermittent nature of availability, the low availability in much of the world due to cloud cover, and the relatively dilute nature of the solar radiation that reaches the earth's surface. However, mature technologies exist to overcome these challenges.[18]

Solar energy can be used in two basic ways: 1) Thermal energy can be captured for use for either heating or cooling (absorptive chillers); and 2) solar energy can create electricity using photovoltaic technology.

Solar Thermal Energy Systems

Thermal energy systems are defined as either being active or passive. Active systems use mechanical devices to move energy. Passive systems rely on the principles of physics to move the energy.

A typical active thermal solar system consists of solar collectors to absorb the rays of the sun and convert them into heat, a circulating fluid (e.g., air, water or some other non-freezing liquid) to remove the heat from the collector, a device to move the fluid, and a means of heat storage. Passive systems incorporate similar components but are designed to use natural convection and radiation to move the heat energy where it is needed.

The potential of thermal solar for producing low temperature heat is enormous. Although sales of solar thermal systems have roughly doubled over the last ten years, solar thermal systems still account for less than 1% of the total energy production. Despite the declining costs of installed photovoltaic and wind power systems, solar thermal can be the most cost-effective renewable energy technology.

The best applications for solar thermal energy are in locations with a large demand for low temperature heat that can be used during the day. Examples of these include pool heating, food processing applications, car washes, laundry facilities, and heating of buildings occupied primarily during the day-time. Water heating remains the dominant use of installed solar thermal systems. The countries with the most installed capacity per capita are Cyprus and Israel while China and the U.S. are the countries with the highest installed capacity. China's market is dominated by evacuated-tube solar collection devices while the U.S. is dominated by unglazed, plastic pool heating equipment.[19]

Solar thermal systems also are used to produce electrical power. Examples include the Solar One project near Barstow, California and the Nevada Solar One project, both located in desert areas of the western U.S. These systems concentrate solar energy to create steam, which is then used to drive steam turbines. This can either be done by the movement of mirror segments directing the sun's rays onto a central tower or by trough-shaped mirrors which concentrate heat onto a pipe running through each one. Large systems are being built around the world including one being constructed in California that utilizes trough concentrators. It is expected to generate 345MW, a capacity comparable to a mid-sized coal-fired power plant.

Photovoltaic Technology

Photovoltaic technology converts light rays into electricity using materials with special properties. Photovoltaic materials are designed in such a way as to allow light to knock electrons free and through a conductive path. The basic material is generally a semi-conductor with two oppositely charged, or "doped," layers sandwiched together. Silicon is the primary material used in most photovoltaics. Three major categories of photovoltaic production technology are used today: crystalline (wafers are cut from a pure silicon crystal), amorphous, and thin film.

Photovoltaic systems can be tied to the electrical grid or stand independently. "Off-grid" systems generally require a battery bank for power storage. Grid-tied systems can sell power to the electric utilities using net-metering agreements. They utilize electronic devices called inverters to convert the power into an alternating electrical power form that matches the electrical grid. Photovoltaics can provide remote or mobile power needs for communications devices, highway construction warning lights, exterior lighting and applications far away from an existing utility grid. In these applications, the devices are not connected to the electrical grid. Photovoltaic panels or modules last a very long time, typically warranted for 20 years and usable much longer with some decrease in performance. Other components of systems such as batteries may need to be replaced every 7 to 15 years.

Photovoltaic technologies have an EROEI from 15:1 to 30:1 depending on the production methods used. There are material recycling and disposal issues that need to be addressed if hazardous materials, such as cadmium and lead, are used in their production.[20]

Wind Energy

Wind energy has been used for millennia to drive mills and to pump water. Rural electrification in the U.S. brought about a surge in the wind power equipment industry in the early part of the last century, until the federal rural electrification program installed the electric grid across the U.S. nearly eradicated the industry. Today, wind power technology has become a major international market force, often cost-competitive with natural gas produced power. Large turbines, some one megawatt and above, are being connected to the electric grid.

Energy production from wind power increases exponentially with increases in wind speed. As a result, site locations must be carefully considered. With an 80% decline in the cost of wind power generation equipment in the last 20 years, wind power is now being produced in 46 of the 50 U.S. states. In Minnesota, a wind farm with 143 wind turbine generators has been developed near the city of Lake Benton.[21]

Wind turbine generators for electric production were once plagued by reliability problems. This has been overcome by new product designs (e.g., lift-type vertical axis machines and vaneless ion generators) and technological advancements (e.g., improved aerodynamics). New materials and improved solid state controls are helping to mature this technology and overcome some of the challenges it faced in the past.

The EROEI of large commercial wind turbines averages 18:1. Sustainability issues related to wind power include wildlife disturbance and physical disturbance of land, such as the need for roadways to access wind farms. Overall, sustainability issues are minor compared to the issues associated with power generation from fossil fuels.

Bio-energy: Biomass Digesters, Ethanol and Biodiesel

Biomass resources can be categorized as being either forest-derived or agricultural-based. A report published in 2005 indicates that sustainably produced bio-energy sources have the potential to offset U.S. petroleum imports by more than one-third by the middle of this century. With relatively minor investment, 1.3 billion dry tons of biomass could be made available annually for this purpose.[22]

The question of EROEI of corn-based ethanol has been a controversial subject, with reputable sources reporting widely variable results. The debate is due to differences in the system boundaries and assumptions being used in these studies.[23]

What energy inputs count? Do you include the energy expended to make the tractors used by the farmers? Do you consider the energy expended by the farmers themselves growing the corn? What about the energy consumed in the production of engineered seed? How do we account for the solar energy that falls on the land where the corn was farmed and the deterioration of the soil quality where the corn is grown? These issues are complex and variable. Farming practices and ethanol production processes are improving, but present EROEI estimates for corn-based ethanol range between 0.7:1 and 1.3:1. There are issues with spiking corn prices and potential increases in GHG emissions due to the use of electricity from coal-fired power plants in ethanol production. Many feel that corn-based ethanol is a short-term, non-sustainable solution, but that it plays a role in developing a market for future cellulosic ethanol in the U.S.

Cellulosic ethanol production refers to the use of feed materials such as grasses and corn stover (stalks and husks) to generate ethanol. While this can be done a number of ways, the most promising method uses enzymes to "digest" pretreated lignocellulosic materials to obtain glucose, a form of sugar. The sugar is then fermented into ethanol identical to the starch-based ethanol created from corn or sugar beets. Reducing the cost of engineered enzymes is seen as the major barrier to economic production of cellulosic ethanol. Currently, there are various projects attempting to scale these technologies to commercially viable production quantities. Cellulosic ethanol processes reduce carbon dioxide emissions by as much as 85% compared to gasoline and have an EROEI approaching 10:1. Cellulosic ethanol also has the obvious advantage of utilizing a feedstock that is indigestible to humans and thus does not compete with food production.

Biodiesel refers to petro-diesel substitutes derived from plants or fats rendered from animal products. The most common sources of biodiesel today are soy, corn, and animal fats. Certain forms of algae contain 50% natural oils, which results from photosynthesis and the trapping of carbon dioxide. The use of this oil to create biodiesel is a very promising emerging technology. Research has demonstrated how shallow algae ponds can utilize either waste water or carbon dioxide from coal-fired power plants to mass produce oil-rich algae.[24] Since crude oil is actually algae and plankton stored and compressed for millions of years, this process bypasses natural processes and offers a way to obtain more use from hydrocarbon combustion byproducts. The EROEI for biodiesel will depend on the feedstock and the processes used, but the currently quoted number is 3:1.[25]

Hydro-power, Tidal Power and Wave Power

Hydro-power, like wind power, has been in use for millennia. This technology captures energy embodied in the motion of water to produce mechanical work or electricity. A typical system utilizes a horizontally spinning turbine much like a boat propeller, connected to an electric generator. Wave power uses a piston device which is moved by the wave oscillation to produce power.

Installed hydro-electric power accounted for 19% of the world's electricity in 2004. Large-scale systems are in use in China, Brazil, Canada, Egypt and the U.S. The potential for additional large-scale hydro-power dams is limited by the vast amount of land required for new dam sites and concerns about the loss of ecosystems in the areas sacrificed.[26] Small-scale hydro-power systems cause very little environmental damage and are being used in locations that lack an existing infrastructure.

Tidal power and wave power systems do not require damming but further observation is needed to determine the impact on aquatic eco-systems. The world's first commercial wave power plant is the Aguçadora Wave Park in Portugal, established in 2006. Prototype tidal power plants are in located in France, Russia and elsewhere. The only operating plant in North America is located on an inlet in the Bay of Fundy in Nova Scotia. Scotland is developing a tidal power facility hoping to use it to generate 10% of its power by 2010.[27] Like hydro-power plants, it can be expected that large scale applications will have associated environmental impacts, yet provide sustainable electrical production solutions.

Geothermal

Geothermal energy, used directly for heat or for electricity generation, has enormous potential. An assessment of the geothermal resources available in the U.S. found that the resource is approximately equal to 2,000 times the country's total primary energy consumption.[28] Worldwide, geothermal is used commercially in over 70 countries, for power and/or heat. It is a proven technology. Essentially, this technology involves drilling into the earth to depths of two to six miles or more, through the earth's crust. Convective heat from the core of the earth as well as from the decay of radioactive isotopes is used to heat water to steam. The process harnesses steam and hot water from deep wells (typically near known hot springs or volcanoes), and generates electricity using turbines.

Electric heat pumps, used for heating and cooling buildings, are referred to as geothermal if they use wells as a heat rejection loop. These systems do not technically use geothermal energy, but rather use the ground near the earth's surface as thermal mass for energy storage.

Geothermal power production has become a booming industry in the U.S. There are 62 geothermal plants operating in California, Nevada, Utah, Hawaii and Alaska that produce 3% of the nation's renewable energy. Approximately 75 more projects are underway in the U.S. that will double capacity to 5,400 megawatts within the next few years.[29] Already, 5% of California's power is supplied by geothermal sources.[30] A steam field in the Mayacamas Mountains north of San Francisco, The Geysers, was one of the first geothermal energy production projects in the country. It has 21 plants that currently generate 750 megawatts of electrical energy, approximately one-fourth of California's total green power production.[31] There is potential that by 2050, geothermal energy could produce as much as 10% of the nation's power, if investment in this renewable resource is substantially increased.[32] New binary production processes, that use water (with temperatures as low as 150°C or 300°F) from geothermal wells to heat a secondary liquid (e.g., isopentane) that vaporizes at lower temperatures, offer the potential of increasing geothermal capacity.[33]

Factors affecting the feasibility and sustainability of geothermal power generation relate to drilling and operations of the systems. The cost of drilling wells to the required depths is the major barrier to using this energy source. Improved drilling methods and materials will lower these costs in the future. New technologies are available to identify prime locations to commercialize geothermal energy. Improved techniques to predict the available energy flow rates once systems are installed are needed to avoid degrading the heat source.

There are environmental impacts associated with drilling and operating geothermal power facilities. Drilling has potential to release trapped sub-terranian gases and to cause seismic disturbances. A small amount of electrical energy is needed to pump water. However, geothermal energy has few negative environmental effects.

Municipal Solid Waste (Landfills) as an Energy Source

In 2005, an average of 4.5 pounds of solid waste was generated by each person in the U.S. daily—almost one ton per year. Approximately two-thirds of that was either deposited in landfills or combusted.[34] There are several opportunities for energy recovery associated with solid waste. Recycling reduces energy used in the production of raw materials for industry. Energy recovery systems can be tied to solid waste incinerators and the heat can be used for either power generation or used directly. Similarly, solid waste can be processed using plasma arc technology and heat can be recovered from that process. Lastly, there is an opportunity to capture methane generated from the organic portion of the solid waste.

Recycling is an important way to reduce energy consumption. Recycling one aluminum can saves enough energy to run a 100W light bulb for 3 1//2 hours. Every pound of steel recycled saves 5,750 kilojoules (5,450 Btus) of energy, enough to light a 60-watt bulb for over 26 hours. Recycling paper cuts energy usage in half and reduces the need for forest products.

Incinerators are used to reduce solid waste volume in localities where landfill costs are high. This is generally an expensive and environmentally questionable practice. In addition to the sometimes toxic increases in air emissions resulting from incinerators, a new and more concentrated waste stream is produced and opportunities for landfills to be "harvested" in the future for recyclables are lost. The economic and environmental fundamentals of these systems are sometimes misinterpreted. Put into perspective, one can see that solid waste incineration is not a source of "renewable energy." Where solid waste incinerators are in use, there are large amounts of energy that can be recovered for beneficial use.

Plasma-arc technology is now being used in a few prototypes to incinerate solid waste but with significantly less local air emissions, virtually eliminating the possibility of toxic air emissions. The by-products of treating solid waste with plasma are a type of clean synthetic gas and a glass-like solid which can be resold as a raw material for ceramic products. Since plasma essentially transforms any material, even anthrax, into benign chemicals, this technology eliminates a major hurdle facing traditional incinerators.

Systems for capturing landfill gas are in place in 425 locations in the

U.S. currently and have potential in at least another 560 locations.[35] This practice involves utilizing the gaseous releases of anaerobic digestion in the landfill for the production of power or for other uses of the gas, such as generating usable heat. While the technology to install these systems is somewhat expensive, these systems have environmental benefits. The gases released are approximately half methane, the principle component of natural gas, and half carbon dioxide. Methane released into the atmosphere contributes to global warming on a per unit basis over 20 times greater than carbon dioxide. A quarter of human-related methane releases are from landfills.[36]

ENERGY TECHNOLOGIES OF THE FUTURE

Nuclear (Fission)

A 2003 study published by the Massachusetts Institute of Technology (MIT) is among the most comprehensive resources for understanding the appropriate role of nuclear energy and assessing barriers and solutions for future use.[37] The study identified four major barriers:

1) High relative costs;

2) Potentially adverse safety, environmental and health effects;

3) The possibility of security risks stemming from proliferation; and

4) Unresolved challenges in long-term management of nuclear wastes.

However, there are substantial resources being dedicated to finding solutions to these challenges.

Another concern in expanding the use of nuclear energy is the limited supply of high-grade enriched uranium ore. The MIT study recommends that a global assessment of the available resources be performed. The assumption used for the study is that a 100 year supply of quality uranium ore is available at the current rate of use, yet information to support this is limited. Lower quality ore can be used in the fission process but the EROEI drops steeply.

The EROEI for power generation using higher grades of ore is difficult to estimate. Little information is available on what the long-term energy costs will be to manage radioactive wastes from spent fuel and decommissioned power plants. Reprocessing and reusing spent nuclear fuel is one solution being explored. Determining the carbon emissions resulting from the production of electricity from nuclear power requires that the entire lifecycle be examined because of the long-term waste management and security concerns. For the short-term, nuclear power generation offers a relatively carbon-free energy source that will be more heavily relied upon in the future.

Nuclear Fusion

The International Thermonuclear Experimental Reactor (ITER) project seeks to investigate and demonstrate the feasibility of energy generation from nuclear fusion. The partners in the project are the European Union, Japan, China, India, the Republic of Korea, the Russian Federation and the U.S. Although there remain major hurdles to overcome, there is an increasing probability that the use of fusion-generated energy will be feasible within the next 40 to 50 years.

Energy generated from fusion is the result of changing the mass and energy state of atomic nuclei by fusing two nuclei. In Figure 5-3, two nuclei (deuterium and tritium), fuse together to form helium, a neutron, and a large amount of energy.

Fusion generates no radioactive waste and in the primary reaction does not produce any carbon emissions. The nature of the reaction lends itself well to providing base load power generation. The major hurdles fusion technology faces today include the difficulty of safely containing large-scale nuclear reactions and the costs involved to commercialize current reactor designs. For now, the EROEI of fusion is under 1:1 due to the energy required to generate and contain the reaction.

ENERGY
CONVERSION
TECHNOLOGIES

Fischer-Tropsch Technology

In the Fischer-Tropsch process, coal or biomass is converted into a synthetic gas, which can then be converted into a

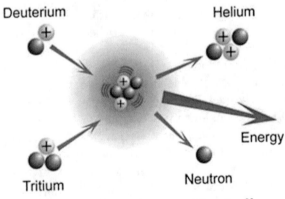

Figure 5-3. Diagram of Fusion Reaction[38]

diesel-like fuel, among other products. The diesel-like fuel is low in toxicity and is virtually interchangeable with conventional diesel fuels.

Although this technology has been available since the 1920s, and coal-to-liquid (CTL) fuels were used in Germany during World War II, it is not widely used today. The only commercial production of CTL occurs in South Africa, where the Sasol Corporation currently produces 150,000 barrels of fuel from coal daily.

Commercialization efforts and research are currently underway to further develop these technologies. Automakers consider CTL and biomass-to-liquid (BTL) fuels viable alternatives to oil, since they do not compromise fuel efficiency or require major infrastructure changes.[39] Commercialization of Fischer-Tropsch technology is likely to be 15 to 20 years away. Potentially, plants could accept a variety of feedstocks, using biomass when it is economically feasible or using coal when biomass is not as available.

CTL is not a sustainable energy source. Creating diesel fuel from coal releases 7 to 10 times more carbon emissions into the atmosphere than deriving it from crude oil. In addition, a sharp increase in coal consumption would accompany commercialization of CTL. This would exacerbate the environmental impacts of coal mining. However, when biomass is used the result is a liquid fuel with significantly lower carbon emissions. The feedstock can be biomass that is predominantly cellulose-based, rather than using sugar-based biomass to create ethanol by current practices.

Fuel Cells and Hydrogen

Hydrogen is not actually a renewable energy resource. While hydrogen is a very abundant element, pure hydrogen is not found in nature. It is only obtained by separating it from compounds to which it is bonded. In the case of water, it must be separated from an oxygen atom. Commercially available hydrogen is stripped from hydrocarbon compounds such as natural gas. In the chemical process of breaking the hydrogen bonds, energy is stored and then released when the hydrogen atom is allowed to bond again with another atom. Various methodologies make this possible but the end result is invariably subject to the Second Law of Thermodynamics—meaning that it is not possible to get more energy out of the process than is used to break the chemical bond. In reality, the energy losses can be as high as 35%.

Fuel cells are an electrochemical energy conversion device. Inside fuel cells, hydrogen and oxygen react and produce electricity, water and heat. A comparison with a more common device, the battery, is helpful to introducing fuel cell principles. Batteries are another type of electrochemical device, where electrical current flowing to them causes a chemical reaction to take place which is then reversed when current is allowed to flow out. Whereas batteries store a charge, however, fuel cells actually take an input fuel (e.g., hydrogen) and output an electrical current. When pure hydrogen is used the byproducts of the reaction are water and heat. When other fuels are used, including diesel, methanol and chemical hydrides, the byproducts are carbon dioxide, water and heat.

There are several types of fuel cells available today and newer designs being researched. Solid oxide fuel cells (SOFC), proton exchange membrane fuel cells (PEMFC), and alkaline fuel cells (AFC) are three available technologies. These technologies operate at different temperatures and differ in size and weight for the same amount of power generated. PEMFC is most suited to use in vehicles due to a lower operating temperature, smaller size and lighter weight. Other technologies are more suited to stationary applications for either back-up power or combined heat and power applications where waste heat used.

Fuel cells are achieving commercial success in the stationary market including some residential systems installed in Europe. Fuel cell applications for vehicles are not yet commercially viable but research and development efforts are continuing. The high costs of fuel cells and the lack of a hydrogen delivery infrastructure are the major market hurdles that will need to be overcome.

One potential use for fuel cells is for large-scale energy storage in conjunction with renewable power sources. One issue with large scale solar and wind power production is the intermittent nature of the resources. Large scale power storage is possible if the power is used for electrolysis to produce hydrogen from water. The hydrogen can then be used to produce power through fuel cell technology. Though there are large energy conversion losses, if this technology makes power storage on a large scale feasible, it could have exciting applications.

ENERGY AND THE DEVELOPING WORLD

We live on a small planet. Yet there is still a massive gap in the way the issues and technologies discussed in this chapter will affect the populations in different parts of the world. When the energy consumed in the poorest nations for industrial uses, transportation and electric power are totaled, it still does not exceed the percentage used by cooking with firewood. Deforestation, along with the associated erosion and destruction of watersheds, are the primary energy issues these people face.

For approximately half of the people living today, utility-provided electricity is still unavailable or economically out of reach. Using automotive batteries to provide minimal amounts of electricity in homes poses safety and environmental challenges. Many communities rely on pollution-generating diesel generators for power. International organizations such as the World Bank are assessing the sustainability of solutions as well as least-cost options to bring power to rural communities in the developing world. The hunger for energy in these populations will require finding sustainable energy solutions or there will be devastating economic, social and environmental consequences.

CONCLUSION

While the world we live in is still a bountiful place, an awareness is building that current rates of fossil energy consumption and disregard for the limits of our ecosystem are not sustainable. Many of the technologies in this chapter are still in their infancy, such as wave power and algae-derived biodiesel. Others, such as wind and solar technology, are mature and are seeing their markets expand as the cost of fossil energy increases along with awareness of global climate change. Some still may prove untenable. Others may be important parts of the future of energy—or may serve only niche applications. Many of the solutions are likely to be small-scale, moving away from the model of massive infrastructure.

The numerous technologies touched upon in this chapter point to an energy future that will be more diverse and challenging. The primary challenge we face is developing and scaling up these energy solutions quickly in order to minimize the environmental and economic consequences of our current path. In the development path of India, China and other nations with fast growing economies, there are significant opportunities for using new technologies rather than the ones we know to be unsustainable. This will require vision, cooperation, intellect and perseverance to achieve.

Endnotes

1. See Price Waterhouse Coopers. United States. www.pwc.com/extweb/service.nsf/docid/FA2588D6FBOE127C852570D600810B03, accessed 5 January 2007.
2. Energy Information Administration, U.S. Department of Energy, *Greenhouse Gases, Climate Change and Energy*, http://www.eia.doe.gov/oiaf/1605/ggccebro/chapter1.html.
3. Energy Information Administration. http://www.eia.doe.gov/emeu/international/energyconsumption.html. *Table 1.5 Energy Consumption, Expenditures and Emissions, Selected Years, 1949-2006*.
4. Energy Information Administration. http://www.eia.doe.gov/emeu/aer/eh/elec.html. *Energy in the United States: 1635-2000*.
5. Nadel, S. *et al* (2004), *The Technical, Economic and Achievable Potential for Energy Efficiency in the U.S.: A Meta-Analysis of Recent Studies*, American Council for an Energy Efficient Economy, from the proceedings of the 2004 Summer Study of Energy Efficiency in Buildings.
6. Arasteh, D. *et al*. (2006, 13-18 August). *Zero energy windows*. U.S. Department of Energy LBNL-60049. http://gaia.lbl.gov/btech/papers/60049.pdf, accessed 22 July 2007.
7. American Council for an Energy-Efficient Economy. (2005, September). *New Windows*. http://www.aceee.org/consumerguide/windo.htm, accessed 21 July 2007.
8. Ibid.
9. U.S. Department of Energy. (2004, 8 July). *Energy solutions for your building*. http://www.eere.energy.gov/buildings/info/homes/buyingwindows.html, accessed 21 July 2007.
10. Canadian Home Builders' Association. *Buying a new home—energy efficiency for savings and comfort*. http://www.chba.ca/NewHome/WhyBuyNew/Energy_Efficiency/index.php, accessed 22 July 2007.
11. Oak Ridge National Lab. (2003, June). *Towards a Climate-Friendly Built Environment*, for the Pew Center on Global Climate Change.
12. United States Federal Energy Management Program, *Fact Sheet: Energy Savings Performance Contracting*, available at http://www1.eere.energy.gov/femp/pdfs/espc_fact_sheet.pdf.
13. Hansen, S. (2006). *Performance Contracting: Expanding Horizons*. Lilburn, GA: The Fairmont Press.
14. Resource Dynamic Corporation (2004, August). *Combined Heat and Power Market Potential for Opportunity Fuels*, for U.S. DOE EERE.
15. National Research Council (2002). National Academy of Sciences. *Effectiveness and Impact of Corporate Average Fuel Economy (CAFE) Standards*, Washington, D.C.: National Academy Press.
16. Southern States Energy Board (2005, July), *American Energy Security: Building a Bridge to Energy Independence and to a Sustainable Energy Future*, section on transportation efficiency at pgs. 109-120.
17. Information about the 25x'25 Initiative can be viewed at http://www.25x25.org/.
18. Philibert, C. (2005). *The Present and Future Use of Solar Thermal Energy as a Primary Source of Energy*, for the International Energy Agency.
19. International Energy Agency. (2006). Solar Heating and Cooling Programme. *Solar Heat Worldwide: Markets and Contributions to the Energy Supply*.
20. Alsema, E. *et al*. (2006, September). *Environmental Impacts of PV Electricity Generation—A Critical Comparison of Energy Supply Options*, Presented at the 21st European Photovoltaic Solar Energy Conference, Dresden, Germany.
21. Data from the American Wind Energy Fact Sheet, obtainable from the American

Wind Energy Association, MC Street, Suite J80, Washington, D.C. 2000.

22. Oak Ridge National Laboratory. (2005, April). *Biomass as Feedstock for a Bioenergy and Bioproducts Industry: The Technical Feasibility of a Billion-Ton Annual Supply*, for the U.S. Department of Agriculture and the U.S. Department of Energy.

23. Shapoiri, H. *et al.* (2002). *The Energy Balance of Corn Ethanol: An Update, 2002*, for the U.S. Department of Agriculture.

24. Sheehan, J. *et al.* (1998, July). *A Look Back at the U.S. Department of Energy's Aquatic Species Program—Biodiesel from Algae*, for the U.S. Department of Energy Office of Fuels Development. p. 9.

25. Visalli, D. (2006 April), *Getting a Decent Energy Return on Investment*, article published on http://www.energybulletin.net/14745.html, accessed 20 July 2007.

26. Wikipedia. *Hydropower*, http://en.wikipedia.org/wiki/Hydropower, accessed 20 May 2007.

27. Wikipedia. *Tidal Power*. http://en.wikipedia.org/wiki/Tidal_power_plants#Resource_around_the world, accessed 1 August 2007.

28. Massachusetts Institute for Technology (2006). *The Future of Geothermal Energy: The Impact of Enhanced Geothermal Systems (EGS) on the United States in the 21st Century*, p. 1-4, for the U.S. Department of Energy.

29. Davison, P. (2007, 7 June). Push for geothermal juice picks up steam. *The Courier-Journal*. p. 3B.

30. Ibid.

31. *Welcome to the Geysers*. www.geysers.com, accessed 18 June 2007.

32. Ibid.

33. Ibid.

34. United States Environmental Protection Agency, Municipal Solid Waste Generation, Recycling and Disposal in the United States: Facts and Figures for 2005, available at http://www.epa.gov/epaoswer/osw/conserve/resources/msw-2005.pdf.

35. United States Environmental Protection Agency, Landfill Methane Outreach Program, http://www.epa.gov/lmop/proj/index.htm, accessed 20 July 2007.

36. United States Environmental Protection Agency, Methane: Sources and Emissions, http://www.epa.gov/methane/sources.html, accessed 20 July 2007.

37. Deutch, J. *et al.* (2003 July)., *The Future of Nuclear Power*, for the Massachusetts Institute of Technology.

38. ITER. *Fusion Energy*. http://www.iter.org/a/index_use_1.htm, accessed 18 July 2007.

39. United States Government Accountability Office (2007, February). *Crude Oil: Uncertainty about Future Oil Supply Makes It Important to Develop a Strategy for Addressing a Peak and Decline in Oil Production*, for the U.S. Congress. p. 60.

Chapter 6

Corporate Sustainability Programs

"Whether the current drive toward corporate sustainability is genuine or mere public re-lations, the fact remains that, in the long run, sustainability will not merely exist as an option: It will and must become an investor-driven as well as physical necessity. The hu-man modus operandi is inherently shortsighted, ever overemphasizing the immediate but as countless indicators have recently demonstrated, issues of sustainability have become very much concerns of the here and now. It is only a matter of time before these concerns garner the scrutiny of the investing public. The earth itself is an asset, and profiting at the expense of this resource without accounting for its depreciation is not a practice that can continue unchecked. The present scale and scope of industry has left an ominous footprint, and only the most arrogant of companies could be reckless enough to believe investors will ignore consequences that now loom just over the horizon."

ARI DAVID KOPOLOVIC, CAMBRIDGESHIRE, ENGLAND
IN *BUSINESS WEEK*, 19 FEBRUARY 2007, P. 19.

"Sustainability is not just cleaning up after yourself. It's not something you take care of after business, it's core. Good business means doing good for the planet. The goal is growth, finding new niches, converting waste into product—and crushing the competition."

NEIL GOLIGHTLY, DIRECTOR OF SUSTAINABLE DEVELOPMENT
AT THE FORD MOTOR COMPANY, 2006 SUSTAINABLE OPPORTUNITIES SUMMIT,
KEYNOTE SPEAKER.

Corporations are getting involved in sustainability! This is an im-portant development that is changing corporate behavior throughout the world. While corporations have always been motivated to manage busi-ness affairs in a manner that improves profitability and extends life of their companies, sustainability programs are now in vogue. Businesses have a history of progressive involvement with the health and welfare of their employees, customers and stakeholders. Corporate social responsi-bility (CSR) programs have been in existence since the early 1970s. CSR is more than corporate philanthropy or providing employee assistance and benefits. According to the World Business Council for Sustainable Devel-

179

opment, "Corporate Social Responsibility is the continuing commitment by business to behave ethically and contribute to economic development while improving the quality of life of the workforce and their families as well as of the local community and society at large."[1]

Beginning in the 1990s, a trend to incorporate sustainability into the mission of corporations began to emerge. There are a number of charters that businesses use to develop their sustainability programs. There include:

- The Coalition for Environmentally Responsible Economies (CERES) Principles (formerly the Valdez Principles) which has been joined by over 80 organizations.[2] Its mission is to integrate sustainability into capital markets.

- The International Chamber of Commerce (ICC) Charter for Sustainable Development.[3]

- The International Standard Organization's (ISO) Standard 14000 and 14001 guide corporations in developing environmental management programs.[4]

Corporations are broadening their social roles. Both the CSR movement and the drive towards sustainability are expanding as a result. Julie Fox Gorte, an investment strategist for the Calvert Group, stated that "All of a sudden, corporate responsibility is an idea whose time has arrived... we're seeing more companies who think it's not just a philosophy, but good for business too" (Iwata 2007).

The importance of corporate leadership cannot be understated. According to Forrest (1996:1), "By taking the lead, corporations can move society toward sustainability far more efficiently and with less turmoil than governments and legislation... companies with the vision to help shape the future will have a much greater chance of recognizing opportunities than those who merely react to the changes." Forrest (1996:2) further explains that organizational sustainability has two commonalities:

1) A goal of conserving irreplaceable resources; and
2) A goal of environmental maintenance.

In an effort to apply the concept of sustainability to business, Dyllick and Hockerts (2002:131-132) defined it "as meeting the needs of a firm's direct and indirect stakeholders (such as shareholders, employees, clients, pressure groups, communities, etc.) without compromising its ability to meet the needs of future stakeholders as well." To this end they suggest that "firms have to maintain and grow their economic, social and environmental capital base." While a company will cease to exist once its economic capital is exhausted, a company can become functionally unsustainable long before this happens. Dyllick and Hockerts (2002:133-134) provide a set of definitions in regard to corporations and their economic, ecological and social sustainability:

- *Corporate economic sustainability*—Economically sustainable companies guarantee at any time cash flow sufficient to ensure liquidity, while producing above average returns for their shareholders.

- *Corporate ecological sustainability*—Ecologically sustainable companies use only natural resources that are consumed at a rate below natural reproduction, or at a rate below the development of substitutes. They do not cause emissions that accumulate in the environment at a rate beyond the capacity of the natural system to absorb and assimilate these emissions. Finally, they do not engage in activities that degrade eco-system services.

- *Corporate social sustainability*—Adds value to the communities within which they operate by increasing the human capital of the individual partners as well as furthering the societal capital of these communities. They manage social capital in such a way that stakeholders can understand their motivations and can broadly agree with the company's value system.

The beauty of these definitions lies in their universal applicability to businesses. Corporations can apply them to their mission while maintaining their responsibilities to their shareholders and stakeholders. They can focus their efforts on resolving problems that meet interrelated criterion. Recently, there has been shift "strongly toward business as a major actor" in implementing sustainability (Dyllick and Hockerts 2002:131).

While the broader definitions of sustainability are relatively new, leaders in the business world have emerged as active proponents. Richard J. Mahoney, a former CEO of Monsanto commented that "Our commitment is to achieve sustainable development for the good of all people in both the developed and less-developed nations" and also that "companies need to rectify the past and provide the technology necessary to serve the people of the world in the future without leaving behind a mess."[5] Corporations have learned that focusing on the economic sustainability of businesses alone is inadequate for their long-term viability and sustainability (Gladwin *et al.* 1995).

Multinational corporations have become commonplace as companies strive to reduce costs and globalize their businesses. By 1993, a United Nations survey of 169 multinational corporations found that 43% had environmental policy statements that committed them to social responsibility in managing environmental impacts (United Nations 1993:22). By 1999 over 130 major corporations were members of the World Business Council for Sustainable Development. As Friedman (2005:293) states so aptly, "... there are going to be a lot of jobs involving the words 'sustainable' and 'renewable'... This is going to be a huge industry in the 21st century."

Globalization of business and industry offers the opportunity to lower product and manufacturing costs by transferring production to countries where production costs are minimized. Milton Friedman pointedly stated that it is now "possible to produce a product anywhere, using resources from anywhere, by a company located anywhere, to be sold anywhere."[6] Moving product components and finished product around the world is costly and energy intensive. Environmental regulations and their enforcement are variable across countries. Global companies are often tempted to seek locations where environmental enforcement is lax.

Today, the objectives of sustainable development have been formalized in the Business Charter for Sustainable Development, sponsored by the International Chamber of Commerce. The charter, launched in 1991, concerns itself primarily with the environmental aspects of health, safety, and stewardship. The objectives commit businesses to improve "their environmental performance in accordance with the principles, to having in place management practices to effect such improvement, to measuring their progress, and to reporting their progress as appropriate, internally and externally."[7] The charter outlines 16 principles that cover various aspects of sustainability including changing management practices, assessing environmental impacts, incorporating precautionary principles, efficiently using materials and resources, allowing technology transfer, being open in reporting, improving facility operations, minimizing wastes, among others. The ICC's efforts have been impressive. Jan Stromblad, Senior Vice-President for Environmental Affairs at ABB stated that "ABB has steadfastly built its EMS (Environmental Management System) around the 16 Charter Principles... there is no better set of guidelines."[8] There are now 2,300 companies that have signed the Charter and agreed to its philosophy and terms.[9]

Berry and Rondinelli (1999) studied the approaches of 38 major multinational corporations (including 3M, DaimerChrysler, Dow, Dupont, UPS and Toyota) and the corporate citizenship programs they engaged in to achieve sustainable development. Corporate citizenship attempts to address social problems using philanthropy, external relationships with stakeholders (Miller 1998:104-108), or other means. For example, Johnson and Johnson (1997:3) strives for "environmental neutrality and resource efficiency consistent with the principles of sustainable development." Berry and Rondinelli (1999:33) found that corporations and the alliances they forge can promote "corporate citizenship for sustainable development" while reducing "the adverse environmental impacts of business operations on local communities." They found that approaches corporations can take include:

1) Developing clean manufacturing and pollution prevention processes and technologies;

2) Exploring environmentally neutral or beneficial products;

3) Conserving natural resources; and

4) Improving environmental conditions around the world.

The new bottom line is that corporations are becoming better corporate citizens, advancing sustainability agendas, and actively promoting their sustainability programs. According to Edwards (2005:50), the "service and manufacturing sectors are shifting from merely meeting environmental compliance standards to realizing the competitive advantage of devising and implementing sustainable business strategies" and "instead of being seen as an impediment to business development, sustainable practices, including ecological, economic and social business concerns, are now seen as business opportunities." This perspective represents a new paradigm in corporate philosophy. This shift is revolutionary. It is a force that is reshaping today's business world.

Yet in the drive to meet corporate economic and financial goals, there remains tension and conflict with corporate *social* goals—including goals associated with corporate sustainability. For example, conserving resources might involve installing more efficient equipment and improving energy efficiencies in existing systems. Once mechanical infrastructure is in place, capital investment is necessary for improvements. Despite the need to improve efficiency, making the investment decision to replace infrastructure can be a difficult one. "Shared savings" programs, whereby improvements are implemented by a third-party, provide one solution for companies. When corporate sustainability goals are in place, decisions can be made and actions taken, that will lead to the ultimate achievement of these objectives.

EXAMPLES OF CORPORATE SUSTAINABILITY INITIATIVES

Corporations are developing and implementing new sustainability programs. UPS is an example of a company that is pursuing a corporate sustainability policy and has a written sustainability plan in place. Their sustainability goal is to "provide optimal service and value to our customers by striving for the highest operational efficiencies and minimizing the impact to the environment."[10] Disney has initiated a number of sustainability programs including a multifaceted energy management program designed to reduce energy costs.

Fortis, which ranks among the top 25 financial institutions in the world, developed its own corporate sustainability statement that establishes the parameters of its program. Fortis did so "not because it is something new to us, but because we believed we needed to embed the wide range of initiatives and activities at our company in an updated vision and integral policy."[11] The Fortis (2004:1) sustainability plan, named "Agenda 2005," included the following commitments:

- "Strengthen internal sustainability networks and increase involvement among our employees;

- Further integrate sustainability into our organization's core processes, such as client acceptance, lending policies, product development and investment policies;

- Publish a separate sustainability report, for the first time in 2005;

- Initiate proactive stakeholder management and engagement;

- Include corporate sustainability in our group-wide purchasing policy."

The Fortis *Corporate Sustainability Statement* also states that "corporate sustainability means conducting business in a responsible manner; achieving sustainable economic growth while anticipating the legitimate interests of our stakeholders; and taking social and environmental responsibility" (Fortis 2004:1).

DuPont has taken initiatives to mitigate global warming. According to Dawn Rittenhouse, Director of Sustainable Development, DuPont has reduced airborne carcinogen emissions by 92% and greenhouse gas emissions by 72% (Iwata 2007). Once heavily dependent on fossil fuels for paints and polymers, DuPont has invested billions in environmentally safe cosmetics, detergents, carpet materials, cellulose based ethanol, and bio-butanol. DuPont, Shell and BP Amoco, along with other companies, have committed to reducing greenhouse gas emissions to a level that is 15% below 1990 levels by 2010 (Speth 2004:186).

There are numerous real-world examples of the initiatives corporations have undertaken. Wharington International, a furniture manufacturer in Austria, developed a new sofa and armchair with a goal of ultimately offering a high quality sustainable furniture line. Called the "Re-Define" project, the idea was to manufacture furniture that avoids toxic chemical release, minimizes components, uses metals with low embodied energy, avoids environmentally damaging solvents, and uses materials that enable the purchaser to make minor repairs.[12] These design changes reduce the potential of obsolescence—increasing the life and utility of their furniture. Avoiding designs that tended to go out of fashion quickly, Wharington International anticipated that the project cost could be recouped from sales in about two years.[13]

Small-scale process adjustments can yield significant cost savings, quickly returning their initial investments. Hitega, a Chilean manufacturer of textiles, set about to improve water management and reduce its energy usage. Improvements included recycling softened water that was used to cool baths, improving softener regeneration processes, and recycling water from air conditioning systems. The company instituted a maintenance program for steam traps and installed screens to reduce suspended solid effluent. The project provided environmental benefits that included "water, energy and chemical conservation, and reduced emissions" –resulting in a reduction in manufacturing costs.[14]

General Electric is pursuing an "Eco-imagination" initiative hoping to double revenue from environmentally clean technologies and products to $20 billion annually by 2010 with investments in "fuel efficient jet and train engines, wind turbine power, energy-saving fluorescent light bulbs, and water purification" technologies (Iwata 2007).

Johnson Controls formats their sustainability programs with a triple bottom line approach to corporate sustainability. This program considers economic prosperity, environmental stewardship and social responsibility as three key performance measures in the decision making process.[15] Their program has included: a) constructing a new Milwaukee headquarters building that achieved a LEED Gold rating; b) launching a comprehensive program to identify energy saving improvements at manufacturing sites in 2006; c) piloting sustainable Energy Education and Communications Programs for employees; and d) instituting programs for recycling and waste reduction (Johnson Controls 2006:17). The company annually publishes a Business and Sustainability Report.

Even "big box" retail companies can change their ways. Wal-Mart's CEO established three new goals for the company in 2005: 1) rely 100% on renewable energy; 2) create zero waste; and 3) sell products that sustain resources and the environment (Worldwatch Institute 2007:155). Wal-Mart CEO Lee Scott was quoted saying, "What struck us was this: This world is much more fragile than any of us would have thought" (Fetterman 2006:1). In 2006, Wal-Mart, now the largest U.S. retail company with 1.8 million employees, launched a multi-faceted sustainability program. The program involves slashing energy use and encouraging its 60,000 suppliers "to produce goods that don't harm the environment and to urge customers to buy green" (Fetterman 2006:1). Since then, it has become the world's largest retailer of organic foods (Worldwatch Institute 2007:155).

Wal-Mart is also the largest private electricity user in the U.S. Actions taken by the company have reduced its electric bill by 17% since 2002.[16] The company recently inventoried and reported its carbon dioxide emissions, saying it annually emits 20.8 million tons worldwide. According to *USA Today* (25 September 2006), Wal-Mart has committed to the following goals:

- "Slash gasoline use by its trucking fleet (one of the largest in the U.S.) and use more hybrid trucks to increase efficiency by 25% over the next three years and double it within 10 years. This will save $310 million a year by 2015.

- Buy 100% of its wild-caught salmon and frozen fish for the North American market only from fisheries that are certified as "sustainable" by the non-profit Marine Stewardship Council within three to five years. That designation means areas of the ocean aren't fished in ways that destroy fish populations.

- Cut energy use at its more than 7,000 stores worldwide by 30% and cut greenhouse-gas emissions at existing stores by 20% in seven years.

- Reduce solid waste from U.S. stores by 25% within three years."

While these goals may seem ambitions for such a large corporation, Wal-Mart is taking other actions as well. These include purchasing refrigerators with LED interior lights, experimenting with wind power and permeable paving, installing skylights in stores, using motion sensor-switches to reduce light levels, forming employee "sustainable value networks," using both biodegradable and recyclable packaging, selling gift cards that use corn-based polylactic acid and requesting products from suppliers that use organic cotton.

Wal-Mart is encouraging its customers to purchase compact fluorescent lamps (CFLs). According to Andy Ruben, Vice-President for Sustainability:

> "… compact fluorescent light bulbs… account for only 5% of light-bulb sales, but at Wal-Mart we are redoing our aisles to make CFLs more visible, mostly at eye level. Soon 20 to 30% of the light bulb sales will be CFLs. We have a goal of selling 100 million CFLs in 2007, more than double what we did last year. That will save our customers $3 billion on electricity. It will save 700 million incandescent bulbs that never have to be produced. It will prevent 20 million metric tons of CO_2 from being released into the atmosphere."[17]

Wal-Mart, along with Herman Miller and the Knoll Group, committed themselves to paying higher prices for lumber products that were produced using sustainable practices (Hawken 1993:155).

Many of these initiatives also reduce Wal-Mart's costs. Reducing excess packaging in one line of toys lowered shipping costs by $2.4 million and saved the equivalent of one million barrels of oil (Fetterman 2006:1). It is likely that additional savings are possible by redesigning packaging and decentralizing product manufacturing. Despite the recent efforts of Wal-Mart, the idea of shipping goods halfway across the planet (to capitalize on lower unit labor costs), from point of manufacture to point-of-sale, is antithetical to the concept of sustainability. The company is considering alternative production techniques that provide consumers with products that can be manufactured closer to where they are needed.

In the last year, Wal-Mart has opened what it calls its new "High-Efficiency Supercenters." These stores in Kansas City and Rockton, Illinois use 20% less energy by including innovations in their design, and more efficient HVAC systems, water systems, lighting components, and construction materials (Davis 2007:D4).

The Cincinnati-based Kroger Company is installing skylights in its stores to reduce electrical requirements for aisle lighting, replacing incandescent fixtures with more efficient fixtures, and redoubling maintenance efforts to reduce energy usage. According to Rodney McMullen, Vice-Chairman of Kroger, "Given our size, we have a certain responsibility to the environment and to being a good citizen... we take everything we save with energy reductions and we reinvest it with the customer in lower prices and improved service" (Davis 2007:D1). The food store chain has over 20 full time energy technicians who work on boosting energy efficiency. Since 2000, their energy conservation program has reduced energy usage by 20% companywide (Davis 2007:D4).

Utilities are joining the sustainability revolution. Many are aggressively pursuing incentive programs within their service territories. These programs are typically related to energy production and consumption. They may offer incentives to their customers to reduce energy use, develop alternative energy production sources, or design unique programs promoting the efficient use of the energy they supply. Many utilities are considering "green" power sources rather than "brown" power sources such as coal. According to Brian Ward of Palo Alto Utilities, "The whole philosophy of the utility is just to get greener."[18] The National Renewable Energy Laboratory has estimated that there are now 750 utility companies that offer green power programs.[19] These custom programs are related to sustainability.

Utilities have active programs that are broad in scope and have international implications. Puget Sound Energy (PSE), is a regulated supplier of electricity and natural gas that serves northwestern Washington State. In addition to offering its customers a green power program, PSE has a sustainability policy directed at greenhouse gas mitigation. The policy statement recognizes that increased atmospheric concentrations of greenhouse gases contribute to climate change. It identifies "near-term strategies" that PSE (2007:1) intends to explore and implement:

- Pursue a diverse energy portfolio mix of resources that includes renewable generation;

- Work with partners in the utility industry, state government, and national government to explore and evaluate opportunities to reduce greenhouse gas emissions;

- Develop a strong energy efficiency program;

- Support the advancement of scientific understanding of climate change;

- Support a market-based national system (e.g., "cap" and "trade" or carbon tax) or sub-national system that covers a large enough area to prove cost-effective and useful;

- Call for the removal of barriers and disincentives to the advancement of the aforementioned recommendations (e.g., governmental facilitation of transmission from renewable energy projects); and

- Advocate government incentives to develop renewable generation and other greenhouse gas reducing technologies.

PSE's policy is to implement cost-effective measures that mitigate or offset greenhouse gas emissions while maintaining diversity in its energy portfolio.

Duke Energy's first sustainability report was published in 2006. Duke, a regulated U.S. utility, serves customers in North Carolina, Ohio, Indiana, northern Kentucky, South Carolina, and Central and South America. Duke defined its business approach to sustainability as one that creates "long-term shareholder value by embracing opportunities and managing risks deriving from economic, environmental and social developments."[20] Duke has committed itself to increasing investment in energy efficiency, installing pollution control equipment that will reduce sulfur dioxide and nitrogen oxides by 70%, and partnering on economic development projects. According to Roberta Brown, Vice-President of Sustainability and Community Affairs, "our approach is characterized by research and analysis, integrating sustainability into our business, focusing on what matters most, and establishing clear accountability, goals and objectives."[21] Duke Energy's recent actions include: 1) signing a 20-year purchase agreement for 100 megawatts of windpower; 2) testing diesel-electric service trucks; 3) establishing an energy efficiency department; 4) converting a 450 acre site into a wildlife conservation area; and 5) committing to invest $3 million annually in greenhouse gas reduction projects.[22]

Non-utility companies are becoming players in electrical production. The largest bio-energy producers are not traditional utility companies. Surprisingly, the companies are International Paper, Weyerhaeuser, Koch Industries, and Georgia-Pacific Corporation. What these companies have in common is that they generate electrical power from their waste streams. Their electrical production is relatively small and costs can be more than power purchased from their utilities. Weyerhaeuser generates up to 18 megawatts at its plant in Springfield, Oregon using cogeneration turbines.[23] The fuel used is wood waste, a byproduct from pulp manufacturing operations. Waste disposal costs are reduced.

Corporations are going solar. According to Google's David Ratcliffe, Vice President of Real Estate, "Google is committed to advancing green technology throughout the workplace... If we can dispel the myth along the way that you can't be green and profitable at the same time, that's another benefit."[24] Their new headquarters campus in Mountain View, California is designed with the largest rooftop solar collector system in the U.S.—rated at 1.6 Megawatts. One of the drivers for this large installation was to protect the company from the potential volatility of energy costs and achieve predictable pricing for their electrical use over the 30-year project life. The project used 9,200 solar panels, taking advantage of federal and state solar energy incentives and is projected to have a simple payback of only 7 ½ years. The annual electrical energy produced is 2.6 million kWh, reducing annual electrical costs by $393,000, and CO_2 emissions by the equivalent of 3.6 million lbs.[25]

SUSTAINABILITY AND THE AUTOMOBILE INDUSTRY

When it comes to sustainability, the automotive industry is getting into high gear. Once resistant, the industry is actively incorporating the philosophy of sustainability into its management and manufacturing processes. In 2007, General Motors became the first automobile manufacturer to join the U.S. Climate Action Partnership, "breaking a long stalemate over rising emissions from the transportation sector."[26] Subaru has an assembly plant that is the first in the U.S. sends no wastes to landfills. The plant recycles 99% of its wastes and uses the remaining 1% to produce electricity.[27] The 832 acres site northwest of Indianapolis is designated by the National Wildlife Federation as a wildlife habitat, home to deer, coyotes, bald eagles and Canadian geese.[28] Tom Easterday, Senior Vice President of Subaru Indiana stated that the company's mantra is to "Eliminate the environmental risks of our operations."[29]

The energy efficiency of vehicles and the types of energy they use has an important impact on sustainability. If vehicles can not only use less fuel (especially gasoline) but also use fuels that are more environmentally benign, sustainability can be more easily achieved. More efficient gasoline powered vehicles are now available. Automobile manufacturers are changing their products, product components and reengineering their vehicles. To meet changing consumer demand, hybrid (gas-and-electric powered) vehicles, electric vehicles, hydrogen vehicles, natural gas vehicles and bio-diesel vehicles[30] have found their niche in the automobile industry. Toyota, BMW, Mitsubishi Motors and Ford provide examples.

Toyota's corporate philosophy is to seek harmony with people, society and the global environment while creating prosperity through the manufacture of automobiles.[31] Their corporate motto is to "only use what you need, when you need it, in the amount that you need."[32] This motto is about resource management, just-in-time inventory systems, and waste reduction. Toyota has been remarkably successful with waste reduction initiatives. Since 2000, hazardous wastes that are shipped to landfills have been reduced by 40%, and other wastes have been reduced by 11%.[33]

Beginning in 1997, Toyota led the industry with the production of hybrid cars. Its cumulative sales of hybrid vehicles have totaled 347,000 in Japan and over 700,000 elsewhere in the world.[34] How did this happen? At a time when Detroit was selling gas-guzzlers and the hybrid appeared uneconomical, Eiji Toyoda (then CEO of Toyota) was told by his concept planners that there was no market for hybrids. His comments were, "Go back and take another look at the hybrid... and this time don't look at the economics."[35] Today, the Primus midsize sedans average 22 kilometers per liter (55 mpg) in combined city and highway driving.[36] Sales are expanding and it is the third best-selling U.S. passenger car. In 2006, Toyota began manufacturing a hybrid model of its Camry sedan in Georgetown, Kentucky.

BMW has redesigned its automobile components and assembly processes with recycling in mind. The company is very concerned with what happens to vehicle components at the end of their usefulness. Their automobiles are designed to maximize the life of the vehicles. BMW has constructed a disassembly plant to recycle old vehicles and has experimented with bar-coding parts to identify material components and to provide instructions on future recycling and reuse (Hawken 1993:72).

Mitsubishi Motors is among those companies that have published a corporate sustainability program. One of the aspects of their environmental sustainability plan is to incorporate Design for the Environment (DfE) processes to minimize the environmental loads of automobile parts and components throughout their life cycle. The plan has four primary components:

1) Environmental management;
2) Expanding recycling programs;
3) Prevention of global warming; and
4) Prevention of environmental pollution.

To improve environmental management, the company is expanding green procurement policies and promoting ISO standard 14001. This certification defines a step-by-step process that assists organizations in the development of environmental management standards. It consists of a process that involves establishing general program requirements, environmental policies, planning, implementation, corrective action, and management review.[37] One of Mitsubishi's goals is to achieve a 95% recycling rate by collaborating with local governments, promoting the recycling of product and materials, reducing emissions of byproducts, and using water resources more effectively.

Mitsubishi Motors complies with targets for domestic and EU fuel economy standards, plans to reduce vehicle CO_2 emissions by 20% by 2010, and promotes the development of HFC 134A-free air conditioning equipment. HFC 134A is a refrigerant gas that has no ozone layer impact and offers a comparatively low global warming potential (USGBC 2003:171). Mitsubishi Motors promotes fuel cell research and development, is developing compressed natural gas hybrid-electric vehicles, and plans to expand the production of vehicles that achieve high fuel economy coupled with low exhaust emissions.[38]

Ford Motor Company's CEO, William C. Ford Jr., spearheaded a $2 million project to place a green roof that captures rain water on its 4 hectare (10 acre) complex at Rough River Michigan (Engardio 2007:64). Meanwhile the company kept manufacturing high mileage SUVs. According to Andrew Winston, a corporate environmental strategist at Yale University, "Having a green factory was not Ford's core issue. It was fuel economy" (Engardio 2007:64).

Regardless, Ford exceeded its government-required fleet fuel-efficiency standards for its 2006 model year. Ford has been involved in vehicle research and testing of vehicles that use alternative fuels, such as ethanol and hydrogen. In 2007, Ford created a new position of Vice President for Sustainability, Environment and Safety Engineering. Ford CEO Alan Mulally was recently quoted as saying, "Green is good business."[39] Ford is subsidizing a test of its hydrogen-fueled busses in Orlando. The primary exhaust emission from hydrogen-fueled vehicles is water. Since the first hydrogen fueling station has opened in Florida, the state is paying $250,000 each for eight hydrogen-powered buses valued at over $1 million apiece.[40] The busses "will ferry customers, tourists, and employees at Orlando International Airport, the Orange County Convention Center and other tourist spots throughout central Florida."

U.S. automakers still have a long road ahead of them in order to capture green-oriented consumers. While GM actually sells more vehicles that get over 12 kilometers per liter (30 mpg) than any other manufacturer in the U.S., Larry Burns of GM conceded that, "We didn't appreciate the image value of hybrids… we missed that."[41]

DEVELOPING A CORPORATE SUSTAINABILITY PLAN

Step #1 – Redefine the Corporate Mission

To be successful, a corporate sustainability program must not only include planning, but go beyond this—to implementation. Implementing a program requires a cultural change within the corporation. It is important to align corporate goals with those of sustainable development. This is a challenging task—and to be successful, corporations must assess operations and implement changes. Implementing changes in businesses involves risk—and business risks must be managed. A high degree of creativity is needed to redesign a business to meet profitability requirements in the short term, while pursuing corporate citizenship responsibilities and long-term sustainability goals.

Sustainability programs often require a change in the direction of corporate philosophy. They refocus the corporation's goals on social and environmental responsibility, and on implementing those goals with long-term solutions. This requires nothing more than a redefined vision of how a corporation sees itself in a future context. Corporate sustainability programs require rethinking basic values, expanding these values to all levels of the organization, implementing plans to achieve sustainability goals, and conducting periodic reviews and adjustments to ensure that results are achieved.

Motivations to develop and carryout corporate sustainability programs vary widely. Many corporate executives firmly believe that sustainability is a core concept that must be aligned with the corporation's long-term goal structure. Alternatively, programs can be driven by altruistic beliefs, a perceived need to appear "green-like" to customers, investors, or other stakeholders, or simply because competitors are involved. The fact of the matter is that even when corporations are motivated, sustainability programs are something new—something most executives have not been trained to comprehend and implement. Not all corporate executives have the skills and resources for such a task. There is also a tendency among some executives to consider sustainability as only another reporting requirement.

Corporations have environmental and occupational legislative mandates that demand adherence to environmental regulations. Corporations may have existing social responsibility programs and also engineering programs to implement technological improvements such as efficiency and waste reduction activities. However, there are often no regulations that require corporations to be sustainable, or have sustainability programs, and rarely are they required to make operational changes such as reducing energy use. Regardless, sustainability programs are voluntary efforts that can demonstrate to stakeholders that the corporation is serious about social and environmental responsibility.[42]

When developing a corporation's sustainability program, it is important that the corporation's Board of Directors and highest level managers actively participate, offer support, and provide direction. According to Barbara Krumslek, CEO of Calvert Investments, "The concept of sustainability expresses a simple truth that today's business organizations—including mutual funds—should conduct themselves in a manner that meets the needs of the present without compromising the ability of future generations to meet their own needs... at Calvert, we believe that all investing must ultimately become sustainable investing—and that all businesses must eventually develop sustainable business models."[43]

Many corporations begin the path to sustainability by inventorying programs, procedures and processes that are already in place. Often managers of large corporations, especially multinational ones, are unaware of many of the successful sustainability-related activities happening in their organizations. Even without a sustainability program *per se*, most corporations already have existing internal programs in place—yet many of these have not been structured within a sustainability agenda. The purpose of inventorying existing programs is to determine what programs and activities are already functioning within the company, and assess how they are linked to sustainability. This is a base line assessment process—taking stock of the strengths and weaknesses of existing programs.

Executives are often amazed at how much they learn about their organizations from this process. The internal programs in place might include those that address employee safety and health, energy management, environmental mitigation, waste management, internal recycling, product design, purchasing standards, regulatory compliance, etc. Inventorying existing activities and programs can yield surprising results. Dincer and Rosen (2004:5) provide a compelling argument that there are four dimensions of sustainability:

1) Societal sustainability;
2) Economic sustainability;
3) Environmental sustainability; and
4) Technological sustainability.

Any corporation can begin planning for sustainability by identifying existing programs and grouping successful ones within these four categories. The most relevant programs will impact more than one category. The assessment will identify programs that 1) already align with a corporate sustainability agenda; 2) or that need to be reworked; and 3) some that clearly do not align with a sustainability agenda and may need to be phased out.

After an initial assessment is completed, the policy components are developed into a corporate sustainability agenda. This process involves: 1) redefining the corporate vision/mission; 2) redefining corporate goals and objectives; 3) identifying programs to implement process and operational changes; and 4) establishing sustainability reporting procedures. *Policies* establish the framework. Programs implement the identified policies.

To repeat: Corporations often begin with a reassessment of their vision, goals and existing policies. Redefining the corporate vision or mission requires planning for the long term despite compelling tendencies that bias management decisions toward short term results. Redefining the corporate vision/mission involves the following:

- Assess industry trends. Research how other companies in the same industries and markets are addressing sustainability issues. Study which approaches and solutions have succeeded and which ones have not.

- Assess internal policies and programs to determine how they align with industry trends and how they shape the long-term mission of the company.

- Negotiate and obtain stakeholder input and consensus.

- Modify the corporation's vision/mission statement to incorporate the "language" of sustainability.

Whole Foods, a retailer of organic foods, has a vision statement called *Sustainability and Our Future* that begins with, "Whole Foods Market's vision of a sustainable future means our children and grandchildren will be living in a world that values human creativity, diversity, and individual choice. Businesses will harness human and material resources without devaluing the integrity of the individual or the planet's ecosystems. Companies, governments, and institutions will be held accountable for their actions."[44] The company plans "to implement this new vision of the future by changing the way we think about the relationships between our food supply, our environment, and our bodies."[45] Their approach uses a comprehensive strategy and involves all levels of employees.

Step #2 – Redefine Earlier Goals

The next step in developing a corporate sustainability program involves redefining corporate objectives. This can be accomplished by:

- Providing clearly stated objectives,

- Developing a plan for implementation,

- Establishing milestones and negotiating dates when the objectives will be met,

- Allocating resources to implement the stated objectives, and

- Instituting standardized approaches to assess and track achievements (e.g., measurement and verification) toward sustainability targets and goals.

It is important that goals and objectives support the mission of the corporation. It is helpful to establish comprehensive, high-level goals that have broad implications. Intel has an energy efficiency goal to "reduce our energy use on an average of 4% a year from 2002 to 2010, which will amount to a 30% overall reduction during that period."[46] Intel diversifies its energy resources by being the largest purchaser of wind energy in Oregon. Establishing goals and objectives that provide opportunities for systemic solutions—those that resolve multiple problems simultaneously—is one means of leveraging results. For example, improving interior lighting systems can increase light levels, reduce energy and maintenance costs, reduce waste, and have a beneficial environmental impact while simultaneously improving employee job satisfaction.

After these initial steps, there are management and operational actions that need to be taken to implement the new or revised sustainability agenda. These will vary across corporations as a function of their values, missions and resources. Examples of these might include:

- Address environmental pollution caused by the corporation by developing an environmental management plan.

- Provide a program of assessing production waste streams and redesigning processes to reduce waste.

- Establish a corporate recycling program for bottles, aluminum, plastics, paper, cardboard and scrap from processes.

- Audit corporate energy use and institute a set of interventions that will reduce energy usage in production and in support services.

- Establish a green energy purchasing program.

- Develop and implement a water conservation plan.

- Inventory refrigerant gases and phase out the use of refrigerants that greatly contribute to global warming or ozone depletion.

- Change equipment and material purchasing policies to increase the use of environmentally green products.

- Incorporate life-cycle cost analyses into procurement policies and process design.

- Review equipment maintenance procedures and make necessary changes to optimize equipment performance.

- Incorporate green building technologies and alternative energy solutions into the design and construction of new buildings and major renovation projects.

- Assess and reconfigure product lines.

- Integrate a plan for corporate community development and volunteer efforts.

Step #3 – Implement the New Sustainability Plan
This is accomplished by administrators, management, employees and stakeholders using dedication and teamwork. After the sustainability program has been launched, implementing and reporting actions must:

- Assess, verify, measure and track results.

- Periodically report to the public, investors, and other stakeholders on the progress made toward meeting sustainability goals.

- Develop a promotional campaign to advertise the activities and successes of the sustainability program.

Describing the process of establishing a sustainability program is easier than implementing one. Executives within an organization must develop a vision, set goals that align with that vision, create policies that take a longer term view, and devise programs that set an agenda as to how to position the company in the marketplace years in the future. Without "sustained" support from management, corporate sustainability programs, like many other long-term initiatives, can become marginalized. Lack of such support dooms sustainability programs to languish after being launched. On the other hand, corporations that want to seriously incorporate sustainability as their core mission can take initiatives to successfully develop, integrate, and manage their sustainability policies.

Since each corporation is unique, one key to success is to customize its sustainability efforts to match the needs of the company. The most successful sustainability programs respond to the business, social, environmental, and resource needs of the corporation, its stakeholders, its customers, and the communities it serves.

SERVICE INDUSTRIES

As the need for corporate sustainability programs continues to grow, diverse service sector companies are being formed to satisfy emerging markets. Examples include opportunities for architectural and engineering firms that specialize in sustainable construction, advertising companies that promote green corporations; companies that develop green energy such as wind farm developers, and financial institutions that invest in green companies. There are companies that market "green" products, such as certified wood or certified marine products. Some companies specialize in the management and recycling of construction wastes. Others provide and monitor high-tech computer systems to manage energy use.

New technologies require new software. These include software that tracks and analyzes energy consumption, software that provides measurement and verification analysis, and software that schedules and programs energy and water consuming devices in facilities.

Entire markets have developed for new business opportunities such as eco-tourism, land conservation and landscape restoration, reducing waterway pollution, re-colonizing endangered species, and developing educational programs and seminars on emerging "green" technologies. The global market for carbon emissions trading in 2006 was roughly $25 billion, doubling since 2005.

Perhaps one of the most interesting concepts involves energy services companies (ESCOs). These companies provide a wide range of customized energy and water-related solutions for industrial, commercial, institutional, and governmental organizations. ESCO services include establishing energy and water conservation projects, providing alternative energy supplies, upgrading building mechanical and electrical systems, and making building envelope improvements. ESCOs guarantee savings that result from the projects they implement. The services ESCOs provide are directly linked to sustainability.

A consulting industry has arisen from the idea that corporations need advice as to how to appropriately implement sustainability programs. SustainAbility, a firm founded in 1987, advises corporations on the opportunities and risks associated with corporate responsibility and sustainable development. According to their website, by "Working at the interface between market forces and societal expectations, we seek solutions to social and environmental challenges that deliver long term results."[47] They assist corporations (e.g., Pfizer, Dow, Baxter, DuPont and Shell) by performing research, establishing benchmarks, developing environmental surveys and producing final reports. The firm publishes a newsletter called *Tomorrow's Value*, prepared in partnership with the United Nations Environment Program and with Standard & Poors, which benchmarks corporate sustainability advances. The report ranks corporate sustainability based on what companies are actually doing to meet their sustainability goals.

CHALLENGES TO IMPLEMENTING
CORPORATE SUSTAINABILITY PROGRAMS

To complicate the process of implementing sustainability programs, corporations have day-to-day threats and emergencies which require immediate action, often diverting attention from long range goals. Generating business opportunities and maximizing profit margins are often at the top of corporate management's agenda. Executives can view new programs as being associated uncertainty, involving potential risks, and generating additional overhead and costs. Counter to the trend of corporations and multinational companies for whom sustainability agendas are meaningful, many businesses fail to focus their attentions on sustainability. These companies will ultimately suffer from higher operating costs and lower margins, and will have difficulty attracting quality employees.

Cost-center managers and their accounting systems often discount the long-term values of sustainability programs. Treating utility expenses as fixed overhead costs, rather than as a manageable variable cost, is a common mistake. Another has to do with the way utility expenses are charged to cost centers. Many companies fail to charge utility expenses, such as energy and water, directly to the divisions that actually consume them. Basing capital investment decisions solely on initial cost and failing to perform a component life-cycle cost analysis almost invariably backfires.

Some corporations take initial steps toward sustainability programs without fully implementing them. This may be due to inadequate skills, the enormity of the task, or a perceived lack of resources. Corporations on the path to sustainability quickly find that individual actions such as building a signature green building, reviewing an employee safety program, or changing corporate reporting standards are inadequate for implementing a sustainability program. While each of these might create public relations opportunities, each is just a beginning.

There is a plethora of case examples for corporations of what not to do when instituting sustainability solutions. Engaging in sustainability efforts half-heartedly results in a waste of corporate resources and yields fruitless results. Using program financial goals as the primary criterion in the decision-making process is a problem, a clue that the company is on an unworkable path.

Recently, a large distiller of spirits in U.S. decided to improve its corporate image by instituting energy and environmental improvements. The company wanted to reduce its costs, and it set a goal of reducing energy usage by 10%. To accomplish this, a request for qualifications for a performance contract was advertised. However, management of the process was delegated to a temporary employee—who did not understand how performing contracting works.

To encourage responses from ESCOs, the distiller stated that projects with return-on-investment ratios (ROI) as low as 8% would be acceptable. As this seemed achievable, there were approximately a dozen ESCOs that expressed interest and many responded. Three were selected for final consideration. Prior to the development of their final proposals, the three short-listed ESCOs were informed that the company had decided to change the project's financial requirement to a 2-3 year payback. While such returns are possible when targeted technologies are selected, this goal is rarely achievable for comprehensive energy and water savings initiatives.

This was a major change in project financial requirements. At the time, the distiller's price-to-earnings ratio yielded investors less than 5%. Changing the project financial goals was an obvious shortcoming, demonstrating a lack of serious commitment on the corporation's behalf. To date, the distiller has not implemented any of the recommendations made by the ESCOs, regardless of the savings available.

BENEFITS OF A
CORPORATE SUSTAINABILITY PROGRAM

Some corporations choose the path of sustainability as a means of improving their corporate citizenship. But in order for these benefits to accrue, they must take sustainability seriously.

Sustainability programs yield tangible benefits which include the potential of reducing operating costs and expenses, especially recurring costs. Lower operating costs accrue from energy and resource savings, labor savings, fewer potential liabilities, and insurance costs. Additional savings comes from lowered expenses of manufacturing, packaging, waste management, and transportation. Maintenance costs can be reduced or better managed. Maintenance improvements decrease equipment malfunctions and lower the probability of unscheduled production stoppages. "Right-sizing" and reducing run-times for energy-intensive equipment, such as electric motors, provides immediate savings in energy costs. Savings from increased employee productivity are often one of the greatest benefits of sustainability improvements.

Sustainability can provide intangible benefits, which include changes in the quality of life of employees, and in the perception of the company by stakeholders. Quality of life improvements include better indoor air quality, greater employee comfort, reductions in environmental contaminants, better employee health, less absenteeism, and greater employee retention. There are opportunities to publicize the success of a corporation's ability to achieve its sustainability goals.

The potential for liability can be reduced. According to Engardio (2007:52), "Embracing sustainability can help avert costly setbacks from environmental disasters, political protests, and human rights or workplace abuses—the kinds of debacles suffered by Royal Dutch Shell PLC in Nigeria and Unocal in Burma." Royal Dutch Shell was beset with problems stemming from environmental damage in the Niger Delta in the 1990s which led to their ceasing operations. In Burma in the 1990s, Unocal knowingly used forced labor to construct the Yadana natural gas pipeline—and a lawsuit about this was not settled until 2004.

Stakeholders and investors are becoming more interested in sustainable companies. This creates marketing opportunities. Solar collectors or a "green roof" at a corporate facility not only provide alternative energy, but also serve as a visible advertisement that the company is interested in sustainability. Investments in companies involved in sustainability are increasing. Assets of mutual funds that invest in companies that meet social responsibility criterion increased from $12 billion in 1995 to $178 billion in 2005 (Engardio 2007:56). Institutions with $4 trillion in assets weigh "sustainability factors in investment decisions" (Engardio 2007:56).

CONCLUSIONS

Corporations are taking the lead in sustainability—but this involves much more than improving business margins and profitability. Corporations must redesign their policies when they seek to become sustainable—by focusing on the economic, ecological and social aspects of sustainability. Within the last decade, corporate social responsibility programs have grown in popularity. There are a number of charters that are used to guide businesses in their sustainability efforts. Examples cited include CERES, the ICC Charter for Sustainable Development and ISO Standard 14000.

Corporations are very creative in their sustainability efforts. Corporations are changing their vision/mission statements, revising policies, and incorporating sustainability goals. This is happening in the retail, manufacturing, automobile, banking, oil production and utility industries. Fortis, a financial services company, developed its *Corporate Sustainability Statement* to address business responsibility, economic growth, and social and environmental concerns. With its new sustainability agenda, Wal-Mart is making changes that impact product packaging, transportation, buildings and infrastructure, and the products it sells. Toyota, BMW and Mitsubishi Motors are examples of corporations in the automobile industry that have sustainability initiatives. DuPont has a sustainability policy to curb global warming. Duke Energy developed its first corporate sustainability report in 2007.

Corporations can begin developing sustainability plans by performing an initial assessment that inventories existing corporate programs. The steps involved in developing a successful plan include: 1) redefining the corporate vision/mission; 2) redefining corporate goals and objectives; 3) selecting and implementing process and operational changes; and 4) establishing sustainability reporting procedures.

Sustainability programs make business sense—they help save corporations money. Sustainability efforts by corporations include reducing wastes, redesigning products, conserving natural resources, improving infrastructure, changing maintenance procedures, and addressing social concerns. A recent survey of energy engineers and managers reported that 48% of respondents worked at companies that have a policy in place to reduce greenhouse gases (Thumann *et al.* 2007). Intel not only purchases green power, but also has a goal of reducing energy use by 4% annually through 2010. Corporations are also pursuing other energy-related initiatives. Some companies are hiring energy executives while others are changing manufacturing processes.

Sustainability is increasingly important to companies. Businesses are viewing corporate sustainability programs as new opportunities—ones that have profit potential. General Electric plans on doubling its revenue from environmentally clean technologies and products—to $20 billion by 2010. As a result of these initiatives, employment is increasing in the commercial, manufacturing, industrial, service and consulting sectors, creating new "green-collar" jobs. Corporations pursuing sustainability offer hope for an enlightened future.

Endnotes

1. Holme L. and Watts. R. *Making Good Business Sense.* As quoted from www.mallenbaker.net/csr/CSRfiles/definition.html, accessed 28 May 2007.
2. For information about CERES see www.ceres.org/coalitionandcompanies/.
3. The International Chamber of Commerce provides information regarding the ICC Business Charter for sustainable development on its website: www.iccwbo.org/policy/enviropnment/id1309/index.html
4. Information on ISO Standard 14000 and 14001 can be found at http://www.iso14000-iso14001-environmental-management.com/.
5. As quoted in Naisbitt (1994:208-209).
6. As quoted in Naisbitt (1994:14-15).
7. *ICC Business Charter for Sustainable Development.* www.bsdglobal.com/tools/principles_icc.asp, accessed 20 January 2007.
8. International Chamber of Commerce. *About the Business Charter for Sustainable Development.* www.iccwbo.org/policy/environment/id1307/index.html, accessed 16 May 2007.

9. Ibid.
10. *UPS Sustainability Statement.* www.sustainability.ups.com/overview/statement. html, accessed 21 May 2007.
11. Fortis (2004, December). *Sustainability Statement.* www.fortis.com/Sustainability/index.asp, accessed 27 December 2006.
12. From *Business and Sustainable Development: A global guide.* www.bsdglobal.com/viewcasestudy.asp?id-93, accessed 21 January 2007.
13. Ibid.
14. The recycling of dye cooling water cost just $750 but delivered a saving of $400 per year. www.bsdglobal.com/viewcasestudy.asp?id-71, accessed 21 January 07.
15. See www.johnsoncontrols.com.sustainability.asp, accessed 4 January 2007.
16. Carey, J. (2007, 29 January). Big strides to become the Jolly Green Giant. *Business Week.* p. 53.
17. Ruben. A. (2007, 16 April). New stores will use less energy. *Newsweek.* p. 92.
18. Caldwell, E. (2007, 28 June). "How to cover your 'carbon footprints' at home." *USA Today.* p. 8D.
19. Ibid.
20. Duke Energy (2006, 2007), *Sustainability Report.* www.duike-energy.com/environment/sustainability.asp, accessed 10 June 2007. p. 3.
21. Ibid., p. 6.
22. Ibid., p. 8-9.
23. Cashing in. *Eugene Weekly.* www2.eugeneweekly.com/2001/02_08_01/news. html, accessed 16 May 2007.
24. Tribble, S. (2007). Google plans huge solar project. *The Mercury News.*
25. Proceedings of the West Coast Energy Management Congress (2007, June 6-7). *Update on California Solar initiative: Google's 1.6 MW Solar Campus,* Chapter 9.
26. Governor's Office of Energy Policy (2007, 10 May). Business coalition for climate action. *Kentucky Energy Watch.* 8 (19).
27. Subaru (2007, 11 July). A car should last a long time (advertisement). *USA Today.* p 9A.
28. Balogh, S. (2007, 16 February). A groovy trip to Subaru's zero landfill assembly plant. http://groovygreen.com/index.php?option=com_content&task=view&id=341&Itemid=58, accessed 17 July 2007.
29. Ibid.
30. Rudolph Diesel's patent (1893) for his engine was originally designed to operate using a variety of fuels, including peanut oil and coal dust. See *A history of the diesel engine,* www.ybiofuels.org, accessed 20 April 2007.
31. Bremer, B. (2006, September). Toyota's North American energy program: World class energy teams. *Proceedings of the 2006 World Energy Engineering Congress,* Washington, D.C.
32. Ibid.
33. PEW Center. *Waste Management Solutions.* www.pewclinate.org, accessed 16 May 2007.
34. Kageyama, Y. (2007, 8 June) Toyota's hybrid sales pass 1 million. *The Courier-Journal.* p. D3.
35. Webber, A. (2007, 25 July). What are the dinosaurs of Detroit thinking? *USA Today.* p. 11A.
36. Kageyama, Y. (2007, 8 June) Toyota's hybrid sales pass 1 million. *The Courier-Journal.* p. D3.
37. *What is ISO 14001.* www.bsiamericas.com/Environment/Overview/Whatis-sISO14001.xalter, accessed 19 January 2007.

38. Mitsubishi Motors (2005) *Environmental Sustainability Plan,* www.mitsubushi-motors.com/corporate/environment/sustainability/e/details.html, accessed 26 December 2006.

39. Healey, J. (2007, 24 April). Environmental chief at Ford gets bigger role. *USA Today.* p. 4B.

40. Florida opens hydrogen fueling station. *USA Today.* 24 May 2007.

41. Kiley, D. (2007, 29 January) How the hybrid race went to the swift. *Business Week.* p. 58.

42. See the Roberts Environmental Center, *Environmental and sustainability reporting,* Claremont McKenna College website for information on the Pacific Sustainability Index for corporations. www.roberts.mckenna.edu, accessed 7 January 2007.

43. Calvert Investments. (2004, December). *Corporate sustainability report.* http://www.calvert.com/pdf/GRI_Sustainability_2004.pdf, accessed 5 July 2007.

44. Whole Foods. *Sustainability and our Future.* www.wholefoodsmarket.com/company/sustainablefuture.html, accessed 4 July 2007.

45. Ibid.

46. Stangis, D. (2007, 16 April). Being green is just good business. *Newsweek.* p. 90.

47. SustainAbility. *Tomorrows value.* www.sustainability.com, accessed 4 January 2007.

Chapter 7

Local Policies for
Sustainable Development

"...the impersonal forces of structure and the personal volition of agency are not always easy to distinguish... Cities may differ in their underlying structure — in the cards they hold. But political leaders can use that structure — play those cards — in any number if ways. The challenge is to understand what specific economic and political contexts underpin choices regarding development options and how they explain variations in cities."

SAVITCH & KANTOR,
CITIES IN THE INTERNATIONAL MARKETPLACE (1992:31)

This chapter focuses on those policies that facilitate improvements in cities and local governments in an effort to achieve urban sustainability. "Sustainability can provide a qualitative measure of the integrality and wholeness of any given system" (Bell and Morse 1999:103). In regard to cities, Castells notes that "a city, ecosystem or complex structure is sustainable if its conditions of production do not destroy over time the conditions of its reproduction" (Castells 2000:118-122). Local solutions must be sought to foster the health and longevity of the systems that fundamentally support the well-being their localities. Managing the energy use of cities is part of the solution. According to Portney, "Sustainable cities frequently attempt to address energy issues" by directly influencing the city's "consumption of energy." This is done by offering consumers public transportation alternatives and creating home energy conservation opportunities (Portney 2003:95). Promoting an urban ecology that engenders construction of environmentally sensitive developments and energy efficient structures is a valid alternative.

This chapter concerns urban policies that impact urban sustainability—particularly those involving energy—and programs with which to implement them. Cities, though quite different in many ways, have common concerns that are important to urban development. Such concerns are manifested in urban policies and programs designed to reduce energy usage, improve the environment, facilitate interaction and foster cooperation. Since energy and sustainability are linked, this chapter gives examples of how urban areas are reducing energy usage in hopes of achieving sustainability. Concerned cities have an array of activities directed toward achieving these goals, including:

•   Understanding that they have choices and that their choices impact sustainability;

•   Assessing the sustainability policies and practices used by other local governments;

•   Identifying how local governments are pursuing—or not pursuing—goals and programs related to energy and sustainability;

•   Discussing specific environmental and energy related policies that apply to cities and identifying what specific types of programs are being pursued;

•   Using recent research to identify policies that are in effect to manage and reduce energy use in U.S. cities.

The U.S. population is increasingly on the move. Examples include the relocation of populations from south of the U.S. border into the U.S., from the cities to the suburbs, from the Rust Belt to the Sunbelt (Kahn 2000:1), and retirees from the northern U.S. and Canada to Florida and Arizona.

From 2000 to 2006 the country's fastest growing U.S. cities were McKinney, TX (+98%), Gilbert, AZ (+74%), North Las Vegas (+71%), Port St. Lucie, FL (+62%), and Elk Grove, FL (+59%).[1] The cities with the most rapid population declines were Detroit, MI (-8%), Cleveland, OH (-7%), Pittsburg, PA (-7%), Flint, MI (-7 %), and Dayton, OH (-6%).[2] While Chicago and Philadelphia lost population, population gains occurred in Houston, Phoenix and San Antonio.

These migrations have caused changes in the population densities of cities. In some cities, central business districts continue to decline as suburban areas increase in population. This results in decreased urban densities. New Orleans is a startling example of how urban populations migrate due to a natural disaster.

While cities have changed in population and size, the need for resources and energy has increased unabated. American cities depend on inexpensive energy, especially oil (which is costly to transport and store) and electricity (which is normally used the instant it is produced) to drive their local economies. Even temporary loss of energy flows disrupts transportation and communications, causing economic damage and social consequences. Transportation systems, resources, material, products and energy are indeed the lifeblood of modern economies. While cities have benefited from development, this same development has been fraught with hazards, environmental externalities and economic dependency. These concerns have fostered interest in alternative development approaches such as sustainable development.

Newman and Kenworthy (1999:333) believe that developing sustainability in cities can occur by "extending the metabolism approach to human settlements so that a city can be defined as becoming sustainable if it is reducing its resource inputs (land, energy, water and materials) and waste outputs (air, liquid and solid waste) while simultaneously improving its livability." This thinking shares the intellectual lineage traceable to Miller's *Living Systems* (1978) in which cities are categorized as having qualities similar to biological organisms. The ideal of the "sustainable city" has been frequently cited in literature beginning in the early 1990s (Haughton and Hunter 1994; Selman 1996). In practice, tracking progress toward such ends can be challenging. While sustainable development can be a core concept guiding long term changes, alternatively, it can be ignored as a local goal—or it can be adopted and coupled with concerted action.

Portney (2003) defines several principle elements that indicate whether urban areas take sustainability seriously. These elements include among others: 1) involvement in a sustainability indicators project; 2) pursuing smart growth activities; 3) employing land-use planning policies and zoning; 4) using transportation planning techniques; 5) having policies for pollution prevention and reduction; and 6) supporting energy, resource conservation and efficiency initiatives.

The Organization for Economic Cooperation and Development (OECD) has referred to the sustainable development of cities as "the focus for the future" (OECD 1990:36). The OECD further states that "energy efficiency and conservation... are areas where further practical urban research is needed" (OECD 1990:38). Interestingly, much of the empirical research to date regarding cities has been performed not by urban researchers but by policy makers, engineers, and designers who have implemented local solutions for local projects. Sustainability and energy are linked. The presence or lack of a focus on energy policies indicates the degree to which local governments are serious about sustainability.

The city of Seattle has a department named the "Office of Sustainability and the Environment" which maintains an Environmental Action Agenda. The city has established goals for protecting environmental quality and maintaining the quality of life. The agenda contains 22 specific goals that include improving city and residential water efficiency, reducing air pollution, limiting solid waste, protecting urban forests and promoting alternatives to personal vehicle use. Their county government is also involved. King County, in which Seattle is located, entered into a memorandum of understanding with Puget Sound Energy to establish a Resource Conservation Management program in an effort to coordinate strategies among departments, track cumulative savings and communicate results (Moen 2007:31).

There are many examples in the U.S. of successful efforts to initiate energy conservation, efficiency improvements, and alternative energy agendas in cities. Some of these predate the U.N.'s Agenda 21. In Arizona, Paolo Soleri's visionary Mesa City concept, entails construction of community-scale facilities whose "morphology is such to capture and make use of cosmic energy—radiation, winds, water, tides" (Soleri 1971:1). However, Arcosanti, his privately funded town, designed for a population of 5,000, has been under construction for over 35 years.

Other sustainability-related initiatives were developed after Agenda 21. Examples include Santa Monica, California, a city that not only adopted a sustainable city program but also mandated in 1999 that 100 per cent of its energy must come from renewable sources (Beatley 2000:361). The city of Chicago created a partnership with its transit authority, parks and 48 surrounding communities to purchase green power in 2001 and thereafter. The first phase of the project cut greenhouse emissions by 45,700 metric tons and 116,700,000 kWh annually (Gordon and Ode 2005:1). The city of Toronto has constructed a water source geothermal cooling system using water from Lake Erie to serve major buildings and facilities in its central business district, and thus reduce the use of electricity for cooling.

Cities have choices. Since urban areas account for the majority of energy usage, it is clear that our cities must engage in meaningful policy approaches to effectively address sustainability. In reality, changing the behavior and policy direction of regimes in cities is a difficult and complex task.

THE TYPES OF CHOICES AVAILABLE TO CITIES

Cities must operate within the realms of variable and restrictive federal and state legislation, often involving unfunded mandates. Despite such challenges, cities and their "local governments control a remarkable amount of resources—millions of buildings, cars, and dollars of tax revenue—and influence even more through building codes, zoning restrictions and transportation" (Gordon and Ode 2005:4). In 2005, Phoenix, Arizona adopted a new building code that included the 2003 International Energy Conservation Code (IECC). Phoenix now requires buildings to meet specific energy performance criteria in their design and construction.

Cities have choices and the paths they choose will determine their development policies. The policies they establish and pursue invariably impact energy use, either directly or indirectly.

Desai, N. noted that "the sustainable cities of the future will need to develop policies that promote sustainable resource use including energy, land, water and human skills" (Inoquchi *et al.* 1999:239). "City re-

7-1. Street Scene in Kowloon, an Urban Area of Hong Kong

sources are quite variable and that variation influences strategic alternatives" (Savitch and Kantor 2002:309). The forms of energy that cities select are among these choices. Confounding this issues, Wheeler (Legates
and Strout 2000:439) notes, "Exactly what constitutes a 'sustainable city'
is impossible to determine, given the extent to which cities are embedded in the global context." While it may be difficult to determine exactly
what makes a city sustainable, cities do have choices.

Cities create policies. Policies impact the sustainability of cities and
their energy consumption. There are policies and technologies that cities can employ to reduce energy use, and others that can cause inordi-
nate quantities of energy to be consumed. Cities can be planned to be
more densely populated, like Hong Kong, or less densely populated, like
Oklahoma City. Planning agencies can moderate urban heat island effects
with more streetscape landscaping and parklands—or they can require
less landscaping. Planning agencies can provide mass transit—or no
public transit. They can provide incentives for carpools or provide none.
Cities can harden policies concerning energy conservation—or they may
inadvertently cause greater use of energy. Cities can regulate the types of
energy used—or allow any type of energy to be used regardless of environmental consequences. Cities can also choose to permit multifunction,
multiuse zoning—or opt for restrictive single use zoning.

Cities can tax energy to promote energy conservation. They can tax
carbon emissions in hope of improving air quality, or alternatively, provide subsidies to their residents for the costs of energy (and then hide
energy costs in their budgets to prevent public review). Cities can establish energy management programs that provide for the reinvestment of
funds generated by energy savings. An example is the program established by the city of Phoenix in the 1980s that annually funds projects
that have saved millions of dollars in avoided in energy costs.

Cites can provide refuse recycling and establish solid waste management programs. Cities can establish policies for fleet fuel performance
standards and require emissions testing for their vehicles. Cities can
hold memberships in organizations such as the International Council
for Local Environmental Initiative (ICLEI) which promotes sustainable
development. Cities can muster concerted political influence to change
state and federal regulations. Cities can assist in "leveling the playing
field" so that traditional carbon-pedaling energy companies do not receive hidden subsidies that are unavailable to alternative energy companies (Chertow and Esty 1997:146).

Cities have infrastructure that must be managed. Infrastructure influences sustainability and consumes resources. Cities and local governments often control and influence infrastructure decisions that affect energy use, and they can create long term commitments to the forms of energy utilized to provide public services. Cities can initiate city-wide energy audits that provide detailed information on the energy consumption of the "different sectors, which can be used to establish locally relevant targets for energy efficiency action and facilitate predictions of future trends in local consumption/emissions" (Bennett and Newborough 2001:127). Municipal electrical utilities can offer demand-side management programs to reduce usage during peak periods, provide alternative energy options, and provide sites for demonstration projects. Cities have control over the planning processes, allowing them to choose to locate an industry near an alternative energy resource, or where on-site energy generation is possible. They can choose sites for facilities to reduce the use of motor vehicle fuels and improve transit system accessibility.

Cities have broad powers to control local transportation and other infrastructure decisions. More efficient transportation equipment can also be placed into service. Cities can require their fleets to use natural gas or opt for vehicles that use other forms of alternative energy. Cities can expand their existing mass-transit systems. Paris and Washington D.C. have their Metros, Hong Kong its MTR, San Diego has its Trolley, San Francisco its BART and Atlanta its MARTA. Each provides its urban area with an important transportation alternative. Cities can build bicycle pathways and bridges or choose not to.

Cities own buildings. Buildings consume resources. Local governments can choose to make their buildings energy efficient—or they can choose to ignore the energy use of their buildings. Cities can regulate energy efficiency standards for new construction. Cities can require energy studies, renovate and remodel buildings to improve energy efficiency by upgrading insulation, installing new fenestration, and replacing lighting systems. Cities can install solar panels on the roofs of their buildings to generate electricity and hot water.

Cities can track building energy usage, restructure operating schedules to change their facility energy consumption profiles, install energy management and control systems, institute four day work weeks, change occupancy densities, provide energy management staff, hire in-house energy engineers and establish internal "building energy monitor" programs. Cities can modify the landscaping of their properties, capture rain water to irrigate lawns, limit parking availability, provide special parking spaces for small commuter vehicles and electric vehicles, regulate speed limits, provide more energy efficient outdoor lighting, hold assemblies outdoors when weather permits, install bicycle racks and provide improved pedestrian access.

Cities own equipment. Equipment consumes energy. Cities purchase computers, office equipment, refrigerators, air conditioners, desk lamps, motorized devices. Cities can choose to purchase the lowest cost equipment item or pay a premium, if necessary, for equipment that uses less energy. Cities can choose to optimize the efficiency of equipment in order to minimize its energy use and reliability—or they can defer maintenance and cause premature equipment failure. Cities can purchase energy-saving equipment (e.g., with Energy Star™ labels) to reduce energy use—or they can dismiss energy use as a component in the decision making process. For example, cities can purchase an incandescent lamp for $.50 that lasts 750-1,000 hours or a $3-$4 compact fluorescent lamp that provides equal lighting, lasts 8,000-12,000 hours (thus requiring fewer replacements) and reduces energy costs by $10-$25 annually.

Cities purchase energy. Purchasers can make demands of energy suppliers. Cities can decide what type of fuels they would prefer to use and require efficiency standards of their suppliers. They can sometimes negotiate favorable and cost-effective utility rate schedules and establish demand-side performance-based standards. Cities can choose to require that public utilities and their officers not be permitted to contribute to local political campaigns. Cities can purchase "green energy" produced by using alternative fuels. Cities can provide their own green energy by installing solar panels to generate electricity. They can install fuel cells to operate motors, and on-site generators for use during periods of emergency and peak demand. Cities, by the methods they use to acquire energy, can create local employment—or expend dollars to purchase energy supplies which may result in less local employment.

In fact, cities have the ability to design and implement programs that are customized for their locations, suited to their budgets, manageable given their political circumstances and doable given their resources. *Examine the sustainability programs developed by cities and local governments and you will find creative solutions, colorful agendas and a plethora of variations on a theme.*

EXAMPLES OF LOCAL SUSTAINABILITY PROGRAMS

Many U. S. cities, especially in the Sunbelt regions, are developing rapidly. Unlike many cities in the northeast, they are growing at a time when the automobile, rather than the railcar or streetcar, or tram is the principle mode of transportation. There are newly developed suburban areas that are literally anti-pedestrian, lacking sidewalks, curb cuts and signage. While most cities have opportunities to use some sort of alternative energy, such as geothermal, solar, or wind power, many have opted instead for greater reliance upon fossil fuels. Some have grown wealthy from the once abundant reserves of oil and natural gas in their regions.

Portney (2003) has identified numerous cities with established sustainable development programs. A sample of these are provided in Table 7-1.

Innovative programs are offered in other cities as well. Orlando's Orange County focused on environmentally and financially appropriate development within urban service boundaries (Russell 2003:80). Seattle established an Office of Sustainability and Environment that

7-2. Downtown San Diego, CA

Table 7-1. Examples of Local Sustainable Development Programs

Brownsville, Texas: Eco-Industrial Park

Chattanooga, Tennessee: Sustainable Chattanooga

Jacksonville, Florida: Jacksonville Indicators Project

Orlando, Florida: Sustainable Communities

Phoenix, Arizona: Comprehensive Plan, Environmental Element

San Diego, California: Report Card of the Region's Livability

Santa Barbara, California: The South Coast Community Indicators Project

Santa Monica, California: Santa Monica Sustainable City Program

Scottsdale, Arizona: Scottsdale Seeks Sustainability

Stuart (Martin County), Florida: Sustainable Community

Tampa, Florida: Tampa/Hillsborough Sustainable Communities

maintains an Action Agenda that lists the city's goals for maintaining quality of life, promoting environmental justice and protecting environmental quality. Seattle has also instituted an internet accessible car share program called "Flexcar" which allows residents to reserve cars for temporary use. Chicago has planted over 400,000 trees in order to reduce the number of "heat islands" and remove carbon dioxide from the atmosphere (Underwood 2007:72).

Some cities have established "green building" programs while others modify ordinances, reduce energy usage and promote the use of renewable energy. Santa Monica aims to both reduce energy consumption and increase efficiency with its sustainable city program. Santa Monica's solution is to completely switch from fossil fuels to renewable resources (Gallagher 2000). The city of Cambridge, Massachusetts, has announced plans to provide free energy audits coupled with no or low interest loans for energy efficiency improvements (Underwood 2007:71). Similar programs are slated for Boston and four other cities in Massachusetts.

As early as 2002, the city of Los Angeles was purchasing 14% of its electricity from renewable energy sources (Bowman 2003:4). The District of Columbia has an Office of Energy Policy that works to provide energy efficiency programs for its residents, businesses, governmental organizations. It also assists low-income residents with paying their energy and utility costs. Indianapolis utilizes hybrid diesel-electric buses in its public transportation fleet, promotes the use of ethanol in city vehicles, and sponsors the sale of energy efficient lamps.

Cities including Baltimore, Maryland, and Evansville, Indiana, have completed energy savings performance contracts (ESPCs) to replace incandescent street signal fixtures with light-emitting diode (LED) lamps in order to reduce energy and maintenance costs. ESPCs provide a third-party guarantee of energy cost savings during the financed term of the project. The new LED lamps for traffic signals last more than 10 times longer than incandescent lamps and use less than 20% of the energy. New York City replaced 80,000 incandescent lamps at over 12,000 intersections, realizing a savings of $6.3 million annually on a $28 million initial investment (Underwood 2007:70).

Austin, Texas, established the nation's first city-wide "Green Building Program" in the early 1990s. Austin now offers a ranking system for buildings that uses energy efficiency as an element in the assessment process (Portney 2003:96). The city has a program for residential energy efficiency, another that provides commercial energy management services, and has established partnerships with its utilities. One of Austin's utility programs controls the run cycles on 35,000 air conditioning systems to reduce peak electrical demand (Morgan 2005:28-24).

Austin has been successful and continues to be a leader in energy efficiency. Engineering professionals took notice—the city attracted two international World Energy Engineering Congresses. In addition, the impact of Austin's city-wide energy conservation program allowed its city-owned utility to avoid construction of a planned 500 megawatt power plant that would have been needed as early as 2000 (Underwood 2007:71)

The city of Metro Louisville in Kentucky merged its city and county governments in 2000. A partnership developed shortly thereafter that included the local government, Jefferson County Public Schools and the University of Louisville called "The Partnership for a Green City." The group also formed a purchasing consortium to improve the effectiveness of purchasing "green" products and combined resources to "reduce, reuse and recycle waste ranging from paper, plastic, and aluminum cans to obsolete electronics, scrap metal and corrugated cardboard" (Ramsey 2005:5). Each institution dedicated itself to developing and implementing a sustainability program. For its part, the city committed to adding 2,000 acres of parkland, expanding its recycling program and conserving energy (Daeschner 2005:7).

COMPARING CITIES AND THEIR POLICIES

Why are Sunbelt cities and their sustainability policies particularly important today? Sunbelt cities are significant centers for urban population growth and development in the U.S., generally outpacing their non-Sunbelt counterparts. Sunbelt cities have become centers not only of urban growth, but also as focal points for new development and investment. Many of these cities are leaders in planning policy. Urban development creates the opportunity to establish policies that promote, monitor and direct both the form and location of growth in urban areas. With new investment and construction, cities can select from a range of newly available technologies as they grow their cities. Their policies will ultimately affect not only the present design of their cities but also their future energy usage and the sustainability of their urban areas.

So why compare varied and seemingly incomparable cities and their policies? Commonalities are often less elusive than they would seem. Each city has a distinctive physical form and culture that can be observed in the collage of its streets, the nature of its structures, and the culture of its people. These factors and others contribute to its aggregate urban form. Christian Norberg-Schiltz noted, "we have over and over again shown that a form only has meaning within a system of forms, and that the idea of independent meaningful forms is a misconception." Yet the issues of population growth and urban development permeate the lore of cities, creating needs for common solutions. Creative legislation dealing with energy and environmental issues is being developed, instituted and tested in states such as New Jersey, California, and Florida. However, despite differences in policy, there are commonalities and patterns in policy—as there are in form.

The cornucopia of policies available to achieve reductions in energy use and improved sustainability include: 1) energy management programs; 2) local policies; 3) organizational participation and memberships; and 4) programs designed to improve the environment. What follows is an examination of each of the selected indicators and a discussion of which policies cities, their utilities, and their local governments have chosen to adopt. For this assessment, the largest 25 cities located in the U.S. Sunbelt (listed in Table 7-2) will provide an example of the variations in their sustainability policies. Several of these policies derive from programs that are sponsored or co-sponsored by the federal government or by local utilities. However, it is emphasized that the selected policies are of a local nature and require local initiative to implement and sustain.

The first category of policy indicators considered are those concerned directly with energy policy. Three measures are selected to assess efforts by a city government and its primary local utility to reduce energy usage within the community. These measures include:

1) The prevalence of certain utility rebate programs;

2) City government programs to manage internal energy use; and

3) Local programs that support energy conservation in general.

LOCAL ENERGY MANAGEMENT POLICIES

Policy #1—Utility Rebate Programs

Cities sometimes own their local utilities, or have a financial interest in them. The city of Memphis has the largest multi-service, municipally owned utility in the U.S. More often local utilities are privately held. Regardless, cities often act in partnership with their local utilities in some manner. Energy Management Policy #1 deals with whether or not the primary local electrical utility serving the city offers rebates for facility lighting upgrades, building envelope improvements, new construction, or other related electrical improvements. These types of rebate programs are a form of incentive for local customers to implement physical improvements to their facilities that impact electrical energy usage. They represent a form of cooperation and commitment between the primary electric utility and its customers. Utility rebate programs may be supported by equipment manufacturers who may discount the costs of critical hardware components, thus partially subsidizing the cost of implementing these programs. Electric utilities tend to favor programs that lower total electrical demand, especially during peak demand periods. Customers, on the other hand, tend to favor improvements that reduce maintenance costs, improve efficiency and directly lower their utility costs.

Electricity is sold in markets such as Phoenix that are regulated and have specifically defined utility service territories, and also in deregulated markets such as those in California where customers are empowered to choose their electric providers. In either case, utility programs are seldom totally independent of governmental intervention and regulation. Most are subject to state regulation and the scrutiny of public service commissions.

Utilities frame their local policies by supporting (or not supporting) energy conservation with their advertising and promotion of rebate programs. Interestingly, a full range of rebate program activity was found among the 25 sampled Sunbelt cities. Utility rebate programs are important as they not only suggest that energy conservation is being promoted by the utility but also imply a modicum of cooperation between the utility and its customers in order to achieve a mutually beneficial solution. The idea of cooperation and partnership was found to be a key principle of sustainability (Beatley 2000:17). So is the belief that corporations must accept responsibility for stewardship by mitigating the environmental consequences of their actions (Presidents Council on Sustainable Development 1996).

Utilities serving urban areas in southern California and Florida typically offer a wide range of utility rebate program options from which their customers may choose. Their programs also involve the implementation of multiple energy reduction technologies. On the other hand, cities in Oklahoma and Virginia offer no or few utility-sponsored rebate programs. A total of 11 cities (44%) in the sample of 25 cities have utility rebate programs. 14 cities (56%) lack such programs. Utilities serving the more populous Sunbelt cities have a greater propensity to offer rebate options. Four of the five largest cities (Los Angeles, Houston, San Diego and San Antonio) have utility rebate programs. Of the five smallest cities sampled, only Miami has a utility-sponsored rebate program.

Policy #2—Energy Efficiency Programs

Cities own many and varied municipal facilities. Local energy Management Policy #2 gauges whether or not the city government has an internal energy efficiency program. The types of buildings owned by cities include courthouses, office buildings, fire and police stations, sewage treatment facilities, emergency action and preparedness centers, libraries, public health facilities, training facilities, subsidized housing, etc. These facilities collectively consume significant amounts of energy. City administrators and managers may view energy use as an unavoidable but manageable cost, an uncontrollable overhead expense, as inconsequential, or of concern only if publicly scrutinized. It would seem logical that cities adopting sustainable policies would be concerned with the costs and impacts of energy in their buildings and facilities. The actions taken by local administrations would likely be manifested in policies that would support energy efficiency improvements such as the installation of energy saving technologies, building envelope and architectural improvements, equipment replacement, and adoption of building standards among others. To these ends, partnerships such as performance contracts might be considered (Hansen and Weisman 1998; Roosa 2002:323-325). The idea of civic engagement and the ethic of institutional stewardship have been linked to improving sustainability (Presidents Council on Sustainable Development 1996).

Local Energy Management Policy #2 (Energy Efficiency Programs) is important. Does the city government feel energy conservation and energy efficiency in its own buildings is important enough to warrant attention? In this sample of 25 Sunbelt cities, 19 (76%) have initiatives to improve energy efficiency in public buildings while only six (24%) lacked such programs. What is striking is the wide range of approaches that cities have chosen to employ. Policies implemented by the sample of major Sunbelt cities include:

- Hiring a city energy manager and implementing recommended improvements to manage and reduce energy use;

- Establishing a written energy policy for government owned buildings;

- Mandating the use of "Green Building" construction techniques or incorporating standards such as those required by Leadership in Energy and Environmental Design (LEED)[3] for new construction;

- Requiring energy analysis surveys of city-owned buildings to determine economically appropriate actions and alternatives to reduce energy use;

- Installing centralized energy management and control systems (EMCS) to automate energy reduction strategies and to control energy-intensive equipment and energy-consuming devices;

- Participating in packaged programs such as the U.S. Department of Energy's "Rebuild America Program" which is geared toward implementing broadly based, integrative strategies and oriented toward policies that reduce facility energy use while lowering the costs of energy;

- Having a departmental division in a city government for Energy Conservation and Management (e.g., San Diego);

- Using Energy Saving Performance Contracts (ESPC) as a vehicle for facility improvements. ESPCs rely with third-party financing, which is subsidized by energy savings and cost avoidance, and which must be verified by measurement and verification procedures.

The most popular policy effort among the sampled cities (with seven cities participating) has been the adoption of the principles and requirements of the U.S. DOE-sponsored "Rebuild America" program. Rebuild America is a "network of community-driven voluntary partnerships that foster energy efficiency and renewable energy in commercial, government and public housing programs" that "work to overcome market barriers that inhibit the use of the best technologies" (U.S. Department of Energy 2003). Among those participating are the four largest Sunbelt cities (Los Angeles, Houston, Phoenix and San Diego). Phoenix has budgeted over a million dollars annually through 2005 to directly fund capital intensive energy conservation improvements. Dallas, Austin and Long Beach have adopted Green Building or LEED construction standards for city owned buildings. San Diego's Environmental Services Operations Station administration building has carports in its parking lot with rooftop photovoltaic panels which generate 91,500 kWh per year (city of San Diego 2002:1). This is more than enough to supply 100% of the building's electrical energy requirements.

With over two-thirds of the city below sea level, New Orleans has concerns about rising sea levels which threaten to displace its urban residents. Long before Hurricane Katrina, New Orleans adopted a unique policy to reduce the threat of global warming. Greenhouse gas emissions have been profiled and municipal emission reduction targets have been mandated through 2015. Their research revealed that municipal buildings were responsible for approximately 35% of the CO_2 emissions released from municipal operations. Mitigation measures, justified by energy savings, were implemented in a number of buildings including city hall, the court complexes, the public library, police headquarters, the airport and others. Measures include mechanical system upgrades, installation of energy efficient lighting systems, tree planting, installing new traffic lights, establishing building energy codes for city buildings and measures to reduce the urban-heat island effect. Despite this epochal and precedent-setting policy initiative, it is obvious that the actions of one city will not resolve the problems associated with global warming. The destruction caused by Hurricane Katrina neutralized the effectiveness of the policies created by New Orleans.

Atlanta's energy conservation program exemplifies programs that offer tangible and measurable financial returns. The city of Atlanta scheduled policy workshops, performed utility rate assessments for over 600 municipal accounts, performed facility energy audits and developed an internal employee energy conservation program. Within one year the city had projected savings from these initiatives of nearly $500,000 (City of Atlanta Online 2003:1). The city has established a policy goal of reducing energy consumption by an additional 10% by 2010 and has appointed an Energy Conservation Coordinator (City of Atlanta Online 2003:1).

Policy #3—Local Utility and Government Programs

Policy #3 gauges policy support by both the primary local utility (excluding rebate programs) and by the city government for energy conservation, energy efficiency and alternative energy programs. This measure asks if both the utility and the local government are active in promoting these programs. Are the goals of both cooperatively directed to achieving reductions in energy use?

Local energy conservation efforts can be supported by citizen actions, organizational support, corporations, utilities, local governments, other governmental bodies or by other means. Local participation and involvement are central to the idea of sustainable cities (Bell and Morse 1999:124). Local governments and utilities have the primary economic means and leadership infrastructure to direct the orientation of community energy policies. The value of Policy #3—Local Utility and Government Programs—can be diminished or enhanced by the fact that in deregulated electrical utility markets, urban residents may have the ability to select from a number of electricity providers.

Energy Management Policy #3 measures the combination of city policies and expanded utility programs supporting energy conservation and alternative energy programs (excluding rebate programs). This can show that there is a broad base of leadership support for these local initiatives within the community. As a result, this local policy program suggests that a level of partnership exists or, as a minimum, the primary utility and the local urban government are working individually or jointly toward a common goal: *reducing the use of energy from conventional sources.* This reflects cooperation and considerable goal alignment between the utility and the local government. These programs often include options for alternative energy. The expanded use of renewable fuels has been identified as a sustainability indicator (Newman and Kenworthy 1999:19).

Utilities can provide a range of policies to support local energy conservation and alternative energy for customers in their service territories. For the 25 sampled cities, it is surprising that only six primary electrical utilities provide no energy conservation or alternative energy programs. Of the 19 cities with such programs, principle utilities in four cities provide assistance with energy studies or "energy audits."

A total of 12 primary utilities serving our 25 selected Sunbelt cities offer programs that allow their customers to opt for the purchase of electricity generated by alternative energy. These programs vary based on local feasibility. Some require an incremental fee for the "green power" purchased, or a financial contribution to support alternative power generation. (Green power is loosely defined as electrical energy generated from renewable resources such as wind, solar, geothermal or biomass).

Green power programs are in vogue and are being branded and marketed. For example, Los Angeles has a green power project to generate 5,000 Mwh annually from micro-turbines using landfill gases. Both Los Angeles and San Diego provide self-generation incentives. The Nashville Electric Service Company offers a program called the "Green Power Switch™" which consumers can purchase 150 kW blocks of electricity generated from renewable sources such as solar, wind and methane gas. JEA, the utility serving Jacksonville, has an incentive program to promote residential and commercial installation of solar photovoltaic and thermal energy systems. Utilities in San Antonio ("Windtricity™") and Oklahoma City ("OG+E Windpower") have programs to develop wind turbine generators. Tucson Electric Power (TEC) has a program called "GreenWatts™" which provides consumers with the option to purchase electrical power generated from methane gas.

Among the most extensive local commitment by a major utility is that of Southern California Edison (SCE), serving Long Beach, which not only has a self-generation incentive program for private installations of alternative energy systems but also generates 20% of its electrical energy from alternative energy sources (Southern California Edison 2004). Including the additional 8% of SCE electricity generated by means of traditional hydroelectric generation, SCE produces an amazing 28% of its total electrical energy production from renewable resources.

An assessment of major cities in the Sunbelt shows the types of programs in place. Among the local governments in the Sunbelt, 14 of 25 cities offer policies and programs to support energy conservation or alternative energy, providing incentives for sustainable technologies for new construction in their local communities. Seven cities offer financial support for community projects involving new buildings that incorporate green building technologies. These programs directly improve the urban ecology. The city of San Antonio has established the "Metropolitan Partnership for Energy," involving the city government and the community at large. The partnership has organized an energy council, educational programs, facility and infrastructure improvements, equipment conservation measures, fleet conservation standards and procurement requirements. Tucson and Las Vegas are among those Sunbelt cities that have adopted building code requirements for energy-efficient construction.

Other cities are less committed. While the city of El Paso has a program, the city budget indicates that only $7,500 is allocated annually. Fresno's energy policy provides only for weatherization assistance for the homes of senior citizens. 11 cities among those sampled lack any active energy conservation program, alternative energy policy, or support for similar local initiatives. Table 7-2 below provides a summary of selected local government and utility energy policies that are presently being promoted.

Combining the results from utility and government policies and programs, research in 2003 found that: 1) a total of 13 cities (52%) meet the requirements and have established policies or programs supported by both the primary electric utility and local government; 2) eight cities (32%) have policies supported by either the primary electric utility or the local government; and 3) only four cities (16%) have no policies supported by either the primary electric utility or local government. Table 7-3 summarizes the results of the research in the category of energy policy indicators.

Among the cities that have programs supported by both the primary electric utility and local government are four of the five largest Sunbelt cities when ranked by population: Los Angeles, Houston, Phoenix and San Diego. Among the cities that have no programs are three of the five with the lowest population: Virginia Beach, Mesa, and Tulsa. Table 7-3 indicates that six of the selected Sunbelt cities satisfy all three categories of energy poli-

Table 7-2. Local Energy Management Policies

Sunbelt City	Sponsored by Local Government	Sponsored by Primary Utility
Los Angeles	Green Building Initiative, Green LA	Green LA (Solar, Vehicles)
Houston	Rebuild America, LEED Program	Support of Energy Studies
Phoenix	Yes, capital improvement projects	Solar Partners Program, EV Charging
San Diego	Green Building, Rebuild America	ESPC Std Contract, Self Generation
Dallas	None	Support of Energy Studies
San Antonio	Metro Partnership for Energy	Windtricity Program
Jacksonville	None	Green Energy Program (solar)
Austin	Green Building Program	Green Choice Program
Memphis	None	None
Nashville/Davidson	None	Green Power Switch Program
El Paso	Yes	Renewable Energy Plan
Charlotte	None	Green Power Program
Fort Worth	None	Support of Energy Studies
Oklahoma City	None	OGE Windpower
Tucson	Model Energy Code	Greenwatts, Sunshare Programs
New Orleans	None	None
Las Vegas	Model Energy Code	Green Power Program
Long Beach	Yes	Self-generation Incentive Program
Albuquerque	Yes, included in 1994 strategic plan	Sky Blue, Renewable Energy Plan
Fresno	Senior Citizen Weatherization	Support of Energy Studies
Virginia Beach	None	None
Atlanta	Yes	None
Mesa	None	None
Tulsa	None	None
Miami	Green Building	Demand Reduction "On Call"

cies. These cities are Los Angeles, Houston, San Diego, Austin, Las Vegas and Long Beach. On the other hand, three of the cities, Charlotte, Oklahoma City and Virginia Beach, have none of these energy policies in effect.

Two forms of local policies are discussed next. The first considers sustainable development as a local policy goal. The second lists creative efforts to improve local transportation systems.

SUSTAINABLE LOCAL POLICIES

Policy #1—Sustainable Development Programs

This policy asks the question: Is either sustainable development or urban sustainability a primary goal of the city? Ancient southwestern

Table 7-3. Local Energy Management Program Indicators

Sunbelt City	Utility Rebates	Policies For Buildings	Locally Supported Programs
Los Angeles	Yes	Yes	Yes
Houston	Yes	Yes	Yes
Phoenix	No	Yes	Yes
San Diego	Yes	Yes	Yes
Dallas	Yes	Yes	No
San Antonio	No	Yes	Yes
Jacksonville	Yes	Yes	No
Austin	Yes	Yes	Yes
Memphis	No	Yes	No
Nashville/Davidson	No	Yes	No
El Paso	No	No	Yes
Charlotte	No	No	No
Fort Worth	Yes	Yes	No
Oklahoma City	No	No	No
Tucson	No	Yes	Yes
New Orleans	No	Yes	No
Las Vegas	Yes	Yes	Yes
Long Beach	Yes	Yes	Yes
Albuquerque	No	Yes	Yes
Fresno	Yes	No	Yes
Virginia Beach	No	No	No
Atlanta	No	Yes	No
Mesa	No	Yes	No
Tulsa	No	Yes	No
Miami	Yes	No	Yes

monuments such as Mesa Verde and Chaco Canyon show that cities in the southwest have been abandoned as a result of factors that included environmental mismanagement and changes in local environmental conditions. Many southwestern mining towns grew to become boomtowns, only to go bust and ultimately become ghost towns. Perhaps, even today, we are growing new throw-away cities. If sustainability is not on the urban agenda and not an identified goal, then it cannot be achieved.

As noted previously, the concept of sustainability has many interpretations. Urban sustainability refers to a somewhat idealized model of urban development which addresses a wider set of concerns about urban growth, patterns of urban development, and issues that arise from urban development. The broad use of the concept of sustainability began after 1992 with the Rio Declaration on Environment and Development (later referred to as Agenda 21). As a result, considering sustainability in local policy-making is a relatively recent phenomenon.

There are multiple definitions of sustainability, and these definitions are subject to a wide range of interpretations. There are four principles of urban sustainability in the European Community's (EC), Sustainable Cities Agenda: 1) the principle of urban management; 2) the principle of policy integration; 3) the principle of ecosystems thinking; and 4) the principle of cooperation and partnership (Beatley 2000:17). These principles can serve as a model for addressing sustainability in U.S. cities.

Local Policy #1 gauges only whether or not sustainable development is a stated urban goal. When assessing city-by-city policies, the interpretation of sustainability was limited to policies that were relevant. For example, if a city's only stated "sustainable" policy is to "maintain a sustainable tax base" then the term was judged to be misapplied, and sustainability was not regarded to be an actual urban goal.

A total of 7 cities (28%) have already established sustainability as a primary urban goal. These cities are Jacksonville, El Paso, Long Beach, Albuquerque, Atlanta, Mesa and Tulsa. Long Beach is atypical in that its *2010 Citywide Strategic Plan* identifies "becoming a sustainable city" as a primary strategic goal. In both Atlanta and Tulsa, urban sustainability is a primary administrative goal, strongly supported by their mayors. In 1985, Jacksonville initiated its *Quality of Life in Jacksonville* plan (Portney 2003:213). The pledge of Atlanta's city council president to "create an efficient, vibrant and sustainable city" includes an energy conservation initiative began in 2002 (City of Atlanta Online 2003:1). Mesa, Arizona has a statement in its *Vision of the Mesa 2025 General Plan* to support the city "as a sustainable community in the 21st century."

Seven additional cities (28%) have identified programs to support sustainable building policies, have demonstration projects underway, or have established land use requirements to promote sustainable development. San Antonio supports community revitalization with a goal that includes "sustaining a strong urban system" (City of San Antonio 2004b). In city of Tucson, a developer-originated project for the new community of Civano aims to build a model sustainable community. "The goal of the Civano project is to create a new mixed use community that at-

7-3. Centennial Park in Atlanta, GA

tains the highest feasible standards of sustainability, resource conservation and development of Arizona's most abundant energy resource –solar—so that it becomes an international model for sustainable growth" (City of Tucson 1998:1). In Los Angeles and Austin sustainable building programs or guidelines have been established for new construction.

While 14 cities (56%) have identified sustainability as an objective, the rest of the selected Sunbelt cities (11 or 44%) have not established it as a goal. The reasons for this are unknown. Perhaps having a goal of "being sustainable" may be unimportant, or not a priority, or counter to the goals of the urban regime. Perhaps sustainability is being considered but has not yet been implemented. Among these cities are Houston, Dallas, Fresno and Las Vegas. (It is interesting that these cities are among those dealing with rapid urban population growth).

Other cities, including Charlotte and Fort Worth, have programs with a development policy based on a "smart growth" agenda. Many of the policies associated with smart growth agendas complement sustainable development initiatives.

Table 7-4. Sustainable Local Policy #1, Sustainable Development Programs

Sunbelt City	Is Sustainability as a Local Policy?
Los Angeles	Yes, sustainable building program
Houston	No
Phoenix	Yes, found in land use plan
San Diego	Yes, goal of Environmental Service Department
Dallas	No
San Antonio	Yes, specific program goal
Jacksonville	Yes
Austin	Yes, established sustainable building guidelines
Memphis	No
Nashville/Davidson	No, excluded in planning mission statement
El Paso	Yes, included as goal in city mission statement
Charlotte	No, focus is on "smart growth"
Fort Worth	No, focus is on "smart growth"
Oklahoma City	No
Tucson	Yes, Adopted Sustainable Energy Code
New Orleans	No, stated as goal of utility
Las Vegas	No
Long Beach	Yes, sustainability is the primary urban goal
Albuquerque	Yes, Sustainable Community Development
Fresno	No, excluded as goal in planning mission
Virginia Beach	No, excluded from vision statement
Atlanta	Yes, administrative goal of Mayor's office
Mesa	Yes, included in General Plan for 2025
Tulsa	Yes, administrative goal of Mayor's office
Miami	Yes

Policy #2—High Occupancy Vehicle Lanes

Are Sunbelt cities seeking creative ways to increase the use of existing highway infrastructure? Traffic movement through urban areas is often impeded by the sheer number of vehicles on the roadways. Highway transportation planners devote endless hours to designing highways that expedite the flow of vehicles. Vehicles stopped on highways due to traffic congestion increase energy use, contributing to urban pollution, and wasting the time of commuters. Traffic congestion in the U.S. created travel delays that consumed billions of gallons of fuel (Brown 2006:3). Recurring disruptions in traffic flow ultimately lead to increased calls for the construction of additional roadways and alternative means of transportation.

At issue is the number of vehicles that are used to commute daily to urban areas, especially during peak periods, with only the driver on board.

Transportation planners in many cities of the Sunbelt and elsewhere have realized that if vehicle occupancies could be increased, especially during peak periods, a reduction in the number of vehicles on roadways would result. To this end, the idea of high occupancy vehicle (HOV) lanes has gained popularity, primarily on intra-urban interstate highways. The decision to construct or designate HOV lanes is influenced by city transportation planners, state and local government officials, and federal funding restrictions. By increasing vehicle occupancy, and reducing total traffic, the costs of constructing additional freeway lanes could be controlled. One obvious goal of HOV lanes is to provide incentives for car pooling and in some cases to cause a switch to alternatively fueled vehicles. Increasing the use of carpools has been identified as an indicator of sustainability in cities (Newman and Kenworthy 1999:19).

Lack of HOV lanes may simply indicate that highway traffic congestion is neither prevalent nor considered problematic during peak periods. On the other hand, the presence of HOV lanes may indicate that the city is exploring innovative solutions to ease vehicular traffic problems, reduce vehicular energy use, and minimize urban air pollution.

HOV lanes are also known as "carpool lanes" or "white diamond" lanes (referring to the large diamonds painted in the center of the lanes). In the U.S., there are now over 4,020 kilometers (2,500 miles) of HOV lanes that collectively move over 3 million commuters each day (North Carolina Department of Transportation 2004). In these lanes, tractor-trailers are not permitted. There are varying types of HOV lanes, almost all constructed or assigned since 1980. Many HOV lanes operate continuously while others are used only during peak commuting periods. Some cities allow bus traffic in HOV lanes. Some high occupancy lanes require vehicles to have a minimum of two occupants, others require three. While enforcement is uneven, operating a vehicle in HOV lanes without passengers can result in hefty fines. The minimum fine in Los Angeles is over $250.

There are recent developments in the use of high occupancy vehicle traffic lanes. Florida law now allows low emission (alternatively fueled) vehicles to use high occupancy vehicle lanes without regard to the number of passengers in the vehicle. In Georgia, both low emission and hybrid vehicles can use HOV lanes regardless of the number of occupants in a vehicle.

Table 7-5 summarizes the research concerning local policies that include sustainability as a city goal and the present availability of HOV lanes.

Table 7-5. Sustainable Local Policy #2, HOV Lanes

Sunbelt City	HOV Lanes
Los Angeles	Yes
Houston	Yes
Phoenix	Yes
San Diego	Yes
Dallas	Yes
San Antonio	Yes
Jacksonville	No
Austin	No
Memphis	Yes
Nashville/Davidson	Yes
El Paso	No
Charlotte	No
Fort Worth	No
Oklahoma City	No
Tucson	No
New Orleans	No
Las Vegas	No
Long Beach	Yes
Albuquerque	No
Fresno	No
Virginia Beach	Yes
Atlanta	Yes
Mesa	No
Tulsa	No
Miami	Yes

Of the 25 Sunbelt cities, 12 (48%) presently have operational high occupancy vehicle lanes. However, application varies widely. The Los Angeles HOV system has over 290 kilometers (180 miles) of HOV lanes. A 56 kilometer (35 mile) long section on I-405, opened in 1998, constitutes the longest continuous stretch of HOV lanes in the U.S. The Los Angeles system has been extraordinarily successful, moving over one million passengers per day and boasting a daily reduction of 30,000 commuter delay hours. Current plans are to extend the Los Angeles HOV system by an additional 491 kilometers (305 miles) of lanes.

Houston provides 119 kilometers (74 miles) of lanes for HOVs with yet another 171 kilometers (106 miles) to be added. Atlanta has 63 kilometers (39 miles) of HOV lanes with an additional 105 kilometers (65 miles) proposed or under construction. Miami currently offers 66 kilometers (41 miles) of HOV lanes with an additional 26 kilometers (16 miles) of lanes proposed. Memphis has only 13 kilometers (8 miles) of lanes reserved for HOVs with no present plans to extend the system.[4]

The remainder of the selected cities (13 or 52%) lack HOV lanes. Charlotte, which is among this group, is planning to install HOV lanes in the near future. Charlotte's plans include construction of 190 kilometers (118 miles) of lanes through 2023 (U.S. Department of Transportation 2004).

Cities concerned about sustainable development and like issues are associating both regionally and internationally, with other cities that have similar policy agendas. Such interaction provides opportunities to learn from the experiences of other cities that have similar policy agendas, fostering the development of best practices. Three organizations in the U.S. offer support: 1) membership in the ICLEI; 2) participation in the Clean Cities program; and 3) being an Energy Star™ partner.

ORGANIZATIONAL PARTICIPATION POLICIES

Policy #1—Local Membership in ICLEI

The International Council for Environmental Initiatives (recently renamed the ICLEI—Local Governments for Sustainability) has a membership of more than 430 local governments—all with the goal of implementing sustainable development. Members of the ICLEI can tap databases and resources that deal with energy services, sustainable transportation, environmental information, case studies of energy projects, educational publications and related local government initiatives. These local governments represent over 300 million people worldwide.[5]

Membership in the ICLEI indicates a policy commitment to sustainable development. According to their website, the ICLEI provides services to "a worldwide movement of local governments to achieve tangible improvements in global environmental and sustainable development conditions through cumulative local actions" (ICLEI 2003). The ICLEI serves as a "clearinghouse on sustainable development by providing policy guidance, training and technical assistance, and consultancy services to increase local governments' capacity to address global challenges" (ICLEI 2003). The organization helps local governments "generate political awareness of key issues, build capacity through technical assistance and training, and evaluate local and cumulative progress toward sustainable development" (ICLEI 2003).

Membership in the ICLEI indicates strong support for sustainable development by urban administrators and their communities, and their desire to cooperate both regionally and internationally with other cities pursuing similar agendas. Of the 25 cities in the sample, only six (24%) are ICLEI members. Those cities are Los Angeles, San Diego, Austin, Tucson, Atlanta and Miami.

The 19 remaining Sunbelt cities (76%) did not hold ICLEI memberships in 2003. Yet more cities are becoming involved in the organization. Some non-member cities have supported certain ICLEI sponsored activities or individual projects. Recently the ICLEI cosponsored the *U.S. Mayor's Statement on Global Warming 2003* which was signed by mayors of 155 U.S. cities. This statement stresses that, "The scientific community is very clear in its warning—we must act now to significantly reduce greenhouse gas emissions below current levels or we will quickly reach a point at which global warming cannot be reversed." It concludes with a request to the federal administration that as "mayors responsible for the well being of our communities, we urge the federal government to maintain, enhance and implement new domestic policies and programs that work with local communities to reduce global warming pollution." Among those who signed the U.S. Mayor's Statement on Global Warming 2003 were mayors from the Sunbelt cities of Long Beach, San Diego, Albuquerque, Las Vegas, Austin, Dallas, Fort Worth, Houston, San Antonio, New Orleans, Atlanta and Virginia Beach. (Many of these mayors represent cities that are not currently active ICLEI members).

Policy #2—Local Membership in the Clean Cities Program

In the U.S., the Clean Cities program supports public and private partnerships to deploy alternatively fueled vehicles (AFVs) and to build the infrastructure needed to support them. The Clean Cities Program promotes greater use of AFVs, with the ultimate goal of reducing the use of petroleum-based transportation fuels and consequently improving the environment. The U.S. Department of Energy (USDOE) is the primary sponsor of the Clean Cities Program. Beginning with its first national conference in 1994, the Clean Cities annual national conferences have been held recently in the Sunbelt cities of San Diego (2000) and Oklahoma City (2002). An impetus for this program is the growing awareness that national energy security will be improved by reducing dependence on imported oil.

Membership in the Clean Cities Program, while voluntary, is not automatic. Certification indicates that the local government is supporting broader use of AFVs, has developed a coalition with multiple participants, and is instituting programs that promote AFV use and development. Membership demonstrates that a city is concerned about energy security and wants to reduce pollution emissions from vehicles.

Cities such as Fresno, a Clean Cities member located in the San Joaquin Valley of central California, are well aware of their problems with ozone, smog and diesel fumes. The natural topography of the San Joaquin Valley, consists of 64,750 square kilometers (25,000 square miles), and creates the nation's largest urban air basin (Walters 2004:3-4). An estimated 60% of its air pollution is caused by uncontrollable sources. Air pollution has resulted in 10% of the Valley's population being afflicted with chronic breathing disorders, and 16% of Fresno County's children have asthma (Walters 2004:3-4). Given such conditions, AFVs may offer long term solutions to improving air quality by reducing vehicle emissions.

According to their website, "The mission of the Clean Cities Program is to enhance our nation's energy security and air quality by supporting public and private partnerships that deploy clean-burning AFVs and build their associated fueling infrastructure" (U.S. Department of Energy 2004). The goals of this program include: 1) to have one million AFVs operating exclusively on alternative fuels by 2010; and 2) alleviate the use one billion gallons of gasoline equivalents per year by using AFVs by 2010 (U.S. Department of Energy 2004).

Clean Cities uses a voluntary approach to AFV development. It cooperates with coalitions of local stakeholders to help develop the lagging AFV industry. "The program thrives on strong local initiatives and a flexible approach to building alternative fuels markets, providing participants with options to address problems unique to their cities, and fostering partnerships to help overcome them" (U.S. Department of Energy 2004). Certification requires that a locally targeted AFV program be developed from coalitions among local governments and other public and private entities. Membership strongly suggests that the city is exploring and investigating alternative transportation system policies and options in addition to subsidizing the use of vehicles that consume fossil fuels.

As a result of central government support for the Clean Cities Program and modest incentives, the sampled cities are quite active in this program. In fact, 19 of the 25 Sunbelt cities (76%) are members either directly or though membership in a local coalition. However, 6 of the 25 Sunbelt cities (24%) do not participate. Those cities that are not part of the Clean Cities program are: Jacksonville, Memphis, Nashville, Charlotte, New Orleans and Mesa.

Evidences of the Clean Cities initiative include actions by Southern California Edison, which serves Long Beach and operates a fleet of 300 electric vehicles (EVs). Electric vehicles are considered to be 90% cleaner than gasoline powered vehicles. In addition, they generate 80% fewer greenhouse gas emissions while using a quarter of the energy that would be used by a similar gasoline powered vehicle (Southern California Edison 2000:3). Phoenix has 10 stations where AFVs can purchase compressed natural gas or bio-diesel fuels (*Valley of the Sun Clean Cities Coalition 2004*). Tulsa's program, which combines state incentives and low interest loans, has helped place over 2,000 alternatively fueled vehicles on its roads.

Participation in the Clean Cities Program in the Sunbelt is expanding. Prior to Hurricane Katrina, the New Orleans Regional Planning Commission was developing a local clean city program (in 2001) to expand the use of alternatives to gasoline and diesel fuel as part of their greenhouse gas abatement program. Their ultimate goal was to apply for Clean Cities Certification. According the USDOE-sponsored *Clean Cities* website (www.ccities.doe.gov), Nashville and Charlotte also plan to become clean cities; both have local coordinators assigned to that end. Beginning in late 2005, tax credits for the purchase of hybrid gasoline-electric vehicles have been provided in the Energy Policy Act of 2005, offering additional incentives for the purchase of alternatively fueled vehicles.

Policy #3—Local Energy Star™ Partnership Program

A city can be a local governmental Energy Star™ partner—and so can its manufacturers, retailers, utilities, builders, and other governments, among others. While partnership is voluntary, there are commitments to which members must agree. Organizations must: 1) sign a memorandum of partnership committing them to continuous improvement of energy efficiency; 2) measure, track and benchmark energy performance; 3) develop and implement a plan to improve energy performance; and 4) educate staff and the public about the partnership and achievements of the program (USEPA 2004). For urban governments, being an Energy Star™

partner permits them to use the label to support equipment purchasing decisions, to assist in developing energy planning strategies, and to make design decisions concerning facility improvements.

Energy Star™ also has a voluntary labeling program that started in 1992, and is cosponsored jointly by both the U. S. Environmental Protection Agency (USEPA) and the U.S. Department of Energy (USDOE). The focus of the program concerns buildings and the energy-consuming equipment they use. Office machines, mechanical equipment, lighting systems, electronics, appliances and other products can be labeled, indicating that they are "energy efficient." The Energy Star™ label has been extended to include new construction including homes, commercial structures and industrial buildings. According to its website, "Through its partnerships with more than 7,000 private and public sector organizations, Energy Star™ delivers the technical information and tools that organizations and consumers need to choose energy-efficient solutions and best management practices" (USEPA 2004). Energy Star™ also offers a building energy-performance rating system which has been used for over 20,000 buildings throughout the U.S. By leveraging private and governmental partnerships, Energy Star™ has proven to be one of the most cost-effective programs sponsored by the U.S. government.

On the other hand, meeting the partnership requirements suggests a modicum of commitment and can be viewed by some administrations as being costly to support and implement. Specifying energy efficient equipment might be associated with higher initial costs. A full-time energy engineer might be required to baseline energy targets and establish goals. Partnership requirements might also be viewed as potentially intrusive for city administrations that consider it politically undesirable to advertise ever-increasing expenditures for energy. City administrations agreeing to measure and track energy performance, and then implement a plan to improve energy performance, may be subject to public scrutiny should they fail to meet these objectives. The perceived political risk might cause some administrators to avoid adopting the Energy Star™ program. As a result, many cities continue more vulnerable policies including unmanaged exposure to fluctuating energy costs and availability.

Due perhaps to these and other considerations, only 10 of the 25 sampled Sunbelt cities (40%) have become Energy Star™ partners. These cities are Los Angeles, Houston, San Diego, Dallas, Fort Worth, Tucson, Las Vegas, Albuquerque, Atlanta and Miami.

The Clean Cities program is the "most popular" of these membership-based programs with 19 cities participating. ICLEI membership is the least popular with six cities. Five cities (Los Angeles, San Diego, Tucson, Atlanta and Miami) participate in all three programs. Six cities (Jacksonville, Nashville, Memphis, Charlotte, New Orleans and Mesa) participate in none of these programs.

Consider the case of Mesa, which is among those cities that has established sustainability as a primary urban goal, yet has not chosen to participate in any of the three programs—Clean Cities, EnergyStar, and ICLEI. New Orleans, a city that had committed itself to reducing greenhouse gas emissions, also has not participated in these membership-based organizations. On the other hand, Las Vegas, Houston, Fort Worth and Dallas are examples of cities that have not adopted sustainability goals, but happen to be both Clean Cities members and Energy Star™ partners.

Table 7-6 summarizes the membership and participation of the selected Sunbelt cities in the ICLEI, the Clean Cities Program, and the Energy Star™ Program.

Next discussed are two local programs designed to improve the environment: 1) curbside recycling programs; and 2) brownfield redevelopment programs. Curbside recycling programs are locally promoted and instituted, while brownfield redevelopment programs have been federally supported and sponsored.

LOCAL ENVIRONMENTAL PROGRAMS

Program #1—Curbside Recycling Programs

Recycling solid waste significantly reduces waste streams and lowers energy usage. Waste recycling is a verifiable indicator of sustainability in cities (Bell and Morse 1999:99) as is managing municipal waste streams (Newman and Kenworthy 1999:19; OCED 1998:38; United Nations 2001:272-275). Local Environmental Program #1 gauges whether or not the city has a curbside solid waste recycling program. When glass, paper and aluminum are recycled and then remanufactured, energy use is substantially reduced. Many consumer-based recycling programs in the U.S. vanished after World War II. They reappeared in the 1970s and 1980s, sponsored by community action groups, schools, and other civic organizations, with material collected at the door or accepted at "drop-off" collection centers. Boy Scout or Girl Scout organizations were among those sponsoring "paper drives." Proceeds from collecting bottles, newspapers and aluminum cans supported the participating organizations.

Table 7-6.[6] Local Policy #3—Participating in Membership-based Programs

Sunbelt City	ICLEI	Clean Cities	Energy Star
Los Angeles	Yes	Yes	Yes
Houston	No	Yes	Yes
Phoenix	No	Yes	No
San Diego	Yes	Yes	Yes
Dallas	No	Yes	Yes
San Antonio	No	Yes	No
Jacksonville	No	No	No
Austin	Yes	Yes	No
Memphis	No	No	No
Nashville/Davidson	No	No	No
El Paso	No	Yes	No
Charlotte	No	No	No
Fort Worth	No	Yes	Yes
Oklahoma City	No	Yes	No
Tucson	Yes	Yes	Yes
New Orleans	No	No	No
Las Vegas	No	Yes	Yes
Long Beach	No	Yes	No
Albuquerque	No	Yes	Yes
Fresno	No	Yes	No
Virginia Beach	No	Yes	No
Atlanta	Yes	Yes	Yes
Mesa	No	No	No
Tulsa	No	Yes	No
Miami	Yes	Yes	Yes

Reduced incineration of urban wastes coupled with declining land-fill space led to increasing public demand for more recycling, eventually causing cities to establish drop-off collection centers.

Today, curbside recycling programs are typically managed by cities either directly or by their subcontractors. The programs involve both drop-off collection points and curbside pickup of items such as newspapers, magazines, telephone books, aluminum and tin cans, corrugated cardboard, plastic beverage containers, glass and other materials. These programs are successful in reducing the quantity of wastes entering urban landfills, and in providing raw materials for new products. Typically, recycling containers are provided to households and the bins are placed at a nearby curb for collection on a weekly or biweekly basis. Often the recycling containers themselves are manufactured from recycled plastic (roughly 25%). Curbside recycling programs, rare in the 1970s and 1980s, became increasingly popular beginning in the 1990s.

Houston began its curbside recycling program in 1990, with 27,000 homes participating (City of Houston 2004). Today the program has grown to 140,000 homes receiving biweekly service (City of Houston 2004). San Antonio's curbside recycling program was initiated in 1995 and was fully implemented in 1998. Today, the city collects over 2,000 tons of recyclables monthly (San Antonio Community Portal 2004a). One of the shortcomings of these programs is that they are generally limited to single family residences, and multi-unit dwellings with less than two to four apartment units. Multifamily dwellings including condominium developments are often excluded. These policies reduce the effectiveness of curbside recycling programs.

The Miami-Dade County curbside recycling program is among the largest in the U.S., serving nearly 300,000 homes (Miami-Dade County 2004). Collection is provided to single-family homes, duplexes, triplexes and cluster homes. Collection is weekly and includes monthly collection of household batteries.

Recycling programs, once a rarity, have become commonplace and are now accepted as a mainstream solution to reduce the wastes going into city landfills. Curbside collection of recyclables, in particular, has become nearly ubiquitous among major Sunbelt cities. A total of 24 (96%) of the cities in the selected sample now provide some form of household curbside collection, making this the most successful program in our selected Sunbelt cities. El Paso is the only city not presently offering curbside collection services. However, El Paso continues to offer and maintain an extensive drop-off program for recyclable materials.

Program #2—Brownfield Redevelopment Programs

How have cities established programs to redevelop unused or abandoned commercial and industrial sites? Are they actually being redeveloped—or are new industries being directed to previously undeveloped (greenfield) locations, thus giving further impetus for urban de-densification? Local Environmental Program #2 provides insight into which cities are exploring brownfield redevelopment as a partial solution to urban expansion. A city scores in this category if it has initiated and continues to maintain a brownfield redevelopment program. An increase in the level of urban redevelopment compared to new development indicates improved sustainability in cities (Newman and Kenworthy 1999:20).

Brownfields are abandoned or underused urban industrial sites that remain undeveloped due to real or perceived environmental contamination (Andrews 1999:249). Reusing these sites can be confounded by the potential, often unknown, costs for environmental remediation and by possible liabilities due to improper cleanup attempts. As a result, investors see the potential of brownfield redevelopment as having unknown liability exposure and likely unquantifiable financial risks. These risks are not unfounded as it is not uncommon for the remediation costs to exceed the market value of these sites. Brownfield sites may often remain idle for a number of years and become eyesores, reducing the actual value of adjacent and nearby properties. The types of properties that tend to qualify as brownfields might include demilitarized armament manufacturing sites, abandoned automobile parts manufacturing sites, underutilized harbor facilities, gas stations with leaking storage tanks, abandoned oil tank farms, idle chemical plants, steel smelting facilities, unused railroad structures and transportation corridors, and the like. There are an estimated 400,000 to 600,000 brownfield sites in the U.S., yet only a few thousand are classified as Superfund sites. In many cases, these properties are located in relatively accessible locations near the perimeter of the original central cities, and often have existing, sometimes usable, utility infrastructure. Some believe that the amount of brownfield or idle industrial land within an urban area is inversely proportional to urban sustainability (Maclaren 1996a:13).

One means of recycling brownfield properties is to mitigate or eliminate the environmental contamination and make the properties available for redevelopment via new construction. Another is to creatively reuse the existing facilities. The goal of urban brownfield programs has been to turn eyesores into redevelopment opportunities that provide amenities and employment opportunities.

The U.S. Taxpayer Relief Act, signed in August 1997, included a tax incentive to expedite the cleanup and redevelopment of brownfield sites. This allowed full deductibility of business expenses for brownfield redevelopment in the year in which the costs for cleanup were incurred, eliminating the requirement that such expenses be capitalized. The Small Business and Liability Relief and Brownfields Revitalization Act provided liability protection for purchasers and landlords, and authorized increased funding for state and local programs to assess brownfields. An EPA funding program that promotes brownfield redevelopment resulted in demonstration projects in cities such as Las Vegas (1998), Albuquerque (2000), and Long Beach (2000).

To provide additional incentives for redevelopment, states including Tennessee used superfund legislation to limit or exempt cleanup liability for qualified sites. California has used the Expedited Remedial Action Reform Act of 1994 for the cleanup of 30 hazardous substance release sites, including a number in Fresno and Los Angeles.

In addition to state and federal incentives, cities began to support brownfield redevelopment by first offering educational opportunities, developing inventories of sites, absorbing survey costs, providing zoning flexibility, designating public funds for infrastructure development (e.g., roads, sidewalks, utility upgrades), developing grant programs, and establishing revolving funds for remediation of contaminated sites. In some cities, brownfield sites have been designated as "Enterprise Zones" in order to attract investment. Enterprise Zone status provides relief from certain types of taxation such as state and local sales taxes, an incentive for businesses to locate operations in these areas.

The combined impact of brownfield redevelopment has led to the "filling in" of cavities in the urban structure with new development. Benefits of these programs include urban re-densification which results in a more compact urban structure, reduction in the use of "greenfields" for new urban development, creation of a new tax base for the host city, and often a reduction in commuting. The demand for "leap-frog development" can be reduced. Existing structures on brownfield sites can often be improved and reused.

Many of these brownfield redevelopment programs are extensive in scope. Albuquerque, as an example, has identified approximately 1,214 hectares (3,000 acres) as brownfields in designated Metropolitan Redevelopment Areas (USEPA 1999:1).

Despite being a relatively recent opportunity, brownfield redevelopment programs have become the norm in major Sunbelt cities. For the 25 selected Sunbelt cities, a total of 19 (76%) have established brownfield redevelopment programs or demonstration projects. In most cases, brownfield redevelopment programs are being institutionalized, refined and expanded. Tulsa has an assistance program for ten selected brownfield redevelopment sites, has initiated environmental assessments, and is exploring financial incentives for them. Table 7-7 summarizes the findings regarding curbside recycling programs and brownfield redevelopment programs in selected cities.

Table 7-7. Local Environmental Programs #1 and #2

Sunbelt City	#1 - Curbside Solid Waste Recycling	#2 - Brownfield Redevelopment Programs
Los Angeles	Yes	Yes
Houston	Yes	Yes
Phoenix	Yes	Yes
San Diego	Yes	Yes, includes development incentives
Dallas	Yes	Yes
San Antonio	Yes	Yes
Jacksonville	Yes	Yes
Austin	Yes	Yes
Memphis	Yes	Yes
Nashville/Davidson	Yes	Yes, began in 1999
El Paso	No	No, task force only
Charlotte	Yes	No, indicated as strategic objective
Fort Worth	Yes	No, under development since 1999
Oklahoma City	Yes	No, applied for EPA funding
Tucson	Yes	Yes, began in 1997
New Orleans	Yes	Yes, began in 1995
Las Vegas	Yes	Yes, began in 1998
Long Beach	Yes	Yes, began in 2000
Albuquerque	Yes	Yes, began in 2000
Fresno	Yes	No
Virginia Beach	Yes	No
Atlanta	Yes	Yes, began in 1997
Mesa	Yes	Yes
Tulsa	Yes	Yes, began in 2000
Miami	Yes	Yes

Of the six (24%) remaining cities that have not established brownfield redevelopment programs, four of the cities have initiatives underway for future projects. El Paso has an assigned brownfield redevelopment task force. Oklahoma City has applied to the EPA for funding of an initial demonstration project. Fort Worth is developing a program. Charlotte has a stated strategic objective to develop a program in the near future. Only Fresno and Virginia Beach have no programs under development.

SUMMARY

Cities have a wide range of choices that impact sustainability and the ways that they can active it. These choices include the types of policies and programs cities can pursue, how they tax their constituents, how they manage infrastructure, and how they purchase energy and energy-consuming equipment.

A comparison of policies used by Sunbelt cities has been provided. In all, ten specific policies or programs and their attributes were discussed in detail. These policies and programs are summarized in Table 7-8.

Table 7-8. Local Energy Policies

Local Energy Policies
- Local utility offers rebates for lighting, building envelope improvements, new construction, and/or electrical improvements
- City has programs to improve energy efficiency and energy conservation in city-owned buildings
- Programs, both utility and local government supported, are in place to support local energy conservation or alternative energy initiatives

Local Policies
- Sustainable development identified in written policy as local goal
- Car pool lanes available on interstate highways

Local Organizational Membership Programs
- International Council for Local Environmental Initiative (ICLEI)
- Member of the U. S. Department of Energy Clean Cities Program
- Energy Star™ partner

Local Environmental Programs
- Curbside solid waste collection and recycling program
- Policies in place promoting brown-field redevelopment

Local sustainability initiatives are being pursued by 25 Sunbelt cities. In this chapter, specific environmental and energy related policies were discussed, and the various policies that cities have put in effect were identified. The ways that many Sunbelt cities are managing and reducing energy use has been considered. These policies and programs focus on local development, organizational memberships and environmental activities. There are variations in ways these measures are defined and placed into practice.

Sunbelt cities vary broadly in their selection and application of policies. Cities such as Los Angeles, San Diego and Miami aggressively pursue multifaceted policies and focus their resources and agendas accordingly. On the other hand, most Sunbelt cities are more selective and limited in their policy choices. While membership in the ICLEI is rare, curbside recycling programs are nearly ubiquitous. Cities such as Oklahoma and Charlotte are among those that have chosen to adopt few of the policies considered. Policies such as utility rebate programs, and those that support local energy conservation efforts and Energy Star™ memberships, are more often adopted by cities with larger populations.

Also discussed in this chapter are examples of cities that use energy policy and energy conservation goals in their agendas as a means of achieving sustainability. Long Beach has local government programs and utility-sponsored self-generation incentives. Atlanta's program includes an internal energy conservation initiative. New Orleans has a policy to combat global warming and was actively pursuing a facility improvement program prior to its near destruction by hurricane Katrina. Tucson is developing a sustainable community based on use of solar energy. Mesa has created a planning agenda for sustainability and has established an energy conservation program for city owned buildings.

The policies and programs considered in this chapter suggest that city agendas are in flux and that some of them relate to sustainability goals. In light of the recent advances that cities in the U.S. (and also overseas) have been making in their agendas, "sustainable city development today has become common-place in the set of policy goals given high priority by the political elites in various countries" (Low *et al.* 2000:45).

Implementing sustainability measures is a challenging—and critical—task. Local policies and programs must create opportunities to meet long term sustainability goals, if they are to succeed.

Endnotes

1. Phoenix climbs to No. 5 in census ranking of cities. *The Courier-Journal.* 28 June 2007. p. 7A.
2. Ibid.
3. According to McGowan (2004:10), "LEED, a design criteria and rating system from

the United States Green Building Council, is becoming a driving force in new construction. The design criteria are comprehensive and encompass every aspect of new construction from sitting to energy efficiency and indoor air quality... LEED includes requirements for optimizing energy performance, commissioning and measurement and verification."

4. Data concerning HOV facilities in Los Angeles, Houston, Atlanta, Miami and Memphis was obtained from the U.S. Department of Transportation's website: hovpfs. ops.fhwa.dot.gov/inventory/inventory.htm

5. Data was obtained from the ICLEI membership website, http://www3.iclei.org/member.htm, accessed 13 March 204

6. Membership information in Table 7-6 was obtained from websites maintained by the ICLEI, the Clean Cities and the Energy Star™.

Notes

Portions of Chapter 7 (Local Policies for Sustainable Development) were previously published in the *Encyclopedia of Energy Engineering* in the chapter entitled "Sustainability Policies: Sunbelt Cities" by Stephen A. Roosa, edited by Barney Capehart. Permission granted by Taylor & Francis Group LLC, a Division of Informa, PLC, Copyright 2007.

Other portions of Chapter 7 appeared in the *Encyclopedia of Energy Engineering* in the chapter entitled "Sustainability Policies: Sunbelt Cities" by Stephen A. Roosa, edited by Barney Capehart. Permission granted by Taylor & Francis Group LLC a Division of Informa, PLC, Copyright 2007.

Portions of Chapter 7 were previously published in the *International Journal of Green Energy*, 2007, 4(2), in the article entitled "Energy Policy and Sustainability in Sunbelt Cities in the United States," pages 173-196 by Stephen A. Roosa. Reproduced by permission granted by Taylor & Francis Group. LLC, a Division of Informa, PLC, Copyright 2007.

Chapter 8

Impact of Sustainable Development Policies on Planning

"The Conde Nast Building (in New York City) contains 1.6 million square feet of floor space and sits on one acre of land. If you divided it into 48 one story suburban office buildings, each averaging thirty-three thousand square feet, and spread those one-story buildings around the countryside, and then added parkland and some green space around each one, you'd end up consuming at least 150 acres of land. And then you'd have to provide infrastructure, the highways and everything else." Like many other buildings in Manhattan, 4 Times Square doesn't even have a parking lot, because the vast majority of the 6,000 people who work inside don't need one. In most parts of the country, big parking lots are not only necessary but are required by law. If my town's zoning regulations applied to Manhattan, 4 Times Square would have 16,000 parking spaces."
OWEN (2005:13), QUOTING BRUCE FOWLE, OF FOX AND FOWLE.

Exciting initiatives are now underway to implement sustainable development. Innovative planning strategies are being used to change policies. In Europe several central governments have establish standards that filter through to metropolitan areas and cities to guide local sustainability efforts. But in countries such as the U.S. that do not have a central governmental sustainability plan, *a lack of policy invariably creates more policies.*

In these countries, policy-making defaults to local, state, and regional initiatives. Without established federal parameters, a variety of "grass roots" laws and regulations emerge. Planners who are experiencing pressures for new development must carefully consider their options, often modifying and improving policies that affect regional infrastructures.

How did this come about? Within this array of policies, there are trends and counter-trends. This chapter explores these trends and delves into planning policies, methodologies, their application to localities and cities, and their relationship to energy and sustainability.

8-1. New Residential Development Near Portsmouth, NH

Architects and planners have a long and commendable history of providing visionary alternatives for the future of cities. These include notable examples such as Ebenezer Howard's "Garden Cities" and Frank Lloyd Wright's "Broad-Acre City." Brasilia, the planned capital city for Brazil, was literally carved out of jungle—it too offers a striking vision of the future of cities. In the new age of sustainability, there are now plans for "Green Cities" and "Eco-Cities." The most important innovations are those that apply the basic concepts of sustainability within existing urban structures, offering hope for sustainable cities.

Local planning decisions invariably impact the environment. Some planning statutes in the U.S. (for example, those that segregate uses) can have the unintended consequences of reduced environmental quality. However, environmental regulations such as those regarding air and water are notable examples of rules enforced by the U.S. Environmental Protection Agency (USEPA) *that yield the opposite effect* and have successfully improved environmental conditions. As part of the planning process, Environmental Impact Statements are required for significant land use changes and major construction projects.

Planning decisions regarding development are often made by local governments. Planning paradoxes persist. One is that despite concerns regarding rising sea levels, there is a propensity to construct buildings on the seashores, barely above sea level. And another... with the high costs of land and increased commuting times, we continue to devote more land and resources to highway construction, despite the fact that more creative transportation options are not only available but can be less costly. And yet another... planning standards and zoning regulations in cities often cause more available land to be used for development projects than is necessary, while undeveloped land is at a premium.

There are other examples of local governments that have established their own sustainability policies and are actively implementing them—with negligible or negative results. Given the interest and efforts of local governments, how does sustainable development provide a new guiding vision for urban development and planning? Improving energy use is an important part of long-range urban development efforts. Do local planers consider it in their efforts to achieve sustainability?

The nature and goals of sustainable planning must allow opportunity for a wide range of solutions. According to Davey (2005:27), sustainable planning "is ultimately ensuring that development continues to occur in such a way that it enables both society and nature to gain in the process... this may involve the exclusion of areas from development altogether to ensure conservation of biodiversity which simply can't be replaced through re-vegetation programs. In other cases, it allows development that observes strict sustainability codes of conduct."

Land is a resource. It is costly to adapt for use and even costlier to reclaim. Land reclamation projects that "create" land by filling-in sea areas have been undertaken in Venice, Holland and more recently for the construction of Hong Kong's new international airport. And once virgin land is adapted for a new use, it is often prohibitively expensive, and sometimes impossible, to reclaim it. Strip mining practices are a notorious example.

An important contribution of planning involves the ways land is used. Available land areas that are suitable for development are at a premium. Governmental, industrial, commercial, residential, transportation and utility interests, all compete for rights to develop available land areas. Planning provides tools to allocate the use of land and resources. Planning suggests that there are benefits in establishing standards and in controlling the decision-making process.

8-2. Interstate Highway Intersection East of San Diego, CA

LINKING COMPREHENSIVE PLANNING AND SUSTAINABILITY

Planning also deals with problem solving and policy-making. Decision-making can be based on a rational model, which involves assessing the situation, considering the information available, identifying alternatives, and selecting the best approach *after all matters are considered*. Meyerson's rational-comprehensive model for planning stated that if "planning is designing a course of action to achieve ends, 'efficient' planning is that which under given conditions leads to the maximization of the attainment of relevant ends" (Meyerson 1955:314). Efficiency is a widely accepted goal of institutions and their policy frameworks, and can be defined simply as system output divided by system input. The test of a planning solution using Meyerson's rational-comprehensive planning model is that it can be proven to "be the most appropriate means to desired ends" (Stein 1995:37).

Charles E. Lindblom (now a professor at Yale University) in the 1950s and 1960s was a critic of the rational planning model. He believed that most problems are too complex to be rationally processed. In his view, too much information is required to make a truly rational and comprehensive decision. Information can be time-consuming to acquire and costly to process. Lindblom considered incremental changes to existing policies to be the truly rational method of achieving systemic changes. Lindblom (1965:178) believed that a "...central decision maker would not try for comprehensiveness in his view of the relations among policies but would instead take up in series each of an unending stream of particular problems, dealing with each with a narrow view of the implications of any policy solution, dealing with neglected implications as quite separate problems." His assertion seems even more valid today—as the internet provides us with endless amounts of information to be considered, so much that "analysis paralysis" can ultimately result. With Lindblom's incremental approach "policy analysts... largely limit their analysis to incremental or marginal differences in policies that are chosen to differ only incrementally" (Stein 1995:43). *Incremental means implementing a series of adjustments.* While it is true that individual perceptions of what constitutes an incremental change can vary, the concept is clear enough (Lindblom 1958:298). Incremental changes are easier to make and are often more palatable.

Lindblom's idea supports the belief that comprehensive solutions to open system problems may be difficult to resolve and perhaps be unachievable. Open systems are subject to constant changes. Incremental adjustments are perceived by policy-makers to be less risky. Lindblom's approach seems pragmatic and in accord with the idea that to be implemented, policies such as sustainable development will require cooperation, management, maintenance and perhaps patience. It offers hope that long-term goals can be achieved by a series of incremental advances.

Planning establishes a useful framework which can be used to redesign neighborhoods, cities and regions. As Berke (2002:21-36) observed, "Up to the 1960s, planning had a long and commendable history of visionary ideas for guiding the development of towns, cities, and regions." U.S. examples include towns such as New Harmony, IN and cities such as Charlestown, SC, Washington D.C., and Savannah, GA.

In addition to the design and layout of cities, planning also deals with achieving social goals and allocating resources. In 1963, Lynton Caldwell of Indiana University studied governance issues including the segmentation of U.S. policies dealing with natural resources. He was among the first to propose the environment as the central focus for the integration of policies that use a comprehensive ecological approach to resolve specific problems (Caldwell 1963:132-139). Resource conservation became yet another planning goal. This vision was noted by Perloff in the 1980s... "We must not only conserve our material resources but must also develop them wisely—and create entirely new ones. Planning can contribute usefully to these difficult tasks and, to some extent, is already doing so" (Perloff 1980:182).

Thanks to the work of Meyerson, Caldwell, Perloff and many others, a theoretical basis evolved which links planning policies with the environment as a central focus and further—to the efficiency of planning solutions, and to the conservation of natural resources. Planning can achieve social goals—and planning has become way to achieve sustainability. Sustainable development and social planning are linked. Together they represent a model that establishes new parameters and redefines the future vision desired for cities. Today's local planning decisions have larger implications and broader applicability than in the past.

Sustainable development has evolved to become the ideal way to define goals. Berke (2002:22) noted a growing debate concerning the "role of sustainable development as an overarching guide to planning that is taking place at the international, national and local government levels around the world." Berke validated how the dimensions of this development, including its links to global issues and balanced ecosystems, were being used to guide local comprehensive plans (Berke and Manta-Conroy 2000:21-33).

From a planner's perspective, sustainable development is a means of achieving certain desired conditions: environmentally appropriate buildings, more functional sites and cities, and improved resource allocation. Berke and Manta-Conroy (2000:23) redefined sustainable development as a "planning process in which communities anticipate and accommodate the needs of current and future generations in ways that reproduce and balance local social, economic, and ecological systems, and link local actions to global concerns." Berke further identified six principles of sustainable development that could be applied to planning: 1) harmony with nature; 2) livable built environments; 3) place-based economy; 4) equity; 5) the concept that those who generate pollution must pay; and 6) responsible regionalism (Berke 2000:33; Berke and Manta-Conroy 2000:23).

The importance of this link between planning and sustainable development cannot be understated. Innovative planning strategies are needed to implement social goals—particularly the goals of sustainable development.

PLANNING FOR SUSTAINABILITY

To be successful, planning decisions must address framework issues, policies, and urban design issues in ways that permit a range of possibilities for development. To this end, planning procedures must offer structures and incentives that are environmentally friendly, use resources effectively, and allow sustainability to flourish. But in the past, planning has been used to do quite the opposite—to encourage unsustainable development, environmental disruption, resource waste, and excessive energy consumption.

There are obstacles that inhibit the adoption of long-term strategies for sustainability. Ironically, one is the inertia inherited from past planning decisions. Widely adopted planning schemes for separated uses are often antithetical to sustainable development practices. Kunstler (1994:118) says that "today we have achieved the goal of total separation of uses in the man-made landscape. The houses are all in their respective income pods, shopping is miles away from houses, and the schools are separate from both shopping areas and dwellings." Urban planning of this type fosters a greater need for automobiles by failing to offer enhanced pedestrian environments. According to Duany, "as long as the conventions of real estate development effectively outlaw the construction of mixed-use neighborhoods, developers will find it very difficult to build anything with a sense of community" (Duany,

Plater-Zyberk and Speck 2000:100). Basically, physical separation of uses not only divides communities but also de-links individuals from their very personal sense of community. It can also be anti-pedestrian.

This bias manifests itself in restrictive planning laws, inappropriate zoning ordinances, selective property inspections and enforcement, and inequitable tax assessment procedures. These policies become replicated across metropolitan areas and are often broadly misapplied, ultimately leading to efforts to "suburbanize" inner city neighborhoods rather than restore the urban fabric. A policy mismatch soon evolves and establishes itself as "the" standard. To compound the problem, governments in other localities clone the standard and adopt it—along with all of its embedded faults. The standard creates a regulatory quagmire. These policies manifest themselves in the form of single-use zoning, suburban density and landscaping standards, lot and deed restrictions, parking requirements, and building code requirements—many of which are inappropriate for their locations and troublesome to manage. They fail to consider urban sustainability, inappropriately modify urban densities, and ultimately result in infrastructure that increases urban energy usage.

There is also the propensity to implement similar planning solutions across entire countries. Development that occurs at interstate highway intersections across the U.S. provides an example that has yielded homogenous solutions, resulting in replicated planning problems. There one finds similar fast-food stores, similar hotels and similar gas stations, all designed for retail obsolescence. Each sits on its own plot of land. Each contributes to traffic congestion, a proliferation of signage, and recurring themes of identical site and construction solutions. This lack of differentiation results from a seemingly endless cloning of planning mistakes. Land use is programmed around the idea of juxtaposed sites rather than being focused on the idea of cohesively planned development.

The fact of the matter is that most local planning policies and regulations are causing more problems than they solve. They need to re-considered, revamped, and redefined. Regardless, there are interesting local and regional examples of U.S. progress toward sustainable development. Local programs are often creative and designed to respond to local issues such as improving energy-consumption, permitting brownfield redevelopment, allowing urban homesteading, redeveloping parklands, addressing vacant structures, and expanding historic districts. In order for these approaches to be implemented, it is often necessary to discard regulations, and revise ordinances, or else endure multiple variances each time a creative planning solution appears on the planning commission's docket. To encourage sustainable development, rewriting comprehensive plans is often a necessity.

There are successful examples of planning initiatives which have been customized for local conditions. Comprehensive plans and local ordinances can be redesigned to promote urban homesteading as a means of bringing people back into the cities. Successful homesteading efforts were often privately funded by groups such as the Urban Homesteading Assistance Board, founded in 1973 in New York City. The goal of local governments in Baltimore, Chicago, Wilmington, and Philadelphia was to recover areas that had lost population (Schamess 2006:40). Urban homesteading proved to be a successful solution.

Youngstown developed a new comprehensive plan to address abandoned properties by demolishing vacant buildings and creating green landscapes that reinforce existing transportation networks and waterways. Louisville and Cleveland promoted redevelopment of their warehouse districts, turning eyesores into assets and centers of commercial development. Baltimore, New York City and Long Beach, CA have redeveloped their harbor-fronts, creating tourist attractions.

Today, urban and regional planners are incorporating the language of sustainable development into their comprehensive plans. This is a new phenomenon in the U.S. These successes tend to highlight localized water and sewer improvements, non-urban land conservation, sprawl mitigation efforts, farmland preservation, preservation of open space, and public transportation initiatives. One regional example is the establishment of the Florida Sustainable Communities Network. Other planning initiatives include the Civano Sustainable Community Project, the Lake Tahoe Regional Plan, the Grand Rapids, Michigan Plan, the Manchester, Vermont Planning and Zoning Program, along with others in Ashfield, Massachusetts and Chattanooga, Tennessee.

Since sustainable development initiatives often have regional characteristics, and local governments often manage the planning process, initiatives such as these provide a structure that was unavailable in the past, setting the stage for a new regionalism—one that allows local governments to work together to solve common planning problems.

PLANNING FOR SUSTAINABILITY
ON THE OTHER SIDE OF THE ATLANTIC

Sustainable development is not something new in Europe. More than 120 European cities signed the Aalborg Charter in 1994,[1] thus joining the European Sustainable Cities and Towns Campaign and committing them to develop and implement local strategies to pursue sustainability. The June 1997 Treaty of Maastricht incorporated sustainable development as a specific policy objective for European Nations. In contrast to countries such as the U.S. and China, national sustainability plans have been prepared by most European nations (Beatley 2000:15). Even Slovenia, one of the newer EU countries, has adopted a national sustainability plan. Regional and local governmental plans followed suit. By the latter half of the 1990s, there were 1,119 European localities with local Agenda 21 (LA21) initiatives (EC 1997). LA21 plans are local manifestos for sustainability. The Building Act in Finland established "sustainable development as the foundation for land use planning" (Beatley 2000:20). By 2000, almost a third of the localities in Finland were developing Local Agenda 21 plans (Beatley 2000:348). Despite common themes, each sustainability plan was customized and the approaches to instituting sustainability are varied. A systematic process that establishes goals and procedures, called "direction analysis," is being used to implement sustainable development agendas in cities such as Stavanger (Devuyst 2001:230-240). The results of local initiatives are impressive.

Cities in Sweden have almost 100% participation in their LA21 program. Zurich, the world's wealthiest city, fought "a guerilla war against the car and won" causing a decline in car use in sections of the city, a reduction in commuting trips, and an increase in mass transit use (Newman and Kenworthy 1997). The city of Stockholm, constructed urban villages centered on a rail system and reduced annual per capita automobile use by 229 kilometers (158 miles) during the 1980s while simultaneously increasing the use of mass transit (Newman and Kenworthy 1997).

In the UK, former Prime Minister Tony Blair required local authorities to develop their own sustainable development plans by the year 2000. As a result, 73% of local communities in the U.K. embarked on the process of developing LA21 plans (Beatley 2000:347). The Sustainable Communities Plan for the UK earmarked £38 billion with the vision of creating sustainable communities in all regions of the country.[2]

Measures included promoting development on previously developed land, increasing energy efficiency, and regulating water usage. Leicester, Leeds, Middlesbrough and Peterborough responded and have since achieved the UK's coveted "Environmental City" status. The Planning and Compulsory Purchase Act was enacted in 2004, requiring local and regional governments in the U.K. to "contribute to the achievement of sustainable development."[3] The Act required efforts to secure:

- High stable levels of economic growth.
- Social progress which recognizes the needs of everyone.
- Effective protection of the environment.
- Prudent use of natural resources.

Each region of the U.K. is involved in this process. In the Cambridgeshire and Peterborough Structure Plan, the city council and their partners developed a plan with goals that require all new developments to:[4]

- Concentrate development to minimize the need to travel and allow service by public transportation.
- Minimize or reduce automobile dependency.
- Provide high-density, mixed-use developments.
- Use pedestrian and cycle-friendly planning.
- Provide well-designed buildings and attractive green space.
- Make efficient use of energy and resources.
- Site buildings for energy efficiency.
- Use sustainable construction practices and materials.
- Use renewable energy where possible.
- Include a mix of housing.
- Provide water conservation measures.
- Provide recycling facilities.

The goals for Cambridgeshire and Peterborough combine solutions for both building construction and sustainable development. Plans such as these are a significant first step as they recognize how sustainable solutions must be integrated in order to successfully meet developmental needs.

In London, residences generate 44% of total carbon emissions.[5] To demonstrate an alternative, London is building a "zero carbon" housing development at a former Dockland's brownfield. Planned for occupancy in 2010, the 233 homes will use a central power plant that burns wood chips to generate heat, electricity and hot water.[6] Supplemental electrical energy will be generated by solar and wind power. Using a form of net metering, excess power will be provided to the electrical grid when it is available and in times when power is inadequate, it will be supplied by the grid.

TRANSPORTATION AND HIGHWAYS

How do people move about in the U.S.? With the exceptions of buses, airplanes, and limited rail transportation, most travel is by automobile. The country's 226 million motor vehicles transverse over 6.4 million kilometers (4 million miles) of paved roadways, enough to circle the earth 157 times (Brown 2006:3). The U.S. has more automobiles than licensed drivers and 90% of its citizens drive to work in cars.[7] More vehicles cause more congestion, resulting in 3.7 billion hours of travel delay for motorists in 2003 (Brown 2006:3), and creating nightmares for transportation system planners. Suburban residents drive more than city dwellers. Based on a study using 1995 data, suburban households drive their vehicles 31 percent more than their central city counterparts (Kahn 2000:16).

Personal transportation is expensive. A recent study of 28 U.S. metropolitan areas concluded that 20.2% of total income for all households was expended on transportation and that 29.6% of the total income of working families (household incomes from $20,000 to $50,000) was devoted to transportation (Lipman 2006:9).[8] For working families, the expenditures on transportation were more than the cost of housing and equal to roughly half of the total household budget for food. In 2000, actual annual dollar expenditures for transportation were a low of $7,880 in New York City to a high of $10,890 in Atlanta (Lipman 2006:12). In that year, the average price of gasoline was $1.46 per gallon.[9] Today, the total costs of transportation are much greater.

The U.S. has become a "suburban nation" (Duany *et al.* 2000). Of the 20 fastest growing counties in the U.S., 15 are located 30 miles or more from the nearest central business district (Lipman 2006:5). People move to the suburbs for various reasons, including a desire to find more affordable housing near the perimeter of metropolitan areas. The perception of lower costs of suburban life is illusionary. The paradox is that while housing costs tend to decline as the distance to the central business district increases, transportation costs tend to increase, offsetting any savings from lower housing costs in the suburbs.[10] Recently, a counter-trend has emerged, spurring more residential development in inner-city areas.

Suburban development is increasingly associated with environmental decay and global warming. Citing the 1970 California Environmental Quality Act, a lawsuit filed by the California Attorney General against San Bernardino County claims that the county failed to account for greenhouse gases when it updated its 25 year growth plan.[11] Such actions are groundbreaking. It means that cities and counties may be held accountable for "greenhouse gas omissions caused by poorly planned suburban development."[12] There is no need to eliminate existing suburbs—we just need to find ways to improve them and exercise greater care in how new ones are designed.

Relying on less-dense suburban development causes the use of transportation fuels to increase. Substantial reductions in transportation fuels can result by adjusting population and development densities and providing alternative infrastructure for transportation. There are differences in how cities use energy for transportation. Studies indicate that gasoline consumption could be 20% to 30% lower in cities like Houston, TX and Phoenix, AZ, if their urban structures were configured in a manner that resembled that of Boston, MA or Washington, D.C. (Newman and Kenworthy 1989:24-37, 1999).

Population density plays an important role. Urban compaction has been cited as a means of reducing both energy usage and utility infrastructure costs (Burgess and Jenks 2000:212). Increasing urban densities and designing more compact cities can reduce energy usage. It can be challenging for cities with low population densities to offer incentives for residents to use publicly accessible transportation systems. People in some cities simply fail to use public transportation—the options available are simply too inconvenient. Phoenix, which sprawls across the desert, has a public transportation system—yet it accounts for only 1% of the passenger miles of New York City's (Owen 2005:11). If Phoenix had a population equal to New York City's and transportation systems were not reconfigured, their use would equal only 6% of New York City's total transportation miles.

Suburbanization requires improvements in highway systems serving metropolitan areas. These projects require land and large capital investments. Like highway systems throughout the world, the management of U.S. highways requires constant planning. Highway systems become temporarily dysfunctional due to crashes, maintenance, rush hour commuting, and natural events. In April of 2007, a section of the San-Francisco-Oakland Bay Bridge was destroyed by an overturned gasoline tanker-truck causing one overpass to "crumple onto another" and creating "nightmarish commutes" for area residents.[13]

Highways can be dangerous. Worldwide, highway fatalities are the number one cause of death for drivers from the ages of 18 to 25. Highway safety is a major sustainability concern. Comparing the U.S. highway system to that of Germany, the oldest national system of its kind, the German system remains one of the world's safest. In the U.S. and many other countries, the highway systems encourage de-densification

8-3. Los Angeles Light Rail System—Station in Long Beach, CA

rather than urban compaction.

Alternatives to building more highways include tramways, using more buses, light rail systems and commuter rail. Like new highways, these options can be expensive. Bus Rapid Transit (BRT) is being used in Los Angeles, Adelaide, Australia and Bogota, Columbia. These systems use dedicated bus lanes to shuttle commuters. BRT systems can be a less expensive to construct than light rail systems. Light rail systems typically cost $150 million per kilometer ($240 million per mile) compared to $41 million per kilometer ($66 million per mile) for BRT.[14] William Vincent, general counsel for the Breakthrough Technologies Institute, is an advocate of BRT. He states "You can get the same number of people out of their cars for about a quarter of the cost with BRT."[15] There are exceptions. Boston's new BRT that runs from Logan airport to South Station will cost $497 million per kilometer ($800 million per mile) compared to New York City's Second Avenue subway line that is estimated to cost $1.24 billion per kilometer ($2 billion per mile).[16]

Germany has invested in a high speed inter-city rail system that connects its primary population centers. The French recently opened their

8-4. Railway Station in Frankfurt, Germany

Train à Grande Vitesse (TGV) ("high-speed train") from Paris to Straus-bourg. Trains reach speeds approaching 320 kph (200 mph) on this 405 ki-lometer (252 mile) link. By providing a high-speed alternative to highways, dependence on automobiles is reduced, and less energy consumed.

What about bicycles as an intra-city alternative? Increasing the use of bicycles can decrease traffic congestion and reduce energy consump-tion. Data from the 2000 U.S. Census reveal that only 600,000 people ride bicycles to work and that only 6 million or so ride a bike at least twice a week.[17] The majority of Americans neither own nor ride bicycles. A city or town that is bike-friendly uses less transportation fuels—and its peo-ple are healthier. However, in order to be a feasible option, cities need to make bicycle riding a safe experience and planners must make bikeways available. To this end, a U.S. law passed in 2007, earmarked $1 billion for the development of new bikeways.

REDESIGNING COMMUNITIES

While migration to the suburbs has been a trend that has drained population from cities and caused commercial centers to fall into de-cline, the counter-trend is to revitalize these same urban centers and business districts. Successful examples include the Belmont area of Port-land, OR; the downtowns of Portland, ME, and Portsmouth, NH; Main-Strasse in Covington, KY; and many others. What do they have in com-mon? In these cities residential developments are within walking dis-tance of commercial centers and a mix of housing is available.

At MainStrasse, the city of Covington improved the boulevard's sidewalks, added period street lighting, curb cuts, restored a public fountain, and improved landscaping. An urban park area was upgraded with amenities and a signature bell tower constructed. These improve-ments started the redevelopment of the historic 1870s storefront build-ings and attracted new restaurants, bars and cottage businesses. Busi-nesses are thriving and MainStrasse now hosts annual events including community festivals and local celebrations.

Portsmouth, settled in 1632, revitalized its former historic center cre-ating an entertainment district complete with craft shops, restaurants, a brewery, art dealers and clothing stores. The historic church in the town center was restored, sidewalks were widened and curb-cuts were in-stalled. Residential areas and a new hotel were developed within walking

8-5. MainStrasse in Covington, KY

8-6. Downtown Portsmouth, NH

distance. Result: an enhanced "pedestrian experience." Parking was made available in the city with the construction of a new public parking garage. With sandy beaches nearby, it is a destination that attracts a lively nightlife. Now, Portsmouth's nostalgic city-center has recaptured the essence of community. The downtown area has become the city's vibrant social center—the preferred place to be on a weekend evening.

Cities can also restore entire historic neighborhoods. The U.S. National Preservation Act of 1966 provides for the preservation of buildings and structures that are significant to American history, architecture and culture. Many buildings in historic neighborhoods are irreplaceable. People are looking for amenities in cities that offer "a healthy mixture of residences and businesses that make walking more attractive than driving a car."[18] Restoration of older neighborhoods represents a form of "new urbanism that is in high demand."[19] Cities that have restored their historic neighborhoods include Charleston, SC, Louisville, KY, Savannah, GA, and St. Augustine, FL, among others. In each case, there have been economic benefits that support tourism. Restoration of older neighborhoods preserves the resources contained in their structures. Cities can also reclaim waterfronts, provide transportation alternatives and add green spaces or parklands. In-fill construction, whereby new development is directed to unused urban areas and vacant lots, is an alternative to leap-frog development, which leaves gaps in development. Redeveloping previously developed land adds amenities, increases density, and provides incentives for people to relocate to central areas.

One of Metro Louisville's efforts to add urban amenities was to develop city parks along the Ohio River, near the central business district. This brownfield site originally was the location of an oil storage facility, a metal recycling facility and a junk automobile lot. Today, Louisville's linear parkland includes walkways, public lawns, green spaces, a marina, playgrounds, and fountains. Amenities such as strategically located parkland can support new economic activity. By 2005, Louisville's Waterfront Park had resulted in an additional 5,300 employees in the immediate area, and more than $364 million in investments in new construction and renovation of existing structures (Karem 2005:37).

Finding places to add green-space and parkland can be challenging for growing cities. In Holland, Amsterdam has developed its *Open City* (1996) plan around the concept of creating "green fingers" or "green wedges," areas of green-space that radiate out from the city center. These areas are important for urban ecology and for recreation. The wedges

8-7. Amsterdam's Planning Center

"form part of the various city districts, peripheral municipalities, the province and protected national landscapes."[20] There are several areas in and near Amsterdam that are being redesigned. A nature boulevard is being developed called the "The Green Uitweg" which will create bicycle links to other areas of the community. Diemerscheg is a green wedge that consists of a series of outdoor "rooms" that are networked to easily accessible nature areas and woodlands; they provide recreational facilities for new residential districts such as IJurg.[21] In order to give residents of Diemerscheg a "Green Haven," a nearby island is being transformed into a recreational and nature area.[22] Westrand was formed by two interconnected green wedges, the parkland of Lake Sloterplas and the Brettenzone, considered to be one of Europe's last urban wildernesses. It consists of a varied landscape of meadows, woodlands and marshy strips.

While there are commonalities, "smart growth" is not simply another term for sustainable development (Devuyst 2001:29-30). Smart growth policies are themselves subject to various interpretations. Like supporters of sustainable development, smart growth advocates propose environmental preservation, collaboration, and a variety of transportation options (ICMA 2002). The 10 basic principles of "smart growth" have been identified by the Smart Growth Network as:[23]

1. Providing a mix of land uses.

2. Taking advantage of compact building design.

3. Creating a range of housing opportunities and choices.

4. Creating walkable neighborhoods.

5. Fostering distinctive, attractive communities with a strong sense of place.

6. Preserving open space, farmland, natural beauty, and critical environmental areas.

7. Strengthening and directing development of existing communities.

8. Providing a variety of transportation choices.

9. Making development decisions predictable, fair, and cost effective.

10. Encouraging community and stakeholder collaboration in development decisions.

With the exception of creating walkable communities and a mix of transportation systems, smart growth often misses the larger connection of how to design developments to make them more energy efficient. Smart growth policies are pursued by cites such as Charlotte, NC—a city that lacks a defined urban sustainability agenda. Alternatively, smart growth policies are used in tandem with a pro-sustainability agenda in Austin, TX.

THE NEW URBANISM

There are movements that propose redesigning communities to correct developmental ills. A fundamental problem with planning practices today is the propensity to construct mono-functional developments with an office building in one location, a shopping mall in another and housing somewhere else, all inaccessible to one another except by automobile. A counter-trend is emerging. Planners have begun to develop alternatives to sprawl and single-use developments along with methodologies that mitigate their effects.[24]

Bressi (Katz:1994:xxv) suggests that professional city planners have been on a perilous path, attempting to solve one set of problems, yet inadvertently creating new ones:

> *"For a century, this reformist profession has been guiding urban redevelopment and suburban expansion with the goals of eradicating the crowding, poverty, disease and congestion that threatened to overwhelm industrial cities, and creating a rational, efficient framework for growth that all but rejected traditional patterns of city and town development. The result of these efforts is a metropolitan landscape that is beset by an altogether different set of problems — traffic congestion, poor air quality, expensive housing, social segregation and neighborhoods whose physical character amounts to little more than the confluence of standard development practices and real estate marketing strategies. The New Urbanists are confronting these problems with an energy and creativity that has eluded planners until now."*

In the 1980s and 1990s, the search for a new vision led to The New Urbanism. The New Urbanism indirectly addresses the issues of environmental pollution and urban energy consumption by focusing on the ecological aspects of neighborhoods, minimizing the use of vehicles, and reemphasizing pedestrian accessibility. Peter Calthope favors the New Urbanist belief that entire regions need to "be designed according to similar urban principles" acknowledging that "the city, its suburbs, and their natural environment should be considered as a whole—socially, economically, and ecologically" (Katz 1994:xi). He believes that regional systems "of open space and transit, complemented with pedestrian-friendly development patterns, can help revitalize an urban center" (Katz 1994:xiv). In the design of cities, streets need to be made less wide, housing more compact and variable, mixed-use development encouraged, and more transportation alternatives made available.

The New Urbanism deals with conservation of environmental resources and equity issues by encouraging pedestrian accessibility and limiting the dominance of automobiles in the design of towns and cities. Transit-oriented developments (TODs) or developments that channel growth toward available bus-lines or light rail nodes, are a favored alternative (Katz 1994:xxx-xxxi). The New Urbanism attempts to recapture the urban sense for locale and community by encouraging integrated design principles for site selection, buildings, streets, neighborhoods and districts. According to Leon Krier, an architect, urban designer and key figure in European New Urbanism, "new urbanists consider it self-evident that human settlements should be ecological... settling the deserts, or swelling existing metropolitan areas looks to me in the long term to be ecologically untenable and probably futile" (EuroCouncil: 2003:5). New Urbanists literally discard the planning guidelines upon which past development has been based, and instead offer a few easy-to-comprehend principles that form the planning basis of their new and redeveloped communities.

Notable examples of communities designed around New Urbanist philosophies are available. Kentlands, in Gaithersburg, MD was developed in 1988. The plan of the city offers a mix of residential, office, retail and cultural developments clustered in six distinct neighborhoods. The town attempts to recapture the essence of the American small town. Many of the homes are designed in an American colonial style. Seaside, in Walton County, FL, is a New Urbanist community developed in 1989. Public spaces such as parks, squares, and amenities were first defined— prior to integrating commercial and residential areas. Construction standards use prescriptive parameters rather than proscriptive mandates, allowing individual creativity and variety in the homes that are developed. There evolved a mix of housing types. Pedestrian alleyways and access points create a very walkable community. Automobile access and parking are controlled and residential garages are rare.

Laguna West in Sacramento County, CA, was developed in 1990. It was the first New Urbanist community designed as a Transit-Oriented Development. These communities are designed to create transportation nodes that facilitate accessibility to public transportation. Amenities include a town hall, town center, village green, and a system of neighborhood parks. Architectural styles and home sizes are varied; homes are near streets and sidewalks are provided. Other examples of New Urbanist communities are South Brentwood Village, CA, Mill Bay, BC, and Windsor, FL.

The New Urbanism in Europe is reclaiming districts from the "desertification of urban centers" (EuroCouncil: 2003:8). Consider the reconstruction of the Rue de Lacken in Brussels. In this project, a 1960s metal and glass skyscraper (nicknamed "the Blue Tower") in the core of the city was demolished and replaced with a new mixed-use neighborhood. In order to construct office space, A.G. Insurance (now Fortis) was required by the city to provide housing along with retail space on the first floor. The new 5-6 storey project included a park-like open courtyard, 41 apartments and townhouses, an office complex for A.G. Insurance, 13 shops, and two underground parking lots with 500 parking spaces (EuroCouncil 2003:8).

In Stockholm, Sweden, the brownfield redevelopment of the Jarla Industrial Estate, a 19th Century manufacturing complex now named Järla Sjö, has been redeveloped using New Urbanist concepts. Buildings on the site had been considered to be "economically and technically beyond repair," and in need of demolition (EuroCouncil 2003:18). Ultimately, most of the large brick industrial buildings were adapted for reuse and become part of the project. The development includes condominiums, row-houses, single-family residences, live/work units, and commercial spaces on 10 hectares (25 acres) of land area. A large factory building, formerly a machine shop, was converted into a bowling facility. Public areas include a marketplace, courtyards, public squares, parks, and waterfront walkways. The development was designed so that the industrial structures complement the new construction. Automobile traffic is controlled and traffic flow is directed toward a central loop that terminates at the site's single entrance. A train station, offering access to the rest of the city, is conveniently located at the edge of the development.

Yet another truly European example of a brownfield redevelopment site that incorporates New Urbanist concepts is the Baden Nord project in Baden, Switzerland. At this site, the city of Baden, formed a partnership with ABB (an electrical company that owned the site) and the Federal Railway, to transform the site into a mixed-use development for 2,000 inhabitants with over 300,000 m² (3.26 million ft²) of commercial and industrial space (EuroCouncil 2003:15). This project is an example of the success of mixed-use developments. Residents can live and work in the development with most needed amenities within walking distance.

Reducing urban sprawl and constructing more sustainable communities reduces urban energy usage and creates less pollution. Providing amenities near places that people live can result in decisions to walk rather than to drive. New York City's Manhattan Island provides an example of a densely occupied urban area that offers innumerable amenities within walking distance, and also a mix of transportation alternatives.

8-8. Manhattan Island

What do supporters of The New Urbanism, smart growth and sustainable development all agree on? They want increased urban density, more thoughtful design of urban communities, improved transportation systems, development of green spaces, strategically located amenities, and local participation in the decision-making process. Like The New Urbanism, smart growth advocates want mixed land uses and more compact developments that encourage walkable communities (ICMA 2002). Supporters of sustainability, on the other hand, have a greater tendency to focus on environmental impact, resource allocation, waste recycling and the longer-term implications of today's development decisions.

ENERGY AND PLANNING

As new facilities are constructed to meet the requirements of massing populations, it is important to consider energy in the planning process. Energy is the basis by which many human needs are fulfilled. It formats how human habitations are designed. The concept of sustainable development is being incorporated into global, urban and local planning guidelines—with energy being a primary concern.

Energy and sustainability are closely linked. How do *energy conservation and energy efficiency* strategies support sustainable planning policies in urban areas? The debate intensifies when the broader problems concerning energy, sustainable development and cities come into focus. Prior to the 1990s, planning practices in the U.S. were not focused on these issues. Early planners established policies and supported land use frameworks that have created a landscape of inefficient and dysfunctional neighborhoods and cities. Fundamental issues include:

1) Planning policies and practices in the U.S. have increased per capita energy use in cities;

2) The energy consumption patterns in cities are directly contributing to environmental problems, such as air quality and increased carbon emissions.

3) The increasing costs of energy strains urban budgets, and limits future financial options;

4) Increased energy use makes it difficult for effective sustainable development programs to be implemented in urban areas; and finally,

5) By providing dysfunctional infrastructures, many existing planning policies are contributing to cities becoming unsustainable.

There are a number of alternatives available to planners who aid in the design and location of utility systems that provide power for cities and regions. These alternatives can be categorized as decentralized, local, and regional. The means of generating and distributing electricity provide heuristic examples.

The first electrical utilities were typically *decentralized*, serving a small community or a major manufacturer. Over time, these evolved into *local* utility networks and connections were expanded as cities grew beyond their local boundaries. Coal or hydropower generation was the norm. These networks grew and were later interconnected with other utility companies in their *regions* but power production became highly centralized.

Utilities choose different types of energy for new generating facilities. Their decisions are often based on the availability of fuels and the economics of development costs. Fuels that are most readily available make the first cut in the selection process. Fuels ultimately selected are often the least expensive ones available at the time of design. Power plants, once constructed, are usually designed to use only one fuel. Coal, oil (or oil shale), natural gas and uranium are the fuels of choice. Oil is expensive and must be imported. Natural gas is much less polluting than oil but is also expensive and unavailable in many locations. Atomic energy has waste disposal and continued safety concerns.

That leaves coal. In the U.S., coal is available and comparatively inexpensive. Yet burning coal is carbon intensive, more so than oil, natural gas, or uranium. Mining coal using strip mining or mountaintop removal permanently changes landscapes, leaving regional planners with abandoned sites that are difficult to redevelop.

Coal-fired power plants in the U.S. are responsible for emitting roughly 80% of the total carbon emissions from electrical production. Roughly 150 more coal-fired power plants are slated for construction in the U.S. by 2030. It is difficult to imagine where they will all be located. Elsewhere, more coal-fired plants are planned, especially in China and the EU states, but also in other countries. In March 2007, EU members agreed to a mandatory 20% reduction in CO_2 emissions compared to 2020, yet Germany has 26 additional coal-fired plants either under construction or being planned.[25]

There is a counter-trend. In an effort to reduce additional emissions of greenhouse gases, a recent California law was passed prohibiting utilities from investing in or signing supply contracts for electricity from traditional coal-fired plants. More specifically, utilities cannot purchase power from sources that generate more CO_2 than a typical natural-gas-fired plant.[26] This includes almost all coal-fired plants. Couple this with a U.S. Supreme Court decision in April 2007 that greenhouse gases qualify as pollutants under the Clear Air Act. This ruling basically requires the U.S. EPA to regulate greenhouse gases.

Locating power plants, especially coal-fired plants, is a regional planning professional's worst nightmare. Planning authorities must decide where to place these facilities—while their constituents are saying NIMBY (Not In My Back Yard) and emphatically NOTE (Not Over There Either). People simply do not want to have power production facilities built near them. Depending on how they are designed and what fuels they use, electrical generating facilities may require large tracks of land for their construction, lots of water for cooling purposes, sludge management and storage areas, access to transportation for fuel delivery, corridors for overhead power transmission lines, and additional land for waste collection and disposal. These requirements have provided impetus to recommission previously shutdown atomic electrical-generating facilities in Ohio, Tennessee, and elsewhere—plants that had been taken off-line earlier often due to safety concerns.

While planning decisions are both formally or informally blessed by local planning agencies, large-scale infrastructure planning to provide energy to various sites is usually performed by the utility companies, with oversight by state governments. Decisions are based on estimates of an increased need for electricity. It is the nature of utility companies, faced with supply concerns, to recommend centralized energy production solutions—which also provides them with the highest profit margins. As a result, the *goals of the utility companies* when they plan locations for new plants are not necessarily aligned with the *goals of their communities*.

The cumulative results of planning initiatives can be dramatic. New electrical generation plants, especially nuclear, are expensive to build. In fact, utility engineers seldom know how much they will cost to construct until after they are built. Rarely does anyone really know how much fuel will cost, so most projections of their economic viability are merely speculation based on engineering estimates.

There are decentralized alternatives to dedicating large pieces of prime real estate to central power plants and power transmission facilities. We need to reemphasize the "eco" (for ecology) in future "economic" assessments. In a climate of economic uncertainty, Seattle managed to overturn their regional Power Authority's scheme to construct two new atomic power plants. Their creative alternative involved not increasing electrical production capacity, but instead reducing electrical demand by means of electrical power *conservation*; by instituting efficiency improvements and revising peak load sharing plans (Heywood 1997:197).

One solution is net metering—a technique that allows building owners to generate electricity, selling power back to the utility grid when excess is available, and drawing power from the grid when additional power is needed. Another is solar-electric generation systems, which can be designed to require no additional land by being placed on the roofs and walls of existing buildings. Such systems can generate power during peak summer periods when sunshine is plentiful and air conditioning loads are greater.

The eco-city movement to develop sustainable cities has yielded interesting results in cities such as Ottawa, Toronto and Curitiba, Brazil (Devuyst, Hens and De Lannoy 2001:94-98; Register 2002:125-129). Toronto installed a geothermal cooling system for its central business district that uses water from Lake Erie for cooling. Curitiba has one of the world's largest bus systems, recycles a high proportion of its wastes, and has recycled abandoned buildings into sports and recreation facilities.[27] High temperature geothermal water sources are used to heat buildings in ancient Pompeii, for district heating systems in Paris, and in cities in Iceland, the U.S. and Turkey.

The town of Navarra, Spain, has installed 1,100 wind turbine generators and has the ambitious goal of providing 100% of all electrical energy by renewable energy sources by 2010 (AEE Hungarian Chapter 2001:263). The tiny region around Navarra has benefited from 4,000 new "green energy" jobs in the last decade and now generates more energy from renewable sources than either France or Poland.[28] Malmö, Sweden, a new district with roughly 1,000 residences, "meets 100% of its electricity needs with solar and wind power, gets its heat from sea and rock strata and from the sun, and fuels its vehicles with biogas from local refuse and sewage" (Worldwatch Institute 2007:99).

These examples prove that there are often feasible alternatives for utility planners to consider—options other than constructing new power plants that use fossil fuels.

CONCLUSION

Planning is a process that involves making decisions, solving problems, and creating policies. Planning also involves developing frameworks that can be used in urban design to achieve social goals and allocate resources. Comprehensive planning principles are not always appropriate for implementing sustainable development. Incremental changes over extended periods of time, when directed toward long-range redevelopment goals, can lead to major improvements. In this chapter, examples of sustainability in planning and development are cited, including activities in the U.S., the UK, and Europe.

Planning mechanisms provide tools with which to implement social goals such as sustainable development. Among the planning solutions available, there are paradoxes, trends and counter-trends.

Planning alternatives include urban homesteading, revitalizing historic neighborhoods, creating amenities in downtown areas, redeveloping abandoned commercial and industrial sites, considering in-fill construction, and encouraging mixed-use developments. In Europe, Local Agenda 21 plans have been developed with the encouragement of central governments to guide cities toward implementing sustainable development agendas. The U.S. has no similar sustainability agenda at the federal governmental level, and local and state governments are taking the initiative.

Planning errors have been made in the past. Failure to create opportunities and conditions for mixed-used developments is one of the most obvious. Other local planning policies and regulations have at times created more problems than they have solved. In the past, planning regulations have not been integrated with the concept of sustainability. These regulations need to reconsidered, revamped, and redefined.

Regardless, there are interesting local and regional examples of U.S. progress toward sustainable development. Local programs can be creative and customized to respond to local issues including meeting energy-consumption requirements, allowing brownfield redevelopment, redeveloping parklands, addressing vacant structures, and expanding historic districts. Transportation planners plan highways or railways, housing planners plan new housing and utility planners plan utility systems. Integrating these decision-making processes is important in order to create opportunities for sustainable development.

Rethinking planning approaches for transportation systems is an important part of the solution. Designing subdivisions accessible by only one mode of transportation—automobiles—is counterproductive. The cost of transportation and the costs of owning and operating automobiles is increasing. Among major U.S. cities, family expenditures for transportation are lowest in New York City, a city that offers a number of transportation options. There are alternatives to more and more automobiles and the need to construct more highways for them. The alternatives include expanding bicycle lanes, bus routes, bus rapid transit systems, tramways, and commuter and rail light rail systems. Some of these options can be less expensive to implement and they offer environmental, developmental, and energy use benefits.

Policies that favor and subsidize increased urban sprawl must be aborted. Redesigning neighborhoods, improving communities, and providing creative solutions for new ones can be part of the solution. The

New Urbanism provides a set of simplified design principles that establish parameters for site construction, buildings, streets, neighborhoods and districts. Examples of The New Urbanism can be found in the U.S. and the European Union. Transit-oriented developments can be designed to create transportation nodes that facilitate the use of public transportation. Expanding urban parklands and developing green wedges, such as those in Amsterdam, are solutions available to city planners.

Energy is often mistakenly omitted from many planning decisions altogether. In order to provide sustainable communities and grow our metropolitan areas, conflicts must be resolved, alternatives considered, and solutions planned in concert with local goals and agendas. To be successful, housing, transportation and energy requirements must be programmed simultaneously. Planning solutions provide opportunities to achieve substantial reductions in transportation fuels. This can be achieved by adjusting population densities and providing alternative infrastructure for transportation. Alternatives to building new power plants are worthy of serious consideration. These include considering energy efficiency and developing decentralized alternative energy production facilities.

Endnotes

1. *The Charter of European Cities and Towns Towards Sustainability*, www.ec.europa.eu/environment/urban/aalborg.htm, accessed 30 June 2007.
2. Communities and Local Government. *Three years on from the Sustainable Communities Plan*. www.communities.gov.uk/index.asp?id=1163434, accessed 2 May 2007.
3. From *More about sustainable communities*. www.cambridgeshire.gov.uk/environment/sustainable/sustainable+communities.htm, accessed 25 April 2007.
4. Ibid.
5. Smith. A. (2007, 14 April). The global warming survival guide—43. Move to London's Green Zone. *TIME*. www.time.com/time/specials/2007/environment/article/0,28804,1602354_1603074_1603739,00.html, accessed 21 May 2007.
6. Ibid.
7. Bureau of Transportation Statistics, U.S. Department of Transportation. *Highlights of the 2001 national household travel survey*. http://www.bts.gov/publications/highlights_of_the_2001_national_household_travel_survey/, accessed 4 July 2007.
8. The study was based on 2000 census data. The 28 metropolitan areas in the study represented 47.1 million households, or 45.1% of all U.S. households.
9. Energy Information Administration. www.eia.doe.gov/emeu/steo/pub/special/summogas.html, accessed 26 April 2006.
10. Lipmam, B. (2006:13). His study suggests that for 28 metropolitan areas, the distance at which the increased costs of transportation exceed savings from reduced housing costs is 19-24 kilometers (12-15 miles) from the central business district.
11. Ritter, J. (2007, 6 June). California sees sprawl as warming culprit. *USA Today*. p. 1.

12. Ibid.
13. Associated Press (2007). *Commute nightmare after fiery Bay Area crash.* www.msnbc. msn.com/id/18392984, accessed 30 April 2007.
14. Nesmith, J. (2007, 13 May). Bus transit fans say rail gets too much support. *The Courier-Journal.* p. D1.
15. Ibid.
16. Ibid. p. D2.
17. O'Grady, W. (2007, May). *When bikes rule the road.* Spirit. p. 117.
18. Gilderbloom, J. and Hanka, M. (2007, 13 June). Louisville's new urbanism. *The Courier-Journal.* p. A11.
19. Ibid..
20. Amsterdam City Planning Department (2007). *Aracam Expositie—Green Fingers.* p. 1.
21. Ibid., p. 2.
22. Ibid., p. 2.
23. U.S. Environmental Protection Agency. (2007, 19 January). *About smart growth.* http://www.epa.gov/smartgrowth/about_sg.htm, accessed 2 May 2007.
24. For a listing if the principles of the *Charter of New Urbanism* see Berke (2002:27-28) or http://www.cnu.org/cnu_reports/Charter.pdf.
25. Schaefer, L. (2007, 21 March). Despite climate concerns, Germany plans coal power plants. *Environment.* www.dw-world.de/dw/article/0,2144,2396828,00.html.
26. This is roughly 370 kilograms of CO_2 per Megawatt hour of electricity.
27. Wilderness Lakes. *Curitiba, Brazil—Inspirational Eco-City.* www.gardenroute.com/ waleaf/archives/20040702.htm, accessed 21 July 2007.
28. Lungescu, O. (2007, 9 March). *Navarra embraces green energy.* www.news.bbc. co.uk/1/hi/world/europe/6430801.stm, accessed 1 May 2007.

Chapter 9

Tracking Local Sustainable Development

"Sustainable development is development that improves the long-term health of human and ecological systems. The strategy avoids fruitless debates over 'carrying capacity,' 'needs,' or sustainable end states, while emphasizing the process of continually moving toward healthier human and natural communities. In theory the direction of this process can be agreed upon though participatory processes in which all relevant stakeholders are represented, and progress can be measured by means of various performance indicators."

STEPHEN WHEELER (2000:438)

This chapter covers complex aspects of achieving sustainability in cities, and is divided into three parts:

Part A (p. 280) – Reviews Indicators of Sustainability

Part B (p. 295) – Analyzes the Values of these Quantitative Indicators, and Provides Regression Analysis of Variables.

Part C (p. 327) – Expands on the Information Developed in Part A and Part B, and Asks: "What Can Our Cities Do?"

This chapter considers methodologies used to track progress toward achieving the elusive goals associated with sustainable initiatives. Tracking systems that assess sustainability have been developed for national economies in regard to environmental performance[1] but are rarer for institutions, corporations and local governments. Local governments, in particular, often desire to know how their progress toward sustainability goals compares with members of their peer group.

Methodologies to track and rank governments and organizations in regard to sustainable practices are a relatively new phenomena. These methodologies use qualitative and quantitative variables. All identify indicators of sustainability and provide a means of comparison. Linking measurements of energy usage to variables that measure the impact of sustainability policies is one approach. Quantitative analysis techniques

279

can be used to interpret the differences and draw conclusions concerning the data available.

This chapter explores different indicators of sustainability, considers variables that can be measured, and demonstrates how sustainability indices can be developed and interpreted. How indexes are developed by selecting variables and identifying relationships will be discussed in detail. Statistical methodologies that can be used to bring credibility to the construction of a sample sustainability index will be presented. The index developed will use both qualitative and quantitative variables. A sample of U.S. Sunbelt cities will be used as an example. The data collected regarding them are used to search for commonalities and differences in their demographic patterns, energy impacts and environmental conditions by analyzing quantitative variables. Based on a statistical analysis, selected variables will demonstrate how conclusions can be drawn regarding the sustainability of cities. A comparison will be developed by dividing subsets of the sampled cities into comparable groups. Conclusions will be offered based on information gained from observing the differences between the groups.

PART A: INDICATORS OF SUSTAINABILITY

Measures of energy and sustainability have been most commonly established at the macroeconomic scale. The Organization for Economic Cooperation and Development uses environmental indicators for whole countries that includes air emission intensities (CO_2, SO_2, etc.), waste recycling, and socio-economic indicators such as energy, energy prices, population density, and transportation infrastructure densities, among others (OCED 1998). United Nations methodologies incorporate techniques that divide indicators into social, environmental, economic, and institutional categories (United Nations 2001).

International research into measurement and monitoring sustainability provides interesting initiatives. A joint study by Columbia University and Yale University developed the Environmental Sustainability Index (ESI), which gauged progress toward environmental sustainability for 142 countries. The study fashioned "a set of 20 core indicators, each of which combines two to eight variables for a total of 68 underlying variables" (Columbia 2001). Variables included levels of sulfur dioxide in the air and protection of land from development, creating an index (from 0 to 100) that gauged the health of the environment (Desai 2002:109). The countries with the highest ratings were Finland, Norway and Canada. Another study by Oxford University focused on sustainability in highly consumptive societies (Lafferty and Meadowcraft 2000:337-421).

Sustainability is multi-dimensional. The success of efforts to provide axiomatic solutions has been mixed. Kutzhanova and Roosa (2003:3-5) suggested developing variables along four dimensions of sustainability that included:

1) The degree to which appropriate ecological and environmental measures can be developed;
2) The degree of efficiency of use of natural resources;
3) The extent of management of urban growth; and
4) The extent to which the cultural framework provides equity for current and future generations, while offering the opportunity for the improvement of the human condition.

While cities have provided a unit for study in the sciences, measuring the sustainability of cities is quite rare. Providing non-subjective variables can be challenging. Nijkamp and Pepping (1998:1484) suggested that, "There is not a single unambiguous urban sustainability measure, but a multitude of quantifiable criteria." Portney (2003:240-241) believes that sustainability relates to variables such as energy consumption, transportation, land use, community building and social justice. Portney (2003:240-241) further observes that, "a city might be said to take sustainability seriously when its program is found to produce environmental improvements, and one city's initiative might be said to be more serious than other cities' initiatives if it produces a greater environmental benefit." Portney (2003:240-241) believes that until such comparative research is feasible, the measurement of taking sustainability seriously will "be based more on judgment than on rigorous objective standards." Apportioning methodologies to quantify the contributions of changes in population and per capita consumption of resources are sometimes used to gauge the impacts of resource use (Holdren 1991).

Cities are viewed as ecological organizations (Miller 1978), as a focus for transportation and its impact on sustainability (Newman and Kenworthy 1999), as a means of comparing urban policies concerning sustainable development (Burgess, Carmona and Kolstee 1997), and as a basis to compare sustainable attributes across cities (Beatley 2000). Advantages of studying energy policies in the context of sustainable cities include:

1)   Urban areas are centers of economic activity, and exhibit the attributes of scale and density;

2)   Spatial clustering suggests concentrated energy use;

3)   Cities have greater population densities;

4)   Most production, transportation and consumption activities occur in urban areas;

5)   The city is an institutional decision-making unit; and

6)   The city is a "suitable statistical entry providing systematic data sets on environmental, energy and socioeconomic indicators" (Nijkamp and Pepping 1998:1484-5).

Assessments across cities are varied. A comparative analysis of twelve European cities by Nijkamp and Pepping (1997:1481-1500) studied success factors of renewable energy policies and determined that policies can have a "double-dividend character" in that environmental quality can be improved while reducing the costs of energy consumption.

In Germany, Merkel suggested four categories of indicators. These categories included ecosystems, energy use, economic life cycles and human health (Merkel 1997). In the UK, the government issued a set of 120 indicators of sustainable development in 1996 and extended the list in 1999 to include a broad range of social issues (Lafferty and Meadowcroft 2000:385).

Indicators of a sustainable community are often locally defined. The Sustainable Seattle program employs "bellwether tests of sustainability" that reflect "something basic and fundamental to the long term economic, social, or environmental health of a community over generations" (Sustainable Seattle 1993:4). Energy, fuel consumption, vehicle miles, non-renewable energy use, and renewable energy use are included as indicators. Sustainability indicators represent more than a collage of social, environmental, and economic factors; they also illustrate integrating linkages among their domains (Maclaren 1996:13). Another example of a transportation indicator is the existence of a policy supporting alternatively fueled city vehicles.

A few Sunbelt cities have adopted specific local indicators. The *Report Card of the San Diego Region's Livability* is an example of a comprehensive planning approach, with sustainability listed among the principle objectives. It notes that "the conservation and efficient use of energy will play a very important role in our future if we are to maintain the amount and quality of desired services that energy facilitates" (San Diego Association of Governments 2000). Santa Monica's resource indicators include a city-wide energy usage profile that is tracked annually and compared to targeted goals.

Indices have been proposed to measure attributes of sustainability. The Barometer of Sustainability is one example (Prescott-Allen 2003). Portney provides an index of "Taking Sustainability Seriously" that includes 34 key indicators. Four of Portney's indicators deal directly with energy conservation and efficiency while several others consider indirect measures such as related transportation and environmental issues (Portney 2003). Those that concern energy use directly are:

1)   Instituting a green building program;

2)   Renewable energy use by local government;

3)   Availability of alternative energy to consumers; and

4)   A local energy conservation effort or program.

Other researchers fail to focus on energy variables as key components in measuring sustainability. For example, Maclaren proposes a set of 16 indicators whose goals include "living off the interest of renewable resources." Remarkably, energy is not included among his 16 suggested sustainability indicators (Maclaren 1996a:79-80). Maclaren offers a detailed multi-step approach to sustainability reporting with a typology of frameworks to develop sustainability indicators. He includes domain-based, goal-based, issue-based, sector-oriented and causal categories of indicators. Again energy is not considered to be an indicator (Maclaren 1996b:192-80). However, elsewhere in his article there are examples of other reports that did use energy as an indicator. Examples include the Sustainable Seattle report (1993) and the United Kingdom's Local Government Management Model (1994). Occasionally researchers note the importance of energy as an indicator of sustainability (such as "green power"), but then surprisingly proceed with their research and analysis without using any energy-related variables (Bowman 2003:4-11).

There are quantitative measures available that can be utilized. Energy use by states can be measured in gross expenditures, dollars per capita, units of energy use by energy type, gross energy consumed, energy use per capita, and others. Indicators of energy use in urban areas include transportation miles, commute travel times, number of vehicles per capita, local weather data, population, household size, and area of enclosed space, among others. When central utilities are municipally owned, data on local utility usage are often available.

The extensive use of hydrocarbon energy has spawned a wide range of concerns that are challenging to assess at the local level. Hydrocarbon emissions can be difficult to measure and track due to a lack of quantifiable local data. One solution is to review local action plans looking for qualitative measures. For example, a checklist has been proposed for "city sustainability" using economic efficiency, social equity, environmental responsibility, and human livability as basic evaluation criteria (Newman and Kenworthy 1999:367-372). Even more subjective, local measures of pollution, global warming and climate change provide examples that have been considered and remain untested (Rocks and Runyon 1972:114-120).

Regardless, the valuable contributions of these researchers provide heuristic examples that variables, indicators and, methodologies are available to use in tracking sustainability.

Ranking Cities Using Qualitative Indicators

A preliminary ranking of cities based on qualitative data from Chapter 7 is revealing. This ranking is based on each city scoring a single point for each dimension of each category of indicators, with the possible scores ranging from 0 to 10. For this example, the policies for this ranking are equally weighted. Weighting the policies equally assures that no one policy or set of policies dominates the rankings. On the other hand, programs and policies that might be of greater import or impact may be minimized in value.

For example, two of the cities in the sample, Los Angles, and San Diego have policies in place in each of the ten categories of variables. As a result, both cities achieved the highest possible score of 10. Miami, which lacks one energy policy indicator, received a total score of 9—the only city to receive this score. Charlotte, with only one environmental policy, had the lowest score of 1. Oklahoma City has two policies in effect, with one in the environmental policy category and one in the organizational participation category. The mean number of policies found among the cities (N=25) was 5.8 while the mode was 7. The cities are ranked based on the total scores in Table 9-1.

Analysis of Quantitative Indicators

From the information gained concerning the types of policies available and selected by the cities, it is clear that policies vary widely in both form and application. Cities and local governments may choose to institute sustainability programs for a wide range of reasons. If policies hope to improve sustainability then what kinds of measures of energy use can be used to gauge improvement at the local level?

For an analysis, measures of urban policies can be studied as independent variables, with energy usage as a dependent variable. Urban policies are considered to be a cause that results in changes in energy use (effects), thus impacting urban sustainability. In order to describe the basis for the analysis, a review of the variables considered will be provided.

This analysis initially considers two aggregate measures (per capita energy usage and per capita energy costs) of the primary dependent variable plus the subcategories of total energy use which include the transportation and residential sectors. Data for energy costs, total energy use and sector energy use were obtained from the Energy Information Administration's *State Energy Data 2000*.[2]

Energy Variables

Per capita energy costs (E_C):

This is the statewide average annual per capita cost of energy during the year 2000 in units of year 2000 U.S. dollars. Lower energy costs can indicate greater urban sustainability since access to the energy source is more equitable. Per capita energy costs are an aggregate measure of the results of energy pricing. Since higher costs of energy can restrict availability, cities in states with higher per capita energy costs may be considered less sustainable. Substantially higher per capita energy costs may indicate that resources are in short supply in a given locality or that resources are being diverted from social needs to provide energy.

Table 9-1. Sunbelt Cities Ranked by Rate of Policy Adoption

Sunbelt Cities In the U.S.	Energy	Local Policy	Organizational Participation	Environmental Programs	Policy Score
Los Angeles	3	2	3	2	10
San Diego	3	2	3	2	10
Miami	2	2	3	2	9
Houston	3	1	2	2	8
Austin	3	1	2	2	8
Tucson	2	1	3	2	8
Atlanta	1	2	3	2	8
Phoenix	2	2	1	2	7
Dallas	2	1	2	2	7
San Antonio	2	2	1	2	7
Las Vegas	3	0	2	2	7
Long Beach	2	2	1	2	7
Albuquerque	2	1	2	2	7
Fort Worth	1	1	2	1	5
Tulsa	1	1	1	2	5
Jacksonville	1	2	0	2	5
Memphis	1	1	0	2	4
Nashville/Davidson	1	1	0	2	4
Fresno	1	1	1	1	4
Mesa	1	1	0	2	4
El Paso	1	1	1	0	3
New Orleans	0	1	0	2	3
Virginia Beach	0	1	1	1	3
Oklahoma City	0	0	1	1	2
Charlotte	0	0	0	1	1

Sources: City of Albuquerque: http://www.cabq.gov/; City of Atlanta Online: www.ci.atlanta.ga.us; Austin City Connection: http://www.ci.austin.tx.us/; Austin Energy www.austinenergy.com; City of El Paso:http://www.ci.el-paso.tx.us; City of Fresno: www.fresno.gov; City of Fort Worth: www.ci.fort-worth.tx.us; City of Jacksonville, Florida: www.thecityofjacksonville.com; City of Houston (2004) also www.ci.houston.tx.us; City of Las Vegas (2002); City of Memphis Online: http://www.cityofmemphis.org/; City of Miami: http://www.ci.miami.fl.us/; City of New Orleans: http://www.new-orleans.la.us/home/; City of Oklahoma City: http://www.okc.gov/; City of Phoenix, Arizona: http://phoe-nix.gov/; City of San Diego Antonio Public Works: http://www.sanantonio.gov/publicworks/?res=800&ver=true; City of Tucson (1998, 26 June); City of Tulsa Online: http://www.cityoftulsa.org/; City of Virginia Beach: http://www.vbgov.com/default/; Clean Cities: www.ccities.doe.gov; Dal-las: Dallas: dallascityhall.com; Energy Star Homepage: http://www.energystar.gov/; Hoover's, Inc. (2004); ICLEI (2003) and http://www3.iclei.org/member.htm; Las Vegas Regional Clean Cities Co-alition (2002); Long Beach: www.ci.long-beach.ca.us; Los Angeles: www.lacounty.info; Metropolitan Government of Nashville/Davidson Co.: http://www.nashville.gov/flashpgs/flashhome.htm; Mi-ami-Dade County (2004); Newman, P. (2003); Nevada Power: www.nevadapower.com; Portney (2003); Rake, L. (2004, 21 March); San Antonio Community Portal (2004a, 2004b); San Diego Association of Governments (September, 2000); City of San Diego (2002); San Diego Association of Governments: www.sandag.org; City of San Diego: www.sandiego.com/environmental-services; Southern Califor-nia Edison (2000, 2004); United States Department of Energy (2002, 2003); United States Department of Transportation (2004); U.S. Department of Transportation: hovpfs.ops.fhwa.dot.gov/inventory/inven-tory.htm; United States Environmental Protection Agency (2003, 2004); United States Environmental Protection Agency (1998, May); Valley of the Sun Clean Cities Coalition (2004); Vision of the Mesa 2025 General Plan.

Per Capita Energy Usage (E)

This variable quantitatively expresses the statewide average annual per capita energy use during the year 2000 in common units of energy consumed (million kilojoules per year). Less energy use is accepted as an indication of greater urban sustainability since less is demanded of the system to maintain equilibrium. To the extent that system efficiencies are comparatively equal, lower energy use suggests less pollution emission potential in the state where the city is located. Cities that are in states with higher per capita energy use may be considered less sustainable as more energy is required to maintain equilibrium. Higher energy use suggests potentially greater pollution potential. It also requires energy to produce and manufacture energy, especially fossil fuels.

For example, states such as Oklahoma, Texas, and Louisiana are oil-exporting states and their per capita energy use will be higher in part due to the energy required to extract, produce, process, refine, and transport marketable oil-based products that are shipped to other states. Regardless, per capita energy consumption is an often-used sustainability measure. Reducing per capita energy usage has been identified both as a goal and an indicator of sustainable cities (Maclaren 1996a:45; Newman and Kenworthy 1999:18; United Nations 2001:231-233).

The Energy Information Administration reports the gross energy usage for four economic sectors: residential, commercial, industrial and transportation. For this analysis, subsets of these data, per capita energy usage (E), transportation sector energy usage (E_T), and residential energy use (E_R) are considered the most relevant since many urban policies focus on these sectors.

Transportation Sector Per Capita Energy Usage (E_T):

This variable quantitatively expresses the statewide average annual per capita energy usage by the transportation sector during the year 2000 in common units (million kilojoules per year). This measure of energy use has been found to be linked to urban density and urban land area (Newman and Kenworthy 1999:101-103).

Residential Sector Per Capita Energy Usage (E_R):

This variable quantitatively expresses the statewide average annual per capita energy usage by the transportation sector during the year 2000 in common units (million kilojoules).

Sunbelt Cities

Many of the cities used for this example are among the principle cities and population centers in their respective states. In a number of the Sunbelt states, these cities represent a substantial percentage of their total state population. The use of state averages is considerably legitimized in states such as Arizona and Texas since the cities in these states constitute a major percentage of their state's population. On the other hand, the use of state averages are less representative for states like Florida, since only a small fraction of the state population is represented by the cities in this study. Compiled using data from the 2000 U.S. Census, a summary of the cities involved in this study is provided in Table 9-2, along with the calculated percentage of population which each represents in its state.

Table 9-2. City Population Compared to State Population (2000)

Selected Sunbelt Cities (by State)	Total State Population each City %	Total State Population of the Selected Cities %
Los Angeles, California	10.9	
San Diego, California	3.6	
Long Beach, California	1.4	
Fresno, California	1.3	17.2
Houston, Texas	9.4	
Dallas, Texas	5.7	
San Antonio, Texas	5.5	
Austin, Texas	3.1	
El Paso, Texas	2.7	
Fort Worth, Texas	2.6	29.0
Oklahoma City, Oklahoma	14.7	
Tulsa, Oklahoma	11.4	26.1
Phoenix, Arizona	25.7	
Tucson, Arizona	9.5	
Mesa, Arizona	7.7	42.9
Las Vegas, Nevada	23.9	23.9
Albuquerque, New Mexico	24.7	24.7
Jacksonville, Florida	4.6	
Miami, Florida	2.3	6.9

Source: Calculated using data from the U.S. Census 2000.

The principal cities in the states of Arizona, Nevada, and New Mexico represent roughly one-quarter of their states' population.

The sustainability indicators used for this analysis include energy usage, transportation, environmental impact and demographic indicators, each with selected variables. The variables used provide quantitative data for each of the indicators, along each of their corresponding dimensions.

Indicators of residential energy use sample and gauge both the use of energy in the residential sector and the extent of use of alternative energy. Residential energy use was chosen since alternative energy technologies are commercially available as are packaged residential applications (e.g., solar water heating systems, photovoltaic roofing shingles, etc.). The independent variables selected to measure residential energy usage include *homes heated by alternative fuels* and *single occupant residences*. The values for these variables were compiled using data from the 2000 U.S. Census.

Homes Heated by Alternative Fuel

This is the percentage of homes in each city in the year 2000 heated by alternative fuels (wood, solar, other renewable sources) plus those that do not require fuels for heating purposes. The ability to heat one's residence with renewable energy or to construct a residence that does not require a conventional heating source is inherently more sustainable than one depending on fossil fuels (oil, natural gas, coal) for residential heating. Using renewable energy rather than non-renewable fuels has been identified as an indicator of sustainability (Rennings and Wiggering 1997:25-36).

Single Occupant Residences

This variable is the percentage of residences in the city that have only one occupant. While appliance energy use may vary, an occupied residence requires roughly the same amount of energy for space heating and cooling regardless of the number of occupants. While single-occupant households are often smaller in floor area than residences constructed for larger families, it is logical that more single-occupant households in a city would cause higher residential energy use per capita than would be found in cities with a lower percentage of single-occupant households.

Cities may choose to develop and provide alternative means of transportation for their populations. When it is possible for urban residents to meet all or a portion of their transportation needs by means other than personal vehicles, energy use tends to decline and sustainability can be theorized to increase. This is evidenced by the number of households that have one or no automobiles. In addition, cities can be planned in a manner that allow shorter commute times, thus reducing the energy impact of automobiles.

The independent variables used as indicators to assess transportation include: 1) *travel time to work*; 2) *alternative means of transport*; and 3) *household vehicles*. The values for these variables were compiled from data found in the 2000 U.S. Census. A description of each of these variables follows.

Travel Time to Work
This variable is the mean travel time (one way) to or from work in minutes during the year 2000 for the city being studied. Reducing commute times to and from work has been suggested to be a goal of sustainable cities (Newman and Kenworthy 1999:19). Vehicles that are driven less use less fuel, resulting in the creation of less air pollution during commuting periods. Substantial commute times can be indicative of decentralized development patterns and traffic congestion, thus contributing to increases in vehicular energy use.

Alternative Transportation
This is a quantitative value indicating the percentage of residents in the subject city who used public transport or walked to their places of employment, plus those who chose to work from home (e.g., home office users, stay-at-home employees, telecommuters, among others, etc.) during the year 2000. This indicator provides a snapshot of the size of the non-commuting population, combined with an estimate of the size of the population that does not necessarily require a vehicle simply to go to work. A larger percentage value is indicative of decreased energy usage, as less dependence on personal vehicles is required to generate income. Smaller percentage values indicate that a greater number of people need to use personal vehicles to reach their primary places of employment. Reduced dependence on personal vehicles for commuting purposes improves urban sustainability by reducing energy usage and pollution from vehicles.

The level of alternative transit use is related to sustainability and urban form (Vig and Kraft 2003:62-65). The extent of use of public transportation systems is a sustainability indicator (Bell and Morse 1999:63; Maclaren 1996:45; Newman and Kenworthy 1999:90).

Household Vehicles:

This variable is the percentage of households with no or only one vehicle per household in a city, for the year 2000. While it may be argued that a smaller number of household vehicles suggests lower economic status and a larger number of household vehicles reflects greater household wealth, it is also likely that design of the urban infrastructure provides lesser or greater needs for private vehicles. As a result, higher percentages may indicate greater urban sustainability—residents of urban core areas have less need for household vehicles to accommodate their transportation requirements. Lower percentages indicate a greater need for, and utility of, multiple personal vehicles. Reducing private ownership of vehicles can be an indicator of improved sustainability (OCED 1998:88).

Urban areas are located in places with widely varying climates, climatic conditions may tend to disperse air pollution, concentrate air pollution or have neutral impact. The impacts of pollution vary by type of pollutants, location of source, distribution and impact. Atmospheric pollution demonstrates inefficiencies in combustion processes, such as those associated with carbon-based fuel consumption.

Combustion fuels such as coal and oil are significant contributors to atmospheric pollution. Alternative fuels create no or negligible atmospheric pollution. Cities located in coastal locations with offshore afternoon winds such as Jacksonville, Florida often experience fewer problems with atmospheric pollution than inland cities like Fresno, California which has continuing and extensive problems with atmospheric pollution. The independent variables used to assess air pollution include the *air quality index (AQI)* and the number of days the air is classified as being either *unhealthy or unhealthy for sensitive groups*. The extremity of environmental climate conditions is represented by the variable *cooling degree days*. These variables are:

Air Quality Index (AQI)

This is a mean value of the daily local or regional air quality index (AQI) for the subject city for the year 2000. Higher values for the AQI indicate poorer air quality due to pollution. The indicator ranges from a low value of 0 to a high value of 500 and is used by the U.S. Environmental Protection Agency (USEPA) to gauge and compare air quality across cities and regions. Air quality levels are typically monitored and compiled by state government entities and reported to the USEPA. AQI values ranging from 0 to 50 are considered "good" and pose little or no health risk. AQI values ranging from 51 to 100 are considered "moderate" yet have levels of pollutants that may pose a "moderate health concern for a very small number of people" (USEPA 2003). AQI values above 101 are ranked progressively as "unhealthy for sensitive groups," "unhealthy," "very unhealthy" and "hazardous" (USEPA 2003).

An AQI value in the "good" range indicates that daily air quality conditions are not problematic or are being successfully addressed by local or regional mitigation efforts. Greater urban sustainability is a natural result. While a given city may have an annual mean value in the "good" range, it will likely experience a number of peak days when the air quality is poorer. Atlanta, as one example, had an annual mean AQI value of 61 for the year 2000 with daily peaks reaching index values of 206. Air quality has been identified as an objectively verifiable measurement of sustainability in cities (Bell and Morse 1999:99; OECD 1998:26-31; Sustainable Seattle 1993).[3]

Days Unhealthy and Unhealthy for Sensitive Groups
This value is the total number of days in the year 2000 during which air quality in the city was categorized as either unhealthy or unhealthy for sensitive groups. As a result, the value of this indicator could range from 0 to 365. The fewer the number of days in which air quality is either unhealthy or unhealthy for sensitive groups, the greater the urban sustainability. Cities experiencing a greater number of unhealthy days may be insensitive to the contributing causes of human health care problems, and inattentive to the need to increase mitigation efforts to improve air quality. The actual number of days with "good air quality" has been identified as an indicator of sustainability (Bell and Morse 1999:99; Newman and Kenworthy 1999:18; United Nations 2001: 140-141).[4]

Cooling Degree Days
This variable is a measure of the average total number of cooling

degree days (CDD) experienced annually by each city when compared to a base temperature of 65°F.[5] The cities in the sample experience a mean of 2,385 cooling degree days, ranging from a low of 679 cooling degree days for Los Angeles and a high of 4,361 cooling degree days for Miami. This variable is used as a measure for residential energy usage (E_R) since more air conditioning is typically required for cities that experience greater number of cooling degree days (as cooling degree days increase, air conditioning systems typically use more energy). A summary for the values of *cooling degree days* has been provided in Table 9-3.

Table 9.3. Climate Data for Selected Sunbelt Cities

Sunbelt Cities in the U.S.	Cooling Degree Days (base=65°)	January Average Solar Radiation (langleys)
Los Angeles, California	679	256
Houston, Texas	2,893	275
Phoenix, Arizona	4,189	305
San Diego, California	801	247
Dallas, Texas	2,878	254
San Antonio, Texas	3,038	283
Jacksonville, Florida	2,817	287
Austin, Texas	2,982	283
Memphis, Tennessee	2,187	155
Nashville/Davidson, Tennessee	1,652	160
El Paso, Texas	2,254	338
Charlotte, North Carolina	1,681	200
Fort Worth, Texas	2,586	254
Oklahoma City, Oklahoma	1,907	254
Tucson, Arizona	3,017	318
New Orleans, Louisiana	2,773	214
Las Vegas, Nevada	3,214	281
Long Beach, California	1,186	247
Albuquerque, New Mexico	1,290	312
Fresno, California	1,963	193
Virginia Beach, Virginia	1,612	210
Atlanta, Georgia	1,810	230
Mesa, Arizona	3,798	305
Tulsa, Oklahoma	2,049	254
Miami, Florida	4,361	350

Cooling degree day data for Table 9-3 were obtained from the National Oceanic and Atmospheric Administration (NOAA). Solar radiation data, an important influence on cooling requirements for Sunbelt Cities, were extracted from tables found in Edward Mazria's *Passive So-*

lar Energy Book.

Table 9-3 provides a summary of the climate data for selected cities located in the newly defined Sunbelt. These cities average 2,385 cooling degree days and 259 langleys of solar radiation on a horizontal surface during a normal January.

Demographic Indicators

Demographic indicators have also been selected for this study. Variables measuring urban *population density* and the relative changes in population and urban density over a period of time (a *10-year relative change in population compared to change in density*) are independent variables helpful in measuring changes along these dimensions. The values for these variables were compiled from data in the 2000 U.S. Census.

Population Density

This indicator is the average population density of the city per square kilometer for the year 2000. Higher population densities indicate less urban sprawl, greater centralization of population, more concentrated development, and a greater degree of sustainability. Low urban population densities suggest greater sprawl, decentralized population, less concentrated development patterns and a lower degree of sustainability. As a point of reference, the average population density in the U.S. is just under 30 inhabitants per square kilometer. Increasing population densities has been suggested to be a direct measure of sustainability (Vig and Kraft 2003:62-65) and as goal and indicator of improved sustainability in cities (Newman and Kenworthy 1999:19, 269).

10-year Relative Change in Population Compared to Change in Density

This indicator records the changes in population relative to changes in density over a recent ten year period. The indicator is calculated as the percentage change in population from 1990 to 2000, less the percentage change in population density from 1990 to 2000. If urban population growth is outstripping increases in density then the value is positive and land area (possibly due to sprawling development) is being added to the city. If urban population growth is not outstripping increases in density then the value is negative and the urban area is likely becoming more densely populated. Densification suggests relatively less suburban development. Population growth and density have been identified as demographic environmental indicators related to sustainability (OECD 1998:74; Sustainable Seattle 1993; United Nations 2001:126-127). However, the use of this calculated quantity as a measure of sustainability is novel and remains untested.

PART B: IDENTIFYING VALUES FOR QUANTITATIVE VARIABLES

Part B takes the next step, analyzes the *values* of these quantitative *variables*, and provides a regression analysis for each.

Data for each of the ten variables listed in Part A of this chapter were obtained and tabulated for all of the 25 selected Sunbelt cities. (There are no missing or omitted quantitative data for these variables). The values of the selected dependent variables are as follows:

Per Capita Energy Costs (E_C)
The values for this variable range from a low per capita annual cost during the year 2000 of $1,951 for cities in Florida (Miami and Jacksonville) to a high of $4,638 for cities in Louisiana including New Orleans. It is interesting that per capita energy costs are lowest in California, Arizona and Florida (energy importers) and highest in Texas and Louisiana (energy exporters). The mean value for the 25 sampled Sunbelt cities is $2,661.

Interestingly, Los Angeles, Long Beach and Miami not only have the highest rates of homes heated by alternative or no fuels but also are located in states with comparatively lower rates of per capita energy usage and per capita energy costs.

Per capita Energy Usage (E):
For data collected for the year 2000, this variable ranges from a low value of 250.0 million kilojoules[6] for cites in Arizona (Phoenix and Tucson) to a high of 936.2 million kilojoules for cities in Louisiana (New Orleans). The mean value for the 25 sampled Sunbelt cities is 409.0 kilojoules. The U.S. national average in 2000 was 371.4 million kilojoules.[7]

Transportation Sector Per Capita Energy Usage (E_T):
Using data collected for the year 2000, this variable ranges from a low value of 94.1 million kilojoules for cites in Florida (Jacksonville and Miami) to a high of 213.5 million kilojoules per year for cities in Louisiana (New Orleans). The mean value for the 25 sampled Sunbelt cities is

113.0 million kilojoules. This mean quantity (E_T) is a subset of p*er capita energy usage (E)* representing 27.9% of the mean of *E.*

Residential Sector Per Capita Energy Usage (E_R):
    For data collected for the year 2000, this variable ranges from a low value of 41.1 million kilojoules for cites in California (e.g., Long Beach and Fresno) to a high of 80.1 million kilojoules per year for cities in Oklahoma (Oklahoma City and Tulsa). The mean value for the 25 sampled Sunbelt cities is 64.3 million kilojoules. The mean quantity (E_R) is a subset of p*er capita energy usage (E)* representing 15.7% of the mean for *E.*
    Table 9-4 below provides the values of the dependent variables. This table has been compiled using data from the Energy Information Administration (*State Energy Data 2000*) and the U.S. Census Bureau.
    As might be suspected, cities located in states with lower per capita energy use (e.g., Los Angeles, Tucson, and Miami) also expend fewer dollars per capita on energy. Cities located in states with higher per capita energy use (e.g., Houston, Dallas and New Orleans), expend more dollars per capita on energy. Yet another interesting finding is that during 2000 residents in California consumed only half the energy per capita as residents of Texas. Residents of Louisiana consume 3.7 times more energy per capita than residents of Florida. On the other hand, the wide range and diversity of residential energy use is not surprising given the variability of climates and the variety of designs for residential housing.

    The two selected energy use variables are: 1) the number of *homes heated by alternative fuels* or no fuel; and 2) percentage of *single occupant residences*. The values of the energy usage indicators for the study are summarized below.

Homes Heated by Alternative Fuels
    The values of this variable indicate that few households in the selected Sunbelt cities use alternative fuels or no fuel for heating. The observed values found for this variable range from a low of 0.3 percent for the cities of Charlotte, Las Vegas and Tulsa, to a high value of 7.3 percent for Miami. Other cities with higher than typical values include the California cities of Los Angeles (4.4%) and Long Beach (3.3%). The mean value for the 25 sampled Sunbelt cities is 1.2%.

Table 9-4. Per Capita Energy Cost and Usage Data (2000)

Sunbelt Cities in the U.S.	Per Capita Total Energy Costs ($) (Ec)	Per Capita Total Energy Usage (Million Kj) (E)	Per Capita Transportation Sector Energy Use (Million Kj) (ET)	Per Capita Residential Sector Energy Use (Million Kj) (ER)
Los Angeles	2,098	265.5	95.4	41.1
Houston	3,551	586.6	132.1	67.5
Phoenix	2,059	250.1	95.2	58.2
San Diego	2,098	265.5	95.4	41.1
Dallas	3,551	586.6	132.1	67.5
San Antonio	3,551	586.6	103.8	67.5
Jacksonville	1,951	260.5	94.1	65.3
Austin	3,551	586.6	132.1	67.5
Memphis	2,419	375.9	103.8	76.5
Nashville	2,419	375.9	103.8	76.5
El Paso	3,551	586.6	103.8	67.5
Charlotte	2,404	464.2	92.9	74.3
Fort Worth	3,551	586.6	132.1	67.5
Oklahoma City	2,706	428.2	131.2	80.1
Tucson	2,059	250.1	95.2	58.2
New Orleans	4,638	936.6	213.5	76.4
Las Vegas	2,413	334.3	108.8	66.0
Long Beach	2,098	265.5	95.4	41.1
Albuquerque	2,259	327.9	132.8	53.4
Fresno	2,098	265.5	95.4	41.1
Virginia Beach	2,372	343.5	103.4	74.1
Atlanta	2,416	357.2	112.0	75.9
Mesa	2,059	250.1	95.2	58.2
Tulsa	2,706	428.4	131.2	80.1
Miami	1,951	260.5	94.1	65.3

Single Occupant Residences

The values of this variable range from low of 19.2% for El Paso, Texas to a high of 38.5% for Atlanta, Georgia. The mean value for the 25 sampled Sunbelt cities is 28.9%

Table 9-5 provides a summary of the energy use data identifying the percentage for each variable and its corresponding city. Data for Table 9-5 were obtained from the U.S. Census Bureau, *Census 2000*.

The selected transportation variables are: 1) *travel time to work*; 2) *alternative* means of *transport* to place of employment; and 3) the percentage of households that have no or only one *household vehicle*.

Table 9-5. Summary of Energy Use Variables (2000)

Sunbelt Cities In the U.S.	Alternate or no Fuels for Heating %	Single Occupant Households %
Los Angeles	4.4	28.5
Houston	0.9	29.6
Phoenix	0.9	25.4
San Diego	0.5	28.0
Dallas	0.6	32.9
San Antonio	0.6	25.1
Jacksonville	1.2	26.2
Austin	0.5	32.8
Memphis	0.5	30.5
Nashville/Davidson	0.7	33.8
El Paso	0.4	19.2
Charlotte	0.3	29.5
Fort Worth	0.5	28.6
Oklahoma City	0.6	30.7
Tucson	0.9	32.3
New Orleans	0.7	33.2
Las Vegas	0.3	25.0
Long Beach	3.3	29.6
Albuquerque	0.6	30.5
Fresno	1.1	23.3
Virginia Beach	0.8	20.4
Atlanta	0.6	38.5
Mesa	0.5	24.2
Tulsa	0.3	33.9
Miami	7.3	30.4

Source: U.S. Census 2000.

Travel Time to Work

The values for this variable range from a low of 18.6 minutes for Tulsa to a high for Los Angeles of 29.6 minutes. Other cities with low values include Tulsa (18.6 minutes), Albuquerque (20.4 minutes) and Oklahoma City (20.8 minutes). Cities with longer average commuting times include Long Beach (28.7 minutes) and Atlanta (28.3 minutes). The mean value for the 25 sampled Sunbelt cities is 24.5 minutes.

Alternative Transportation

Values for this variable provide a range of 17% which varies from a low of 5.3 % for both Oklahoma City and Fort Worth to a high of 22.3% for Atlanta. In Virginia Beach, 5.5% of the residents either work at home or have alternative means of transportation to work. Other cities with higher percentages include New Orleans (21.6%) and Los Angeles (17.9%). The mean value for the 25 sampled Sunbelt cities is 9.9%.

Household Vehicles

The majority (greater than 50%) of households in 12 of the selected Sunbelt cities have no or only one vehicle. Cities with the largest percentage of households with no or only one vehicle include New Orleans (69.6%), Miami (68.8%) and Atlanta (66.0%). Virginia Beach has by far the lowest percentage of households with no or only one vehicle, approximately 35.7%. Other Sunbelt cities with low values for this variable include Nashville (41.8%) and Houston (44.5%). The mean value for the 25 sampled Sunbelt cities is 51.8%. In addition, the data indicate that the majority of households in 13 of the 25 cities have two or more vehicles.

Table 9-6 (from U.S. Census Bureau data, *Census 2000*) provides a summary of the data relative to the transportation indicators for the sampled Sunbelt cities. Residents of Los Angeles, Atlanta, and Miami experience longer than average travel times to work, are more likely to avail themselves of alternative transportation systems, and have a higher than average percentage of households with no or only one vehicle.

On the other hand, residents of Tucson, Tulsa, and Fresno have shorter commuting times, are less likely to use alternative transportation, and also have a higher than average percentage of households with no or only one vehicle.

The environmental indicators are represented by two variables: 1) the *air quality index (AQI)*; and 2) the number of *days per year that the air quality is considered to be either unhealthy or unhealthy for sensitive groups.*

Air Quality Index (AQI)

Within the group of selected Sunbelt cities, Jacksonville with its service economy and offshore breezes has an unusually low mean AQI of 0, by far the best air quality in the sampled Sunbelt cities. Cities with the next lowest mean air quality indexes are Miami (37) and San Antonio (38). The city of Fresno, located in the country's largest air basin and noted for notoriously poor air quality, has the highest mean value of 80. Cities with the next highest mean air quality indexes are Los Angeles (72) and Long Beach (72). The mean AQI value for the 25 sampled Sunbelt cities was 50.8 for the year 2000.

Table 9-6. Transportation Variables (2000)

Sunbelt City	Mean Travel Time to Work Minutes	Alternative Transport %	0 to 1 Vehicles per Household %
Los Angeles	29.6	17.9	56.8
Houston	27.4	10.5	44.5
Phoenix	26.1	8.8	48.4
San Diego	23.2	11.8	47.2
Dallas	26.9	8.2	56.9
San Antonio	23.8	8.2	49.5
Jacksonville	25.2	6.8	47.5
Austin	22.4	10.4	50.3
Memphis	23.0	6.6	58.3
Nashville/Davidson	23.3	7.2	41.8
El Paso	22.4	6.5	46.1
Charlotte	25.1	7.9	47.0
Fort Worth	24.6	5.3	49.1
Oklahoma City	20.8	5.3	48.3
Tucson	21.6	9.8	56.0
New Orleans	25.7	21.6	69.6
Las Vegas	25.4	9.4	51.0
Long Beach	28.7	10.0	57.6
Albuquerque	20.4	8.0	46.5
Fresno	21.7	7.5	53.0
Virginia Beach	23.9	5.5	35.7
Atlanta	28.3	22.3	66.0
Mesa	25.9	6.6	47.0
Tulsa	18.6	7.5	51.6
Miami	28.1	17.2	68.8

Days Unhealthy and Unhealthy for Sensitive Groups

A number of cities have low scores for this variable: Virginia Beach (0), Tucson (0), Jacksonville (2), Miami (2), San Antonio (3), and Albuquerque (3). For residents of these cities, air quality is typically healthy for all groups of individuals and air quality (excluding point sources of pollution) is not normally of concern to human health. The city of Fresno experienced the most days (132) in 2000 during which the air quality was either unhealthy or unhealthy for sensitive groups. Cities with the next highest values were Los Angeles (88) and Long Beach (88). The mean value for the sampled Sunbelt cities during the year 2000 was 28.4 days. Table 9-7 summarizes available data concerning the selected environmental indicators.

Fresno has the highest scores in both categories. Fresno experiences unhealthy ambient outdoor air conditions more often than any of the other sampled Sunbelt cities. Los Angeles and Long Beach, both part of

Table 9-7. Environmental Variables (2000)

Sunbelt Cities In the U.S.	AQI Index	Days Unhealthy & Unhealthy for sensitive groups
Los Angeles	72	88
Houston	53	53
Phoenix	68	30
San Diego	59	36
Dallas	49	35
San Antonio	38	3
Jacksonville	0	2
Austin	41	12
Memphis	54	28
Nashville/Davidson	54	22
El Paso	54	11
Charlotte	55	31
Fort Worth	46	20
Oklahoma City	42	7
Tucson	45	0
New Orleans	45	20
Las Vegas	55	5
Long Beach	72	88
Albuquerque	53	3
Fresno	80	132
Virginia Beach	40	0
Atlanta	61	60
Mesa	53	13
Tulsa	45	10
Miami	37	2

U.S. Environmental Protection Agency: www.epa.gov/airnow/.

the Los Angeles Metropolitan Statistical Area (MSA), tie for second place as having the next most unhealthy air. Prolonged periods of poor air quality indicates significant urban air pollution, often caused by the inefficient burning of fossil fuels, vehicular urban transportation systems, and extended periods of stagnant air. Such conditions are often experienced in both the Fresno and Los Angles basins.

Relevant demographic indicators are *population density* for the year 2000 and the *relative change in population compared to changes in population density* (1990-2000).

Population Density

Oklahoma City (282 per km² or 730 per mile²), Jacksonville (323 per km^2 or 837 per mile^2), and Nashville (398 per km^2 or 1,031 per mile^2) have the lowest population densities among the Sunbelt cities in this study. Miami (3,889 per km^2 or 10,072 per mile^2), Long Beach (3,316 per km^2 or 8,588 per mile^2), and Los Angeles (2,868 per km^2 or 7,428 per mile^2) have the greatest population densities. The mean population density for the sampled Sunbelt cities is 1,203 persons per km^2 or 3,166 per mile^2.

10-year Relative Change in Population Compared to Density

All of the sampled Sunbelt cities, with the single exception of New Orleans, experienced increases in population growth between 1990 and 2000. Six of the cities experienced declines (negative changes) in density while the remaining 19 cities experienced increases (positive changes) in density during the period. Albuquerque had the largest negative percentage change in population density (-14.7%) while Las Vegas experienced the largest positive change in density (36.2%). There are five cities among the selected Sunbelt cities that have slightly negative values: Tulsa (-0.5%), Jacksonville (-0.2%), Oklahoma City (-0.2%), Nashville (-0.1%) and Los Angeles (-0.1%). In Dallas, New Orleans, Virginia Beach, and Atlanta, the change in population equaled the change in density, meaning that changes in population are moving in tandem with changes in density.

While this indicator is less than or equal to 0 for nine of the selected Sunbelt cities, it is positive for 16 of the cities. In a number of Sunbelt cities, population growth substantially exceeds changes in population density: Las Vegas (49.0%), Charlotte (38.3%), Albuquerque (31.3%), Tucson (23.7%), San Antonio (22.4%), and Austin (18.9%). The mean percentage change in population (less the mean percentage change in population density of the sampled cities) is roughly 9%, indicating that population growth is generally outstripping changes in density.

It is interesting to compare the data for population density, and the data indicating percentage change in population less the percentage change in population density. For example, the three Sunbelt cities in the sample with the greatest population density in 2000, Miami (3,889 per km^2), Long Beach (3,316 per km^2), and Los Angeles (2,868 per km^2), saw only negligible changes in the percentage change in population less percentage change in population density, ranging from a value of -0.1 to a value 0.9. Likewise, the three cities with the lowest population density in 2000, Oklahoma City (282 per km^2), Jacksonville (323 per km^2) and Nashville (398 per km^2), experienced only negligible changes in the percentage change in population less percentage change in population density, ranging from a value of -0.2 to a value -0.1.

This suggests that these cities are relatively elastic (able to expand their physical footprint in step with population growth), or have to large municipal areas and they did not accommodate population growth. As a result, population growth is being accommodated by increases in population density, as vacant land within the cities is being developed. Table 9-8 data were compiled using information from the U.S. Census Bureau (Census 1990 and Census 2000 data) and summarizes the findings concerning the demographic indicators relevant to changes in population and changes in population density.

Table 9-8. Demographic Indicators

Sunbelt Cities In the U.S.	2000 Population Density Per Sq KM	1990-2000 Change in Population Change %	1990-2000 Population Population Density %	1990-2000 Change - Pop. Density %
Los Angeles	2,868	6.0	6.1	-0.1
Houston	1,166	19.8	11.6	8.2
Phoenix	904	34.3	18.8	15.5
San Diego	1,323	10.2	10.0	0.2
Dallas	1,135	18.0	18.0	0.0
San Antonio	1,085	22.3	-0.1	22.4
Jacksonville	323	15.8	16.0	-0.2
Austin	825	41.0	22.1	18.9
Memphis	921	6.5	-2.4	8.9
Nashville/Davidson	398	11.6	11.7	-0.1
El Paso	811	9.4	7.8	1.6
Charlotte	877	36.6	-1.7	38.3
Fort Worth	615	19.5	14.8	4.7
Oklahoma City	282	13.8	14.0	-0.2
Tucson	1,001	20.1	-3.6	23.7
New Orleans	1,062	-2.5	-2.5	0.0
Las Vegas	1,197	85.2	36.2	49.0
Long Beach	3,316	7.5	6.6	0.9
Albuquerque	1,124	16.6	-14.7	31.3
Fresno	1,380	20.7	14.6	6.1
Virginia Beach	611	8.2	8.2	0.0
Atlanta	1,154	5.7	5.7	0.0
Mesa	1,024	37.6	19.5	18.1
Tulsa	773	7.0	7.5	-0.5
Miami	3,889	1.1	0.8	0.3

Equally interesting is that the five Sunbelt cities (San Antonio, Charlotte, Tucson, Las Vegas, and Albuquerque) with the largest difference between percentage population change and percentage change in population density (ranging from 22.4% to 49.0%) fell into a relatively narrow density range. The population density, ranging from 877 to 1,197 persons per km², is less than the mean population density for the sampled Sunbelt cities of 1,203 persons per km². Population growth is being accommodated by increases in land area suggesting that these cities are more elastic and that newly developed areas are less densely populated than existing developments. This offers evidence that Sunbelt cities with faster growing populations tend to consume more land. While population density is declining in San Antonio, Charlotte, Tucson, and Albuquerque, population density is increasing in Las Vegas.

The data indicate that rapid population growth in cities such as Phoenix, Austin, Fort Worth, Fresno, Las Vegas, and Mesa is requiring both the expansion of land area and coincidental increases in population density in order to accommodate growing populations.

Energy Costs and Sector Energy Use

If policies are to be successful, to what ends do they need to be directed? One-way Analysis of Variance Analysis (ANOVA) regression is revealing. If energy is considered to be too costly, can prices be influenced by the actions cities take? The relationships between the dependent variables *per capita energy costs* (E_C) and the selected independent variables are not statistically significant.

The first regression analysis considers the aggregate measure of energy costs as the dependent variable and local measures of energy usage as the independent variables. Table 9-9 provides a summary of the regression statistics using the dependent variable *per capita energy costs* (E_C) across the selected Sunbelt cities (N=25). The independent variables used for this regression are: *alternative or no fuels for heating, single occupancy households, travel time to work, population density, household vehicles, days unhealthy and unhealthy for sensitive groups, population density, air quality index,* and *10 year relative change in population compared to density* (raw values for these variables are provided in Table A2 of the Appendix).

The regression did not control for other variables. The results of this regression provide a coefficient of determination (R^2) of 34.2%. This is proved to be insignificant ($p > .1$) and all independent variables are also insignificant.

The analysis indicates that urban policies designed to influence changes in these variables are likely an ineffective means of achieving a goal of impacting energy costs. If cities are motivated to implement their

Table 9-9. Relationship of Energy Costs (E_C) to Selected Energy Use Indicators

Multiple R	0.5848
R Square	0.3420
Adjusted R Square	-0.0528
Standard Error	742.8051
Observations	25

ANOVA (Analysis of Variance)

	df	SS	MS	F	Sign. F
Regression	9	4301906.849	477989.65	0.87	0.5729
Residual	15	8276390.511	551759.37		
Total	24	12578297.36			

	Coefficients	Standard Error	t Stat	P-value
Intercept	1654.4402	2632.9865	0.6284	0.5392
Alt. Heating Fuel	-431.8183	290.6551	-1.4857	0.1581
Single Households	-20.7018	43.4359	-0.4766	0.6405
Travel Time	30.1319	76.3859	0.3945	0.6988
Alternative Transport	23.9260	56.6176	0.4226	0.6786
Household Vehicles	26.3104	34.2651	0.7678	0.4545
AQI	-5.2454	17.6572	-0.2971	0.7705
Unhealthy Days	-5.8015	7.8802	-0.7362	0.4730
Population Density	0.2957	0.5889	0.5021	0.6229
Pop./Density Change	-15.6538	13.5035	-1.1592	0.2645

policies by the notion that aggregate energy costs can be influenced, then this analysis proves this belief to be flawed. As market logic would suggest, aggregate costs are not in the long term influenced by these types of local or regional policies. Market economists believe that energy costs are more likely subject to the forces of supply and demand for energy products, and are influenced by external conditions on the national and international scene, rather than by local conditions.

Cities, even a set of large cities such as our 25 Sunbelt cities, are unlikely to influence aggregate energy costs by implementing local policies. For example, a city might increase the number of homes that use alternative heating systems, reduce the number of household vehicles and provide alternative means of transportation, yet such actions *will not* influence aggregate energy costs.

According to *Self* (2000:22) such "evidence suggests that the pursuit of micro-efficiency does not add up to macro-welfare and prosperity" and that the problem may lie "not so much in any disjunction between the two levels of the economy as in errors in micro-economic theory itself." The analysis renders additional evidence that the conservative view in the economic model of a "spontaneous self-regulating system" can be "strongly challenged" (*Self* 2000:22). Taking all of the variables into account, it is clear that local and regional efforts are likely ineffective in regard to aggregate measures without some other influence, such as a broadly based (e.g., world scale) concerted initiative.

Analysis of Energy Use Indicators

Moving beyond energy costs, our analysis now considers energy usage variables. An analysis of variables representing the economic sectors that are most likely to be affected by disaggregated measures of energy use is in order. It is known that the variable *per capita energy use* (E) includes transportation sector, residential sector, industrial sector, and commercial sector energy usage components. Cities attempting to create incentives for economic development can be viewed as supporting business development interests, and are less likely to pursue policies that regulate industrial and commercial sector activities. As a result, they may be reluctant to implement policies that focus on energy usage in these sectors.

Alternatively, certain variables used in this analysis are logically related to the transportation and residential sectors. Energy use by these sectors may have a stronger relationship with certain variables than others. The variables considered are *transportation sector per capita energy usage* (E_T), and *residential sector per capita energy usage* (E_R). Both variables are subsets *per capita energy usage* (E).

To further probe for the potential of a relationship between sector energy use and their indicators of measures of energy use, variables that would logically be related to the energy usage of the transportation sector are used. The independent measures that have been selected are those that are often the focus of local transportation policies. A selected dependent variable measuring total transportation sector energy use is tested against selected independent variables that are measures of transportation usage by means of an ANOVA regression. To this end, the dependent variable *transportation sector per capita energy usage* (E_T) is tested against the independent variables *alternative transportation, travel time to work, population density* and *household vehicles,* again using regression and analysis of variance. Results of this analysis for the selected Sunbelt cities are provided in Table 9-10.

This regression does not control for other variables. The results indicated in Table 9-10 provide a coefficient of determination (R^2) of 38.2% which is significant ($p \leq .05$). This suggests that the energy usage by the transportation sector (E_T) can be explained in part by these four variables. Using the data, the equation for E_T for all cities (N=25) becomes:

$$E_T = 88.7877 + (2.2041 \text{ X } alternative\ transportation) - (1.5846 \text{ X } travel\ time\ to\ work) - (0.0159 \text{ X } population\ density) + (1.1666 \text{ X } household\ vehicles)$$

Next, the dependent variable *residential sector per capita energy usage (E_R)* is tested against the selected independent variables that are measures of residential energy usage using an ANOVA regression. The selected variables are *alternative or no fuels for heating, single occupancy households,* and *cooling degree days.* For this case, the independent variable *cooling degree days* is introduced since residential sector energy us-

Table 9-10. Relationship of Transportation Energy (E_T) to Selected Energy Use Indicators

Multiple R	0.6184
R Square	0.3824
Adjusted R Square	0.2589
Standard Error	22.4880
Observations	25

ANOVA

	df	SS	MS	F	Sign. F
Regression	4	6263.12	1565.78	3.0962	0.0389
Residual	20	10114.19	505.71		
Total	24	16377.32			

	Coefficients	Standard Error	t Stat	P-value
Intercept	88.7877	65.3376	1.3589	0.1893
Alternative Transport	2.2041	1.6353	1.3478	0.1928
Travel Time	-1.5846	2.2099	-0.7170	0.4816
Population Density	-0.0159	0.0072	-2.2168	0.0384
Household Vehicles	1.1666	0.9469	1.2321	0.2322

age in Sunbelt cities is accepted to be a function of changes in demand for indoor space cooling due to the varying local conditions. Table 9-11 provides a summary of results.

While the regression analysis does not control for other variables, the results for the residential sector variables in Table 9-11 provides an R^2 of 32.6% which is significant ($p \leq .05$). This regression suggests that the energy usage by the residential sector (E_R) can be explained in part by these three variables. Using the data, the equation for E_R for all cities (N=25) becomes:

E_R = 29.2799 – (3.1149 X *alternative or no fuels for heating, or no fuels for heating*) + (0.9801 X *single occupancy households*) + (.0043 X *cooling degree days*)

These findings are important as they support the notion that focused local policies that are directed toward influencing these variables can be an effective means of influencing both transportation and residential sector energy usage. However, policies directed toward other measures are likely to be less effective. For example, increases in resi-

Table 9-11. Relationship of Residential Energy (E_R) to Selected Energy Use Indicators

Multiple R	0.5708
R Square	0.3259
Adjusted R Square	0.2296
Standard Error	10.9422
Observations	25

ANOVA

	df	SS	MS	F	Sign. F
Regression	3	1215.38	405.13	3.38	0.0373
Residual	21	2514.39	119.73		
Total	24	3729.76			

	Coefficients	Standard Error	t Stat	P-value
Intercept	29.2799	16.1572	1.8122	0.0843
Alt. Heating Fuel	-3.1149	1.4236	-2.1880	0.0401
Single Households	0.9801	0.5008	1.9572	0.0637
CDD	0.0043	0.0023	1.8505	0.0784

dential energy use are associated with increases in the number of single occupancy residences, increases in cooling degree days and declines in the number of alternatively heated residences. These findings offer evidence that local policy efforts can be productive when they are focused and targeted toward very specific ends.

The Relationships of Policies to Variables

Ranking systems often use both qualitative and quantitative variables in assessing sustainability. The next challenge is to compare and interpret the information gained from the qualitative investigation of policies to the analysis of the selected quantitative data. An analysis is developed in this section that provides insight into the relationships of policy adoption in Sunbelt cities. Based on this analysis, an index ranks cities based on energy related indicators of sustainability by using both qualitative and quantitative data. Sunbelt cities are then divided into two distinctive groups based on rates of policy adoption and selected variables. Commonalities and differences between groups are discussed.

The ten selected qualitative policies discussed earlier show that many of the sampled cities have policies in place that impact energy use and sustainability. Recall that the selected qualitative policies considered previously include: Utility rebate programs, city operated energy efficiency programs, Utility and government program support, Sustainable development policies, Use of high occupancy vehicle lanes, ICLEI membership, Clean Cities designation, Energy Star™ partnership, Curbside recycling programs, and Brownfield redevelopment programs.

It is possible to rank the cities by the number of categories of policies that cities choose to employ. Recall that the cities in our sample of 25 Sunbelt cities have adopted from 1 to all 10 of these policies. Las Vegas, as an example, utilizes seven of these policies. The number of policies employed by each city can be called its *rate of policy adoption*.

While the literature identified aspects of sustainable cities not directly related to energy or environmental factors, in this study, cities with policies in place that respond to all categories of policies can be considered as promoting improved urban sustainability. On the other hand, cities with no or few policies in place either lack a sustainability agenda or may not be earnest in their quest to become sustainable.

Quantitative rankings are similarly achieved. The variables for each of the 25 cities were ranked using the raw values obtained during the research. Table A2 (see Appendix), provides a listing of the raw values for each variable for all of the 25 selected cities. An index provides a possible range of scores for each variable from .0 to 1.0. The city with the lowest quantitative ranking for each variable was assigned an index score of 0.0 for that variable, and the city with highest score for the variable was assigned an index score of 1.0.

Index scores for the other intervening cities were proportionately assigned based on the values of their variables. With a total of nine variables, the lowest index total theoretically achievable was 0.0, and the maximum index total that was potentially achievable was 9.0.[8] All variables are equally weighted in this index. The total score provides a newly created composite variable.

Table A3 (see Appendix) provides a summary of the calculated index values for each of the selected variables including energy costs. All variables in the index are equally weighted. An index value combining all remaining variables yields the variable *index scores* which is comprised of the sum of the index values for the following nine variables: *Per capita energy use (E), Homes heated by alternative fuels, Single occupant households, Travel time to work, Alternative transportation, Household vehicles, Air quality index (AQI), Days unhealthy, and Unhealthy for sensitive groups* and *Population density*.

Using this procedure, the summed values indicate that Miami has the highest index total of 6.74 and that Nashville has the lowest index total of 3.17. The mean index value of the 25 selected cities is 4.0.

Using the rate of policy adoption or total *number of policies* (from 1-10) as the independent variable, and the computed *index scores* to serve as the composite dependent variable measuring energy use, a regression without controlling for other influences provides an R^2 of 14.9%. The regression also indicates that the relationship between the *number of policies* employed and the composite total values of the indexed variables is significant ($p \leq .1$) with $p = .057$. Table 9-12 details the regression statistics.

These regression statistics indicate that policies developed to implement energy reductions and enhance sustainability are associated with higher total index measures. They show that the composite measure along the selected dimensions has a weak but statistically significant relationship to the number of policies selected by Sunbelt cities. In addition, this model provides empirical evidence that there are various means of reducing energy usage and achieving the sustainability goals of cities.

In an effort to create a more functional and credible model, various combinations of the variables available were probed. The five dependent

Table 9-12. Relationship of Policy Adoption Rates to Nine Indexed Measures of Energy Use

Regression Statistics	
Multiple R	0.3863
R Square	0.1492
Adjusted R Square	0.1122
Standard Error	2.3505
Observations	25

ANOVA

	df	SS	MS	F	Significance F
Regression	1	22.29	22.29	4.03	0.0565
Residual	23	127.07	5.52		
Total	24	149.36			

	Coefficients	Standard Error	t Stat	P-value
Intercept	0.4721	2.7135	0.1740	0.8634
X Variable 1	1.3401	0.6672	2.0086	0.0565

variables *alternative transportation, population density, homes heated by alternative fuels, days unhealthy and unhealthy for sensitive groups* and *household vehicles* resulted in the highest coefficient of determination that yielded a significant result. These five variables provide a new, less cumbersome index score. This revised index provides a possible range of scores for each variable from 0.0 to 1.0, with a total possible range from a low score 0.0 to a high score of 5.0. All variables are equally weighted in this index to avoid any dominating influence by one or a set of variables. Table A4 (see Appendix), provides a summary of the calculated index values for each variable and a new total index score for each of the 25 selected Sunbelt cities.

Using the total *number of policies* (ranging from 1-10) as the independent variable measuring policy adoption with the individual indexed values for the five variables (*alternative transportation, population density, homes heated by alternative fuels, days unhealthy, and unhealthy for sensitive groups,* and *household vehicles*) as variables that are indicators of energy use, this regression results in an R^2 of 40.3%. This suggests that the relationship between the independent variable and the values of the five remaining indexed variables is stronger than in the model offered in Table 9-12 due to the higher coefficient of determination. The result is also significant ($p \leq .1$) resulting in $p = .062$. The results of this regression are displayed in Table 9-13.

The formula for *number of policies* that results is provided below:

number of policies = 3.5596 + (4.2471 X *alternative transportation*) + (8.6678 X *population density*) – (3.6771 X *homes heated by alternative fuels*) + (1.3990 X *days unhealthy and unhealthy for sensitive groups*) – (3.6265 X *household vehicles*)

The strength of the statistical relationship among these variables is valuable. These quantitative variables can now be used as a tool to devise a sustainability index.

Ranking Sunbelt Cities

The major cities of the Sunbelt are growing in size and population. They are individual and dynamic, yet are experimenting with an assortment of policy agendas. This study assesses policies that are being employed by Sunbelt cities. A few of the cities (e.g., Miami, San Diego, and

Table 9-13. Relationship of Policy Adoption Rates to Five Indexed Measures of Energy Use

Multiple R	0.6350
R Square	0.4032
Adjusted R Square	0.2461
Standard Error	2.1660
Observations	25

ANOVA

	df	SS	MS	F	Significance F
Regression	5	60.22	12.04	2.57	0.0616
Residual	19	89.14	4.69		
Total	24	149.36			

	Coefficients	Standard Error	t Stat	P-value
Intercept	3.5596	2.2430	1.5870	0.1290
Alternative Transport	4.2471	2.4539	1.7307	0.0997
Population Density	8.6778	4.8305	1.7964	0.0883
Alt. Heating Fuel	-3.6771	4.6488	-0.7910	0.4387
Unhealthy Days	1.3990	2.0720	0.6752	0.5077
Household Vehicles	-3.6265	3.0200	-1.2008	0.2446

Los Angeles) when compared to others in the sample, are seriously concerned about energy and sustainability. They are aggressively pursuing multifaceted policy agendas.

The majority of the selected Sunbelt cities are less concerned about energy and sustainability issues. Their rates of policy adoption are more variable. In these cases, a city might choose to direct its policies more toward transportation solutions than changes in residential designs. Many cities work closely with their local utilities while other cities have a less cordial or even adversarial relationship with their local utilities. Some cities pursue memberships in organizations with common goals while others are less participatory.

There are also Sunbelt cities that have only recently adopted energy and sustainability policies. Additional time may be required before results can be observed and measured. For others cities (e.g., Virginia Beach, Oklahoma City, and Charlotte), concerns about energy and sustainability do not appear to have become part of the public policy agenda. It is also possible that other effective policy agendas are being pursued that respond to the issues of energy and sustainability that are not considered by this study.

Earlier a functional typology was proposed in which various types of policies could be categorized as *energy management programs, transportation system policies, organizational memberships or those directly affecting the urban environment*. These categories offer a readily identifiable set of functional indicators. This shows how selected policies focus on certain sustainability and energy concerns. The utility of the selected variables can be further demonstrated.

By reviewing the qualitative data, it is clear that policies and programs are in place and being used by Sunbelt cities that manage and influence urban energy consumption. Cities indeed have choices. While a variety of initiatives are being implemented, some are more widely adopted than others. Many are intended to manage or reduce energy use in addition to meeting other related objectives. Table 9-14 summarizes the extent of application of the categories of policies used by the select Sunbelt cities.

As Table 9-14 indicates, curbside recycling programs are nearly ubiquitous while membership is the ICLEI is comparatively rare. In Sunbelt cities, programs which provide alternative energy options, expand high occupancy vehicle lanes, and promote brownfield redevelopment are expanding. Meanwhile, utility sponsored rebate programs are in de-

Table 9-14. Policies Used by Sunbelt Cities

Types of Policies N = 25	Cities with Policy	Cities Lacking Policy
Local government and utility policies		
Utility rebate programs	44%	56%
City-operated energy efficiency programs	76%	24%
Local energy conservation programs	52%	48%
Local policy		
Sustainable development as a policy goal	56%	44%
High occupancy vehicle lanes	48%	52%
Participation in membership-based organizations		
International Council for		
Local Environmental Initiatives	24%	76%
Clean Cities Program	76%	24%
Energy Star Partner	40%	60%
Environmental policies		
Curbside recycling program	96%	4%
Brownfield redevelopment program	76%	24%

cline. Some policies, including city operated energy efficiency programs, Energy Star partnerships, and utility rebate programs were found to be more common in cities with larger populations than in the smaller cities. It is likely that more populous cities have larger staffs, expertise, and resources to implement and support policies of these types.

The Role of Energy

Energy efficiency, energy conservation and alternative energy programs are likely to enhance urban sustainability. To assess this possibility, a set of *dependent variables* was considered in relation to selected sets of *independent variables*. It was determined that there was no significant, direct relationship between the values of the dependent variables that measured the total per capita costs of energy. The analysis implies that the local policies of the largest Sunbelt cities are having no discernable or measurable impact on per capita aggregate measures of energy use (E) and energy costs (Ec).

For these two measures of energy use, this hypothesis is proved false in regard to the selected set of independent variables and in all sampled Sunbelt cities. Why is this? It can be theorized that while energy efficiency is having targeted and local success, the impact is limited and is being more than offset by increases in energy costs and usage resulting from by other forces in the economy.

Variables measuring disaggregated energy use for the transportation and residential sectors were each tested against subsets of the selected independent variables. Transportation energy usage was determined to be in part a function of the availability and use of alternative means of transportation to work, the amount of travel time to work, population density and the number of household vehicles. Residential energy use was found to be related to the percentage of homes that used alternative fuels for heating or did not require heating, the number of single occupancy households, and the annual average number of cooling degree days. These relationships between the disaggregated dependent variables and the selected independent variables were shown to be statistically significant. For the two disaggregated measures of energy use, this hypothesis is proved true when considered in relationship to all sampled Sunbelt cities using the selected independent variables.

Developing a Sustainability Index

A combined index can be devised from the qualitative policy scores and the composite index score using the five quantitative variables *alternative transportation, population density, homes heated by alternative fuels, days unhealthy and unhealthy for sensitive groups,* and *household vehicles.* This new index has a possible range of 0.0 to 10.0 and will be referred to as the "sustainability score." This is accomplished by multiplying the policy score by 50% (or dividing by 2) and adding the score to the combined index value of the five selected variables. The resulting sustainability score provides equal weight to both the policies in place and the results of the indexed variables. The following formula defines how a sustainability score for each city can be calculated using this methodology.

Sustainability Score = (*number of policies* X .5) + index value of *alternative transportation* + index value of *population density* + index value of *homes heated by alternative fuels* + index value of *days unhealthy and unhealthy for sensitive groups* + index value of *household vehicles*

As shown in the formula, sustainability score values represent the sum of the weighted policy score and the total of the index scores of the significant variables. Using this formula, the cities can be ranked in order based on their calculated sustainability scores.[9]

The sustainability score for the selected Sunbelt cities ranges from a high of 9.16 for Miami to a low of 1.91 for Charlotte. Other cities that accompany Miami in the upper tier include Los Angeles, San Diego, Atlanta, and Tucson. The mean sustainability score for the 25 cities is 4.9. Phoenix, with a score of 5.11, can be considered a typical Sunbelt city based on the sustainability score. Las Vegas has a sustainability score of 5.4 which is slightly higher than the mean of the sample. It too can be considered as a typical city among those sampled (both Phoenix and Las Vegas are currently dealing with rapid population growth and developmental issues).

The values of the index range for the 25 cities spans 72.5% of the available range. The mean weighted policy score of the 25 cities is 3.0 while the mean index score (using significant variables) is 1.9. This indicates that rates of policy adoption exceed the indexed values.

There are anomalies within the rankings. Miami and New Orleans are examples of Sunbelt cities with index scores that are higher than their corresponding policy adoption rates. This suggests that their policies have been relatively effective. Miami has the highest population density of the sampled cities, has fairly high rates of public transport use, and the highest percentage of homes that are alternatively heated or require no heat. New Orleans has a low rate of policy adoption and the second highest index score among the selected Sunbelt cities. The high index score for New Orleans is due to high rates of alternative transport use and a low number of days with unhealthy air. Among the sampled Sunbelt cities, New Orleans has the highest rate of households with no or only one vehicle.

On the other hand, San Diego is an example of a Sunbelt city with a high rate of policy adoption but a correspondingly low index score. This is due in part to low rates of alternatively heated homes (in part due to climate). San Diego is a commuter city and has comparatively low rates of households with one or fewer vehicles.

Table 9-15 provides a summary of the policy scores, the summed index of significant variables scores for the five selected variables, and the sustainability scores.

Despite the anomalies noted earlier, there are consistencies as well. Cities that have utility rebate programs, city-operated energy efficiency programs, and local programs to support energy conservation (Los Angeles, San Diego, Austin, Las Vegas, and Houston) have an average sustainability score of 7.3, outpacing the average of the other twenty whose average sustainability score was 4.8. The four cities (New Orleans, Virginia Beach, Charlotte, and Oklahoma City) that lack such programs

Table 9-15. Sustainability Ranking of Sunbelt Cities

U.S. City	Policy Adoption Rate (A)	Weighted Policy Score (A*.5)	Selected Index Measures* (B)	Sustainability Score (A*.5)+B
Miami	9	4.5	4.66	9.16
Los Angeles	10	5.0	3.00	8.00
San Diego	10	5.0	1.77	6.77
Atlanta	8	4.0	2.72	6.72
Tucson	8	4.0	2.15	6.15
Long Beach	7	3.5	2.53	6.03
Austin	8	4.0	1.82	5.82
Houston	8	4.0	1.51	5.51
Las Vegas	7	3.5	1.90	5.40
San Antonio	7	3.5	1.82	5.32
Dallas	7	3.5	1.81	5.31
Albuquerque	7	3.5	1.73	5.23
Phoenix	7	3.5	1.61	5.11
New Orleans	3	1.5	3.09	4.59
Tulsa	5	2.5	1.66	4.16
Jacksonville	5	2.5	1.56	4.06
Fort Worth	5	2.5	1.37	3.87
Memphis	4	2.0	1.75	3.75
Mesa	4	2.0	1.55	3.55
Nashville	4	2.0	1.21	3.21
Fresno	4	2.0	1.05	3.05
El Paso	3	1.5	1.46	2.96
Virginia Beach	3	1.5	1.17	2.67
Oklahoma City	2	1.0	1.36	2.36
Charlotte	1	0.5	1.41	1.91

* See Appendix Table A4 for index values of significant indicators and measures of energy used for the sustainability ranking of Sunbelt cities.

have a much lower average sustainability score (2.9).

Cities that participate in related organizational activities are more successful at achieving measurable results than those that do not when measured against the selected indexed variables. Recall that the organizational activities selected were ICLEI membership, Clean Cities designation, and engagement of the Energy Star™ program. Those cities participating in all three organizations (Los Angeles, San Diego, Miami, Atlanta, and Tucson) scored higher on both the average sustainability score (7.4) and the average index score (2.9). Cities that did not participate in any of these organizations (Jacksonville, Memphis, Nashville, New Orleans, Mesa, and Charlotte) had lower average sustainability scores (3.4) and lower average index scores (1.8). Greater participation in organizational membership programs appears to be associated with greater success at implementing these policies.

Comparing Characteristics of Groups of Cities

To further understand the differences between cities, the sampling of 25 Sunbelt cities are grouped into two exclusive categories based on their sustainability score. Group A includes thirteen (13) cities with a sustainability score > 5.0. As indicated by Table 9-15, this group includes Miami, Los Angeles, San Diego, Atlanta, Tucson, Long Beach, Austin, Houston, Las Vegas, San Antonio, Dallas, Albuquerque, and Phoenix. Group B includes the twelve (12) remaining cities with a sustainability score < 5.0. Group B includes New Orleans, Tulsa, Fort Worth, Memphis, Jacksonville, Mesa, Nashville, Fresno, El Paso, Virginia Beach, Oklahoma City, and Charlotte.

Cities in Group A are designated as Type A cities while cities in Group B are designated as Type B cities. Table 9-16 provides a compiled set of average values for the data comparing Type A cities to Type B cities. Type A cities are characterized as having higher sustainability scores than Type B cities which have characteristically lower sustainability scores.

Both Type A and Type B cities have similar rates of change in population density. However, this is where the commonalities end. The data indicate that when the groups are compared to each other, Type A Sunbelt cities have a set of characteristics that are fundamentally different than Type B cities. Both the climate and geography differ. The summer climate of a typical Type A city is warmer and the winters are sunnier than Type B cities. Cities in California are more likely to be Type A cities while those in Oklahoma and Tennessee are more likely to be Type B cities.

Considering demographic measures, Type A cities are more than twice as large as Type B cities in average population. As a result, Type A cities are more likely to be regional centers of commerce and politically more powerful. Type A cities are more than twice as likely to be their state capitals as Type B cities. Rates of population growth for Type A cities are roughly 50% higher than Type B cities. The population density of a typical Type A city is 2.1 times larger than the typical Type B city. Suburban development is outpacing increases in population density in both types of cities. However, suburbanization is occurring at a faster rate in

Table 9-16. Characteristics of Groups of Cities Ranked by the Sustainability Index

Indicator or Variable	Unit of Measure	Type A	Type B
Number of cities	Each	13	12
Policies adopted	range = 0-10	7.9	3.6
Index score	range = 0-5	2.2	1.6
Sustainability score	range = 0-10	6.2	3.3
Population	Each	1,064,376	518,994
Population density	per square Km	1,614	756
1990-2000 population	per cent change	22.3	15.5
1990-2000 population density	per cent change	9.0	9.0
Pop. change—pop. density	per cent	13.1	6.4
Energy costs	$s per capita	2,589	2,740
Total energy use	Kj per capita	378.7	441.8
Residential energy use	Kj per capita	59.3	69.2
Transportation energy use	Kj per capita	111.8	119.2
Alternative heating fuels	per cent	1.6	0.6
Single occupant households	per cent	29.9	27.7
Cooling degree days	base = 65	2,488	2,273
Solar radiation in January	Langley	280	235
Travel time to work	Minutes	25.5	23.4
Use of alternative transportation	per cent	11.7	7.9
Vehicles per household (0-1)	per cent	53.8	49.6
Air quality	Index value	54.1	47.3
Days with unhealthy air	range = 0-365	31.9	24.7

Sources: United States Census Bureau (1999, 2000); Energy Information Administration (2000); United States Environmental Protection Agency http://www.epa.gov/airnow/index.html (2004); Mazria, E. (1979) *The passive solar energy book*; National Oceanic and Atmospheric Administration http://www.cpc.noaa.gov/products/analysis_monitoring/cdus/degree_days/.

Type A cities than in Type B cities.

Type A cities not only have larger populations and faster population growth rates, they also have higher rates of policy adoption. While levels of commitment to sustainability vary, these cities are not moving from crisis to crisis in their approaches. Instead, they appear to be developing and implementing broadly based and flexible policy strategies.

Type A cities have broader administrative infrastructure in place.

This allows them to implement a broader range of policies than deployed by Type B cities. In regard to the ten policies considered in this study, the larger cities have a greater propensity to adopt policies related to sustainability. Policies are being adopted incrementally. In fact, Type A cities have adopted an average of 7.9 policies while Type B cities have adopted an average of only 3.3 policies.

Type A cities are more likely to have locally sponsored energy saving programs, and have their utilities engaged in energy conservation programs in city-owned buildings. By a wide margin, Type A cities are more likely to have established sustainability goals and be Energy Star™ partners. Interestingly, all cities in the sample that are ICLEI members are Type A cities. All cities in the sample that have not adopted Clean Cities plans or brownfield redevelopment programs are Type B cities. Type B cities are more selective in the policies they choose and are more likely to implement those that require fewer resources to implement. The only city that has not implemented curbside recycling (El Paso) is a Type B city. Charlotte, Virginia Beach and Oklahoma City are Type B cities. At the time of this study, none had documented utility sponsored rebate programs, established policies for city buildings, nor developed locally supported energy conservation programs.

Though overall rates are very low for all Sunbelt cities, homes in Type A cities are 2.5 times more likely to be heated by alternative fuels. They have a slightly higher percentage of single occupancy households. Residents of Type A cities are 48% more likely to use alternative forms of transportation to get to work, or work at home, than their Type B city counterparts. Type A cities have fewer households with two or more vehicles. Type A cities, such as Las Vegas, San Diego, and Atlanta are more likely than Type B cities to have light rail or monorail services. However, residents of Type A cities spend a couple of minutes longer on the average during the daily commute to work. Despite their higher use of public transport, Type A cities have more problems maintaining air quality. They experience an average of seven additional days annually when the air quality can be categorized as unhealthy or unhealthy for sensitive groups, than Type B cities.

Residents of Type B cities spend almost 6% more dollars annually on energy than residents of Type A cities. Germane to this study are the rates of actual energy consumption. *Type A cities are simply more energy efficient.* When measured on a per capita basis, Type B cities consume 17% more total energy, 7% more transportation sector energy, and almost 17% more residential sector energy than Type A cities.

In regard to rates of policy adoption only four cities, Miami (Type A), Oklahoma City (Type B), Charlotte (Type B), and New Orleans (Type B), have policy adoption rates that are less than their index scores. The rest of the cities have policy scores that are greater than total index scores.

Type A cities Miami and Los Angeles are leading the pack based on their sustainability scores. Their high rates of policy adoption are coupled with high index scores. These two cities are "talking the talk" and "walking the walk." These cities are investing money, time and administrative equity in an effort to bridge the gap between policies and their results. These cities are arguably focused not simply on the policies themselves but also on how policies can be successfully implemented to achieve the broadest impact.

Other Type A cities, like Austin, Houston, and San Diego have high rates of policy adoption along with correspondingly low index scores when significant variables are considered. It is unlikely that the policies of these cities are succeeding in decreasing energy usage and improving sustainability. One interpretation is that these cities may indeed have policies in place but are "talking the talk" yet are not "walking the walk." Some of the policies in these cities appear to be nominal, possibly inadequately funded but certainly not providing acceptable results. It is also possible that the adopted policies for these cities have not yet had adequate time to make a noticeable impact when measured by the narrowly selected policies and indicators used in this study. The Type B version of this approach is offered by cities like Jacksonville and Fort Worth who publicly "talk the talk." Their rates of policy adoption within their group are above average, yet they have correspondingly below average index scores. Perhaps there are other measures against which their efforts can be measured. Cities like Tulsa and Mesa are actually decreasing actual aggregate energy use.

Type B cities such as Virginia Beach, Oklahoma City and Charlotte have low rates of policy adoption and correspondingly low index scores. These cities are neither "talking the talk" nor "walking the walk." They are out of step with the crowd, heading in another direction, pursuing other agendas, or perhaps focused on policies and solutions beyond the scope of this study.

Theory of Divergence

It also possible to hypothesize that for some of the sampled cities, policies supporting sustainability and measures of urban energy consumption are on divergent paths. A divergence would be indicated by high rates of policy adoption with the intent of reducing energy consumption combined with observable increases in aggregate energy consumption. If a divergence exists, are the identified policies and programs sufficient to find empirical evidence of reductions in aggregate measures of energy use?

Considering the two types of cities, one with high policy adoption rates (Type A) and the other with much lower policy adoption rates (Type B), *do the data find evidence of a divergence?* The estimates for per capita city energy use can be calculated by using the data that included city population, state population, and per capita energy use for the years 1990 and 2000. The formulas used for estimating values for per capita city energy use and total city energy use are as follows:

Per capita city energy use = Total city population ÷ Total population of state X per capita state energy use

Total city energy use = Per capita city energy use X Total city population

Table A5 (see Appendix) provides detailed estimates of per capita energy usage for all cities in the years 1990 and 2000. The tabular results based on this formula indicate:

- In 2000, the average estimated energy use of a typical Type A city (N=13) is 432,400 billion kilojoules per year while the Type B cities (N=12) use an average of 224,500 billion kilojoules per year.

- From 1990 to 2000, per capita energy usage declined in 92% of the Type A cities and in 75% of the Type B cities. For this period, per capita energy usage has declined by an average of 2.4% in Type A cities and by an average of 4.3% in Type B cities.

- From 1990 to 2000, total city energy usage increased by an average of 53,000 billion kilojoules for Type A cities and increased by almost 20,000 billion kilojoules for Type B cities. From 1990 to 2000, total city energy use has increased by 20.4% in Type A cities and by 10.0% in Type B cities.

Cities adopt more policies, and the policies are having an impact as evidenced by the declines in per capita energy consumption. *While per capita energy use is declining, aggregate energy use is increasing.* Due to improvements in equipment, processes and infrastructure, energy efficiency based on per capita measures has improved in 21 of the 25 (84%) cities in our sample. The reductions in energy use are of a magnitude that impact sustainability by decreasing energy use.

While other factors are not being controlled, the evidence is incontrovertible—*a divergence exists*. The broader policies that cities employ are having an impact. If policies were not in place, energy use on a per capita basis would continue to be increasing at a greater pace. Cities can impact sustainability based on per capita measures. They are likely most successful when employing policies that have a direct impact on disaggregated measures of energy use while only indirectly impacting aggregate measures.

Both Type A and Type B cities decreased energy use by fractional annual rates from 1990 to 2000, with Type B cities decreasing energy use more rapidly. Of the 25 sampled cities, all but four (Tucson, New Orleans, Virginia Beach, and Charlotte) experienced declines in per capita energy usage from 1990 to 2000. Tucson is the only Type A city among the four with increasing per capita energy usage.

However, despite the broader deployment of policies designed to impact energy use, total energy use continues to increase substantially. These data firmly support a hypothesis that proposes that there is a divergence between policies and aggregate measures of energy use. In fact, Type A cities, the group with the highest rates of policy adoption, have greater average increases in energy usage than the Type B cities which tend to have lower rates of policy adoption. There are examples within the groups that are illustrative. Tucson, a Type A city with a high rate of policy adoption, has the largest rate of increase in estimated per capita energy usage from 1990 to 2000, and the second highest increase in total energy use among the 25 selected cities. It might be inferred that Tucson is among those cities that has perceived a set of problems and is rapidly putting programs into place in an effort to resolve them.

Among the 25 cities only two appear to have declining actual energy use. Both are Type B cities. Mesa has an average rate of policy adoption among Type B cities yet achieves the largest percentage decline in per capita energy use of the sampled cities. Tulsa has an above-average rate of policy adoption among Type B cities and achieves a slight percentage decline in per capita energy use.

If average per capita energy use is declining and population densification is occurring among the selected Sunbelt cities, then why does total energy use continue to increase? Are there any additional relationships and results that can be gleaned from the data? Development subsidies to provide incentives for suburbanization are a set of policies that have been touted as one of the solutions to population growth for cities. Populations are accommodated by development on the urban periphery. How are these policies related to the resulting energy use of cities? First, one must accept the notion that most all Sunbelt cities in this study have experienced new suburban development in some form between 1990 and 2000. Increases in population density can be assumed to rely primarily on existing infrastructure. When suburbanization occurs, new construction at the perimeter of the city is typically at lower population densities. While suburbs vary in density, let's assume that on the average, the patterns of population densities in new developments across the sample of Sunbelt cities are likely to be relatively constant. This notion is supported by Newman and Kenworthy (1999:287) who believe that "These identical, mechanical suburbs are becoming universal" and become a sprawling "monotonous megalopolis." If suburbs are nearly identical across cities, the resulting total energy usage due to suburbanization is likely to be of a similar magnitude.

Considering the varying influences of climate on energy, buildings in some cities may require more energy for heating and less for cooling while others may require less energy for heating and more for cooling. As a result, energy usage not accounted for by changes in density will result from factors that are related to new development (for ease of discussion, these changes will be called "suburbanization"). If this is true, average changes in suburban energy use will be relatively constant across cities regardless of whether or not they are Type A city or Type B. Recall that the grouping of these cities into types resulted from rates of policy adoption and indexed values of selected variables and not from unexplored variables such as lot size, size of new residences or measures of infrastructure improvements. The following formulas test this relationship for the period from 1990 to 2000:

α = change in Type A city energy use due to suburbanization as a proportion of total change in energy use from 1990 to 2000

ϕ = change in Type B city energy use due to suburbanization as a proportion of total change in energy use from 1990 to 2000

n_α = 13, sample size for Type A cities

n_ϕ = 12, sample size for Type B cities

$$\frac{n_\alpha)]}{\alpha} = \frac{[(\Sigma\ \%\Delta\ \text{Type A population} \div n_\alpha) - (\Sigma\ \%\Delta \text{Type A population density} \div}{\Sigma\ \%\Delta\ \text{Type A Energy Use} \div n_\alpha}$$

α = $(22.29\% - 9.04\%) \div 2\text{-.}40\%$

α = .6495

$$\phi = \frac{[(\Sigma\ \%\Delta\ \text{Type B population} \div n_\phi) - (\Sigma\ \%\Delta\ \text{Type B population density} \div n_\phi)]}{\Sigma\ \%\ \Delta\ \text{Type B Energy Use} \div n_\phi}$$

ϕ = $(15.52\% - 8.96\%) \div 10.03\%$

ϕ = .6540

$\therefore \alpha \cong \phi$

The fact that the calculated ratios for both groups of cities are near-ly equal is important. Regardless of circumstances, suburbanization is a form of development that has a similar impact on urban energy use regardless of policies, measures of energy use, or city type. These equations estimate the relative contribution of suburbanization to increases in the total energy usage of Sunbelt cities. As indicated from the formulas, suburban development is responsible for a substantial portion of the increasing energy use in cities, more than offsetting the declines in energy use resulting from practices, policies and programs that have tended to reduce energy usage.

This analysis finds a divergence between policy adoption rates and increases in energy usage. The discovery and identification of this divergence is an important finding. It identifies suburbanization to be the probable cause of the divergence. The increases in energy use, due to the processes involved in suburbanization, are having a dampening effect on the sustainability of cities. It clearly contributes to increases in the energy use of cities.

Summary of the Analysis

In this analysis, the dependent variable that measures aggregate energy costs (E_C) was considered against an array of variables across a selection of 25 Sunbelt cities. Each of the measures of energy use were defined and described at length. The values of these variables were provided in tables. A series of analyses were performed using ordinary least-squares regression and analysis of variance. Raw values for variables were used in each regression. The results of the regressions on per capita energy costs (E_C) indicated that no statistically significant relationships were identified between this aggregate measure and the selected independent variables. The results indicate that local policies that attempt to influence aggregate energy costs are likely to be ineffective.

The regressions for the dependent variables transportation sector energy use (E_T) and residential sector energy use (E_R) are more enlightening. The measures of transportation sector energy use show a significant relationship to variables such as the number of vehicles per household, the percentage of those using alternative transportation to get to work, travel time to work, and population density. The analysis suggests that ways of reducing energy usage include reducing the number of vehicles per household, supporting programs that provide alternative means of getting to and from work, impacting travel time to work, and increasing population density. These solutions may be feasible if local planning and transportation system policies are modified.

The measures of residential sector energy usage were found to have a stronger and more significant relationship to the variables of alternative or no fuels for heating, percentage of single occupant households, and cooling degree days. The analysis suggests that ways to reduce residential energy use include increasing the number of homes using alternative fuels for heat and decreasing the number of single occupant households. This analysis supports the concept of decreasing the impact of extreme climates on residences. Possibilities include locating residences in areas with less severe climates or providing improved design of residences to control for temperature extremes (e.g., providing controls or building envelope improvements such as insulation), thus reducing residential energy usage. This means that cities need to be more selective in choosing site locations for their facilities and more creative in how their buildings are planned and designed.

The results of the analysis were interpreted as an indication that local policies directed toward influencing transportation or residential sector energy usage are likely to be effective and yield fruitful results if they are directed toward selected measures. It can be theorized that while energy efficiency is having targeted and local success, the impact is limited and is offset by increases in energy costs and usage resulting from other forces in the economy.

The analysis compared the information gained from the qualitative investigation of policies to the analysis of the quantitative data. This approach provided insight into the relationships of policy adoption in Sunbelt cities. For example, it was found that there exists a statistically significant relationship between rates of policy adoption and indexed measures of energy use. A statistical analysis was used to select five quantitative variables. A sustainability index was devised to rank cities based on energy-related indicators of sustainability by using both qualitative and quantitative data.

Sunbelt cities were next divided into two groups based on rates of policy adoption and selected variables. Commonalities and differences between groups were discussed. Cities were designated as either Type A or Type B cities. The groups of cities were determined to have fundamental differences. Type A cities have larger populations, greater population densities and higher rates of policy adoption than Type B cities. Both Type A and Type B cities are reducing their per capita rates of energy consumption. Per capita rates of energy use are declining more rapidly in Type B cities—despite the fact that Type B cities have lower rates of policy adoption. Regardless, total urban energy consumption continues to increase in both Type A and Type B cities. Energy use is increasing more rapidly in Type A cities than in Type B cities.

An analysis of the data comparing Type A cities to Type B cities revealed that there is a common value that represents the increase in energy use due to new development, which was labeled suburbanization. Policies that promote suburban development were found to be offsetting energy reductions achieved by policies that cities have deployed to reduce urban energy use.

PART C: WHAT CAN CITIES DO?

Part C expands on the information gained from the analyses in Part A and Part B and explores the options cities have to achieve sustainability.

What can cities do in an effort to achieve sustainability? What programs are workable and which ones are doomed to fail? A description of the kinds of policies a city might choose to pursue in order to reduce energy use and advance sustainability is in order. To demonstrate the value of developing sustainability indexes, this analysis focused on cities in the U.S. Sunbelt, a roughly defined region of the U.S.

While using less energy is clearly a part of the sustainability solution, most policies used by Sunbelt cities have achieved limited and mixed results in regard to actually reducing total energy usage. For most Sunbelt cities, aggregate energy use continues to increase. This may be due to the increasing pressures from population growth and development, the economics of the marketplace, or to other causes. However, this analysis suggests that other development policies, such as those that support suburbanization, more than offset the impact of policies designed to decrease aggregate energy consumption.

It is unlikely that cities, either individually or in concert, can significantly influence the cost of energy. Efforts to this end are likely to misdirect resources and need to be reoriented. As a result, engaging in targeted energy sector policies that attempt to influence demand for energy seem to be more appropriate for cities and more likely to succeed. The arcane economic relationship between energy usage and costs, suggests that market interventions designed to increase energy costs may be among the more effective mechanisms to reduce aggregate energy usage. Coupling energy taxes at the local level with carefully managed incentive programs to reduce the use of fossil based fuels might be effective. Such programs might involve a tax on gasoline with the proceeds used to improve public transportation systems or a tax on natural gas with the proceeds applied to subsidizing alternatively fueled heating systems for residences. For most cities, the political reality is that tax policies are rarely palpable among voting constituencies. Many existing policies are ossified, dysfunctional, costly, and may present national security implications. The reality is that once infrastructure is in place, maintaining the status quo often becomes a rigidly pursued and defended default strategy. Interventions tend to be postponed.

In many cities, it is likely that urban policies are often undermining intended objectives due to the divergence between increasing energy use and urban sustainability goals. Development policies that promote suburbanization are contributing to increases in commuting, a greater number of households with multiple vehicles, and lower population densities. These factors are contributing to increasing energy usage, offsetting the impact of policies that are intended to reduce energy use. In other cities, policies are in place but they are likely under-funded, inadequately supported by staff, or require additional management, administrative or political resources. In order to better impact energy usage, policies are more likely to be successful if they target energy usage in disaggregated energy consuming sectors such as the residential and transportation sectors.

Type A cities have policies in place and are working on the problem. These cities are expanding rapidly and their decisions concerning how their expansion is permitted to continue are critical to future energy use. Focusing on how development occurs will establish infrastructure that brackets future energy options. Increasing density is part of the solution. Miami is an example of a Type A city that has successfully reduced the number of families with two or more vehicles and achieved a high rate of use of public transportation. Other Type A cities might consider how this was achieved. Reducing the number of single occupant households is an effective strategy for Type A cities. Local ordinances that require excessive floor plan area for newly constructed apartments are problematic. Singles often enjoy less spacious flats. There is a need to expand the number of alternatively heated residences, perhaps by considering greater use of solar and other alternative energy technologies.

Meanwhile, Type B cities are having greater success in reducing per capita energy use than Type A cities and are succeeding with fewer active policies. Aggregate energy use is growing more slowly in Type B cities. These cities are smaller and policies are likely more customized and targeted. Since growth is slower, it is generally more manageable. However, existing programs need to be reviewed and reassessed to determine which are most effective. In addition, Type B cities might want to consider studying the broader policy structure of Type A cities to determine which policies best align with their goals and are the most adaptable to their local agendas. Type B cities might consider improvements such as expanding and upgrading public transportation to more energy efficient systems. Programs need to be established for residences to create incentives to use alternative energy for heating.

As Type A cities demonstrate, ignoring the problem of urban energy use, or stonewalling with established policies of institutional inertia, do not lead to success and will not promote sustainability. The alternative, taking little or no action is exemplified by cities like Charlotte, a Type B city. Charlotte is among those cities that is pursuing a "smart growth" policy agenda. The success of this policy approach for Charlotte is questionable from a sustainability perspective because it has the lowest rate of policy adoption among our sample of Sunbelt cities. Within its group, Charlotte also has the highest rates of increases in per capita energy use coupled with the highest percentage increases in estimated total energy use (see Appendix, Table A5).

The variables considered in this study suggest that policies need to provide disincentives for suburban development patterns. Greater suburbanization can result in longer commutes, increased reliance on automotive transportation, potentially more air pollution, inevitably resulting in increased use of transportation fuels, and decreasing population density. Las Vegas serves as the quintessential example. The city has policies in place yet is suburbanizing at a rapid rate. This type of urban development ultimately yields the greatest increases in urban energy use.

While suburbanization is a development model that is problematic for both Type A and Type B cities from an energy use perspective, there are also exceptions. Mesa is one example of a Type B city that has found ways to reduce total energy use while accommodating a substantial portion of its population growth by limiting suburbanization and increasing population density.

Recommendations and Conclusions

A wide range of policies are being used by Sunbelt cities in an effort to reduce energy use and improve sustainability. This movement is being motivated by a number of issues including energy costs, air quality and public demand. Sunbelt cities are creatively experimenting with policies and are achieving mixed results. The road to policy implementation has been fraught with slick spots, speed bumps and in some cases black hole sized chuck holes that have impeded the process. However, these policies are achieving observable benefits. Most encouraging is that despite increases in aggregate energy use, the per capita use of energy in Sunbelt cities has been declining.

An important finding of this analysis is that urban policies that are focused on the goals of reducing aggregate measures of energy use and costs are likely to yield no measurable results. Policies that claim such idealistic objectives are misdirected. They have no place in the real world of urban systems that are infinitely more complex and inescapably entropic. Even when partially successful, actual results are difficult or impossible to measure. This analysis offers more than a glimmer of hope for cities as it fails to find evidence of "increasing per capita consumption" that Berke (Ko 2000:2) uses as a thesis to conclude that the "idea of sustainable development" is potentially "unfeasible."

Policies that are based on the notion that sustainability can be successfully achieved by using only one approach, such as "can only come from a catalyst—specifically, design and institution of sustainable cities" (Levine, 1990:24) while possibly well intended are shortsighted and limited in vision. The analysis proves that incremental improvements and appropriate policy changes can be effective and successful. In reality, the combined energy-efficiency gains of the 25 sampled Sunbelt cities (when considered as a group) from 1990 to 2000 is adequate to provide all the energy requirements of a comparably sized Sunbelt city. This was achieved with incremental increases in population density (using less land) coupled with reductions in per capita energy usage (being more energy-efficient).

If, as an alternative to the incremental changes that were implemented, a new city might have been constructed. No matter how sustainable the design, energy use would have increased as a result. After all, even very energy-efficient cities use lots of energy. Regardless, the analysis supports sustainable design solutions that are directed toward energy efficiency, energy conservation and alternative energy policies.

Conclusions that can be drawn from the analysis of transportation sector energy use provide suggestions for ways that cities can both reduce energy usage and improve sustainability. The analysis suggests that expanding policies and programs that provide alternative means of transportation to work, reduce the number of household vehicles, increase population density and improve air quality may be keys to reducing transportation sector energy use.

These findings are in opposition to the report of the 1996 President's Council on Sustainable Development which failed to consider policies focusing on the direct use of transportation fuels as a key ingredient to successfully achieving sustainability. In light of the findings in this study, the Council's proposed measurement techniques now appear questionable and unsupportable. For example, the Council suggested that the "transportation sector could be measured" using indicators such as congestion, national security, transportation efficiency and transportation patterns (President's Council on Sustainable Development 1996:54). The Council failed to clarify how transportation efficiency was to be measured, though it can be measured in units such as consumption per unit distance. It was unclear from their report how the Council intended to measure congestion, national security or transportation patterns. Their assessment lacked statistically significant relationships and an analysis that would provide a resolution to these questions.

On the other hand, the findings are in general agreement with the findings of researchers (Inogucho, Newman and Paoletto, 1999; Register 2002; Newman and Kenworthy 1999:102-103) who propose increasing population densities and improving transportation systems. These are programs that can successfully reduce energy usage and improve sustainability if carefully implemented.

When dependable alternative transportation is available and convenient, it is more likely that it will be useful and by default, reduce reliance on automobiles. The types of alternative mass transport available include autobus, monorail and light rail, among others. Cities such as Atlanta, Los Angeles, New Orleans, and Miami lead the pack in offering alternative transportation systems. Increasing the percentage of households that have no vehicles or only one vehicle can reduce the total number of vehicles, causing residents to increase the use of public transportation, thus resulting in reduced energy use. This can be accomplished by increasing density, providing less parking, changing the design of residences, and limiting garage space.

Programs designed to offer enhanced incentives to motivate families to reduce the number of household vehicles can be successful, especially when targeted toward households with two or more vehicles. Finally, policies that reduce the number of days annually that cities experience unhealthy air are indeed linked to transportation sector energy consumption. Since internal combustion engines are a primary cause of air pollution, policies designed to decrease their use are likely to improve air quality. Expanding policies that promote the use of alternatively fueled vehicles are a key. Such policies exemplify programs that are effective in achieving the goals of simultaneously reducing energy use and improving sustainability.

This analysis supports what many energy engineers and design professionals already understand… residential energy use also varies as a result of the climate. However, the amount of energy required is not an inevitable result of climate alone. The analysis supports a call for an urban vernacular with local solutions for residential architecture that integrate within their designs responses to energy use. Locating Sunbelt residential developments on sites with less severe microclimatic conditions, especially in locations that require less air conditioning is one sure way to reduce residential energy use. Carefully changing the design of residences in response to climatic conditions using engineering and architectural design features can also reduce air conditioning requirements.

Supporting policies that reduce the proliferation of single occupancy households can be an effective means of reducing residential energy use. Such policies can also reduce the demand for housing and yield increases in population density. Supporting programs that offer alternative forms of heating may not always reduce the total energy needed but can reduce reliance on fossil fuels, currently the primary means of heating residences. Finally, constructing residences that are larger than actually needed effectively increases urban energy use.

This study also found that cities that participate in organizations, such as ICLEI, Clean Cities or Energy Star™, are more successful at achieving measurable results than those that do not. This may be due to the educational aspects of their programs, the requirements or standards established by the organizations, or other benefits such as the ability to share information of common applicability.

Most importantly, this assessment offers proof that to be successful, policies must be multifaceted, varied, locally adaptable, accessible and creatively deployed. Policies are not universal. What works well in one city may not work at all for the next and may not apply to another. Policies that are either federally or utility sponsored seem to have greater credibility, broader applicability, and are more likely to be adopted and pursued by urban governments. Policies must be focused and supported for long periods of time to be effective.

The trends of this study followed incremental policy changes over a recent period of ten years. It is unlikely that reactionary policy responses and quick fix solutions to immediate concerns (such as energy price spikes and short term shortages) will resolve the issues central to urban energy usage. In order to increase the potential of success, policies must be flexible and supported both politically and financially for prolonged periods. This requires a long term vision… like the vision urban sustainability provides. While there are many paths and routes that cities may pursue in an effort to achieve sustainability, choosing policies that focus on energy related solutions is an approach that offers significant potential for success.

Endnotes

1. One example is the 2005 Environmental Sustainability Index www.infoplease.com/ipa/A0930889.html, accessed 7 Janaury 2007. It is a composite index that uses 21 indicators such as natural resources, pollution levels, environmental management (among others) that characterize sustainability at the national scale. Finland

is ranked at the top of the list, the U.S. is 45th. Yet another example is the Pilot 2006 Environmental Performance Index which can be accessed at www.infoplease.com/ ipa/A0933290.html.

2. Statewide average per capita data was used as no data was available that would provide a specific energy cost and energy usage value, aggregate or economic sector, for each individual city within each state.

3. Data for the variable *air quality index* was found on the USEPA website: www.epa. gov/airnow/.

4. Data for the variable *days unhealthy and unhealthy for sensitive groups* was found on the USEPA website: www.epa.gov/airnow/.

5. Data for the variable *cooling degree days* was found on a website sponsored by the National Oceanic and Atmospheric Administration (NOAA): lwf.ncdc.noaa.gov/ oa/climate/online/ccd/nrmcdd.html,

6. Note: 1 Kilojoule is equal to 0.9472171 British Thermal Units.

7. Data from the Energy Information Administration: www.eia.doe.gov/emeu/aer/ txt/ptb0105.html.

8. To derive an index for the year 2000 which is uninfluenced by 1990 data, the variable *10 year relative change in population compared to density* is not used. The variable *cooling degree days* is also not used for this index as it is accepted that while cities can establish policies for the designs of structures which respond to varying climatic conditions, they have no direct policy control over local weather conditions.

9. The index weighs the policies and selected variables equally. This equal weighting might be considered to be arbitrary. A goal of the index was to provide a set of values that would value both the rates of policy adoption and measures of their impacts. Conclusions based on this comparative process impact the rankings of the cities. This approach limits quantitative dimensions subject to analysis, further narrowing applicability. Other approaches that have potential are: 1) add or eliminate additional independent variables and perform a hierarchical analysis of variance; or 2) add or otherwise modify the number of policies considered.

Chapter 10

Learning from Las Vegas

"It is a virtual miracle that a city like Las Vegas can exist and thrive in the middle of one of the most inhospitable and resource-poor climates in the world with an average annual rainfall of only 4.5 inches. This anomaly makes the concept of sustainable living a very relevant topic for Las Vegas."

FRANCIS BELAND, DIRECTOR OF THE LAS VEGAS SPRINGS PRESERVE,
AS QUOTED IN EARTHTIMES.ORG,
LAS VEGAS' VISION OF CENTRAL PARK
A MODEL OF SUSTAINABILITY, 1 NOVEMBER, 2006.

Corporations, institutions and especially local governments all deal with sustainability issues even if they have adopted no resolutions or policies concerning them. Often sustainability efforts are manifest in how local governments deal with environmental, energy or social sustainability problems.

To further explore how sustainability impacts local governments, Las Vegas is selected for a descriptive case study. Las Vegas might be one of the most unique experiments in rapid urban development in the U.S. Las Vegas is an example of a city that arose from the idea of our manifest destiny rather that from the idea of sustainability. Developed on the belief that cities can prosper when environmental resources are *unlimited*, Las Vegas pierces the heart of sustainability issues in the U.S. Regardless, there are lessons to be learned from Las Vegas. What can be learned from the history of Las Vegas that can be used to assess its future? How can its sustainability be gauged? What in Las Vegas will survive the next 100 years? A deeper understanding of the history of Las Vegas offers clues. And yes, the idea of sustainable development is germinating in Las Vegas.

The underlying causes of population growth, urban development and increasing energy use can all be found in the case of Las Vegas. In addition, the underlying effects of environmental degradation, dislocation and changes in urban infrastructure are recognizable—and often glaring.

Using Las Vegas as a case study yields a greater understanding of how sustainability evolves to take root in local governments. Case studies are "in-depth investigations" that are "based on a great wealth of empirical materials" and are notable "because of their variety" (Hamel, Dufour and Fortin 1993:45). Cases present "the selected observation point for an object of study" with the analysis establishing the "definition of the case" and permitting "an assessment of its generality in view of the results of the analysis" (Hamel, Dufour and Fortin 1993:45). Case studies provide insight into how urban policies are associated with environmental damage to urban areas (Crenson 1971).

Las Vegas is selected for the case study as a result of: 1) demographic factors; 2) geographic conditions and resources; 3) representative local actions; and 4) the availability of specific statistical information that assess its characteristics. This chapter explores Las Vegas in regard to its history, its resources, and its indicators of sustainability.

ASSESSING LAS VEGAS

Las Vegas is typical of many communities that are being pushed into adopting sustainability agendas without having formally adopted one. The city is growing in population. This growth has manifested itself both in sprawling suburbanization and renewed efforts to increase population densities. These are two primary methods of accommodating growth in the Sunbelt. In other ways, Las Vegas is atypical. Population growth has occurred at a faster rate than experienced in most U.S. metropolitan areas. As a result, problematic aspects of rapid population growth and densification are more noticeable in Las Vegas than in most other Sunbelt cities. Like many Sunbelt cities, Las Vegas, which might be the called the nation's casino capital, has a defined economic specialty. Las Vegas is a city that flaunts it its wealth, making it quite difficult to claim that financial resources are unavailable to implement energy saving technologies or policies that concern sustainable development agendas.

While it can be argued that all cities are inherently unique, there are a number of reasons that make Las Vegas an excellent choice for an in-depth assessment. Las Vegas is unlike Los Angeles and San Diego—cities that have adopted many sustainable development policies. It is also unlike Charlotte and Oklahoma City, cities with very low rates of policy adoption.

Las Vegas, due to its rapid population growth, might be considered to be an extreme example as well. This can be interpreted as a reason to study Las Vegas in depth. Being a boomtown, Las Vegas has benefited from the ability to select from a wide range of proven and tested sustainability programs. Since most of the development in Las Vegas has been recent, the impacts of decisions illustrate what is presently occurring in the developing areas of the U.S.

Exploring how Las Vegas evolved and what types of problems were encountered during its developmental evolution is educational. The case study which follows reveals salient aspects of its history, development, environmental issues and energy policies. It reveals that Las Vegas has unresolved energy and sustainability issues.

HISTORY OF LAS VEGAS

Las Vegas, which translates as "the meadows" in Spanish, was built on a fragile desert plain. Like many other cities in the Sunbelt, Las Vegas has developed due to the growth of its service economy while offering unique commercial specialties. "There is no place like it. It is literally a beacon of civilization... astronauts make out the lights of Las Vegas before anything else" (Denton and Morris 2001:6).[1] Rapid growth that sprouted from dollars gambled away has created an expanding set of centrally located service industries. The city was the product of both the automobile and the availability of fresh water that became available when the Colorado River was dammed.

Landlocked, Las Vegas was located on relatively useless land with sandy and alkaline soil conditions. The city grew in the sands of a surreal desert landscape of scrub, sagebrush and cactus. The transformation from empty desert basin to desert town, to one of the fastest growing cities in the U.S., all occurred in a very brief span of time.

People settled the Las Vegas Valley for a variety reasons. These included the dry climate, inexpensive land, low taxes, and legalized gambling. Las Vegas is a 20th century new town and Sunbelt boom town. The city plan of Las Vegas is a set of paved intersecting grids with ample parking lots and what seem to be too many fences.

Nevada was part of the Utah territory prior to statehood in 1864 (Laxalt 1977:37). Clark County, home of the future Las Vegas, was appended to the state of Nevada in 1867. A gold deposit was discovered south of the town and a mining camp developed nearby at Searchlight from 1892-1907 (Laxalt 1977:31). Completion of a railhead in 1906 spawned the incorporation in 1911 of the city of Las Vegas which today encompasses 313 square kilometers (121 square miles). The San Pedro, Los Angeles, and Salt River Railroads intersected the Las Vegas and Tomopah Railroad, linking the city to Virginia City and elsewhere. The city benefited from its dry climate, a nearby fresh water source (Colorado River), the construction of Boulder Dam (beginning in 1931) and the creation of Lake Mead. Businesses developed due to the opportunity for quick marriages, easy divorces, games of chance, prize fights, inexpensive land, and the eventual completion of Interstate 15 (connecting Salt Lake City, Utah, to San Bernardino, California, via Las Vegas).

Lucky for Las Vegas, development was facilitated by an abundance of inexpensive flat land in the Las Vegas Valley over which a grid of streets was easily laid. The city developed on a location which avoided the use of valuable farmland. The soil was easily dug for foundations and sand was readily available for concrete construction.

Since casino gambling is illegal in both California and Arizona, and Reno is far to the North, the Las Vegas metro area has a regional monopoly on gambling and has become one of the premier gambling destinations in North America. To attract high rollers, Las Vegas casinos offer incentives to lure their customer's gambling dollar, and as a result, many gamblers are willing to travel from international locations. Large stakes gamblers are often provided with perquisites such as free meals, transportation, accommodations and other incentives to visit Las Vegas casinos. More than 50 million people journey to it each year (Denton and Morris 2001:7). For Las Vegas, the range of its initial offerings (gambling and hospitality) has expanded well beyond its immediate region with the development of its international airport. This has contributed to explosive growth since the 1970s.

Within Las Vegas, there is competition among the downtown area, the Las Vegas Strip (Las Vegas Boulevard South), and off-strip locations (e.g., the Hilton and Las Vegas Convention Center). Local residents are enticed by advertising, dining, dinner shows and special offers in selecting locations for gambling and entertainment. Due to the large number of casinos, this creates a highly competitive local market and provides a variety of entertainment options for patrons. Given the special nature of Las Vegas as a gaming center for the U.S., economic theories such as "central place theory" and "location theory" both apply to a significant extent to Las Vegas. Central place theorists would argue that such factors created an economy of scale, allowing trade and specialization to occur.

The economic transformation of Las Vegas was caused by a loosely organized growth coalition that promoted Las Vegas as a place that was profitable for businesses (Rudd and Fainstein 1999:122). The city benefited from Nevada's image of being a tax haven. Since taxes collected from gaming provide substantial income, Nevada is a state without corporate, income, inheritance, gift, franchise or inventory taxes. In addition, Nevada is a right-to-work state and attracts a large pool of low wage labor (Rudd and Fainstein 1999:122). By 1996, Las Vegas, along with Phoenix and Albuquerque, was ranked among the top twenty U.S. cities in manufacturing-related job growth (Wolf 1999:70).

The population growth of Las Vegas since 1940 has been phenomenal by almost any North American standard. People who came from other places populate the city. The city of Las Vegas had a population of only 8,422 in 1940, growing to 24,624 in 1950 (Elliott 1973:314). This growth was spurred by construction and operation of a nearby magnesium plant which operated during World War II. Las Vegas officially became the largest city in Nevada in 1960, with a population of 64,405 (Elliott 1973:314).

Two important technological breakthroughs were crucial to the development of Las Vegas. One resident stated that, "The most important developments in Las Vegas history was the Carrier Air Conditioning Company's ability to air condition large spaces, combined with direct flights from Los Angeles" (Denton and Morris 2001:100). With the outside air temperature on summer days reaching 40° to 43°C (105° to 110°F), the importance of air conditioning the city's cavernous casinos, hotel rooms, offices and residences cannot be understated. Given the dry desert climate, year-round occupancy is difficult and uncomfortable without air conditioning and humidification. Direct flights from Los Angeles provided access to a major population center, allowing visitors access to the city without having to endure a grueling drive through the desert.

Despite being born as a gambling Mecca, the city has spawned significant industrial and commercial development, further spurring population growth. The Las Vegas metropolitan area includes all of Clark County, Nevada. Primary incorporated areas include Las Vegas, North Las Vegas, Henderson, Nellis, Boulder City and others. The residents of Las Vegas are varied, multicultural and have a wide range of incomes. Las Vegas has proved to be a magnet for Chinese and Mexican immigrants who provide labor for the service economy. By some measures Las Vegas is the fastest growing city in the U.S. Census data indicate that the population increased from 273,000 to 863,000 from 1972 to 1992. The year 2000 census posts the Clark County metro population at 1,375,765,[2] a five-fold increase in only 28 years. Today, over 5,000 people per month move into the Las Vegas Basin. The most rapidly growing population segments from 1990 to 2000 were the under 5 and over 75 age groups, suggesting that Las Vegas was attracting families and retirees.

Transportation in and around Las Vegas is by automobile and motor coach. The "Las Vegas Strip" is a linear and paved multilane roadway lined by massive theme hotels and casinos. It is crossed at right angles by innumerable intersecting roadways with traffic controlled by traffic signals. It is also crossed in a number of locations by pedestrian bridges which separate walkways from roadways. Las Vegas is among those U.S. cities that never seem to sleep. Its 1995 tourism campaign was named: "Las Vegas—Open 24 hours."

Las Vegas has created distinctive urban forms and international urban images. However, when compared to European cities, Las Vegas seems at once both contemporary and temporal. Its gambling palaces and other buildings unabashedly imitate. The allure of its amenities are carefully designed. One structure recreates a New York skyline, another recreates Camelot, yet another ancient Egypt. There is a copy of the Eiffel Tower, a Roman palace, an Alpine hotel, a reproduction of the Statue of Liberty and a miniature Empire State Building. It is as if Las Vegas were a contemporary open air museum with a rare collection of reconstructed buildings that rank as global architectural icons.

Las Vegas, an international tourist destination in its own right, can offer remarkable hospitality, holding open its doorways and air curtained entrances to all comers who have both leisure time and money to spend. There is widespread cultural approval of risk taking, gambling and nearby legalized prostitution. Las Vegas is all about hospitality. There are more guestrooms in Las Vegas than anywhere in the U.S.—twice as many as in New York City (Denton and Morris 2001:7). Las Vegas also boasts 17 of the 20 largest hotels in the world. Las Vegas Mayor Oscar Goodman recently stated, "We must be prudent on our approach, assuring the preservation of the good name of Las Vegas …which represents honesty, integrity and a fair shake for the gambler" (Goodman 2002b).

The ever present and moderating influence of the Morman Church, combined with notorious personas like Benjamin "Bugsy" Siegl and the initial gangster control of casinos, created an unusual set of influences for a thriving Las Vegas poised for fame. Legalized gambling, the proliferation of the automobile, economical air passenger transport and a new convention center have contributed to the city's accessibility as a center for entertainment. The city has awe-inspiring theme and amusement park-like facilities (e.g., MGM Grand Adventure and the top of the Stratosphere) with roller coasters, water parks and other unique attractions (e.g., the futuristic Borg Invasion and Star Trek Adventure at the Hilton). Those with a zest for new experiences come to Las Vegas to marvel at the digitally sequenced lights of "Glitter City" and the programmed imagery of its grandiose theatrical productions, which thrive by pushing the entertainment envelope to its technological limits. Ultra-lounges and neon illuminated night clubs abound, some with dance floors that have liquid nitrogen clouds to provide "Ice" conditioned air and action lasting into the night. These and other energy-intensive attractions have generated additional commercialization and employment by extending entertainment and leisure activities to gamblers and their families

The city has nurtured auxiliary and unclassified businesses with increases in education, construction and service sector employment. The University of Nevada-Las Vegas, which opened in 1957, has over 24,000 students. Primary employment centers are located either downtown, at government locations or on the Las Vegas strip. The service sector remains the dominating base industry, accounting for 56% of employment in 1993 and 60% by 1999. The growth in the service sector is due in part to the construction of large hotel and casino complexes and amusement facilities on the Las Vegas strip and in the downtown area. This centralization of the "entertainment" districts impacts infrastructure development by placing demands on water, sewer and utility systems. In addition, support services for the expanding population have added to service sector demands. Population data indicate a trend toward decentralization of population from some urban neighborhoods to the suburbs, leaving the central core areas for commercialization to support visitors and tourists. This provides evidence of a developing spatial mismatch due to movement of the residential population to the suburbs.

The city is not without its critics. Parker notes "Las Vegas explicitly advertises itself as a fake neon city, and hundreds of thousands of people flock there every year precisely because it delivers on its promise" (Rudd and Fainstein 1999:6). There is a shopping galleria which recreates a Venetian canal and a Venetian streetscape. Denton and Morris (2001:7) commented, "Off a sham Piazza San Marco, gondolas glide on simulated Venetian canals carved onto the face of the Great American Desert." There is a concert hall at Caesar's Palace (which has its own Apian Way shopping street) designed to look from street level like a reproduction of Rome's Coliseum. Though the 4,000 seat Las Vegas Coliseum has been endorsed by Elton John, ancient Rome's Vespanian and Titus would glare rather than marvel. Yet Las Vegas' bet on casino gambling and tourism has paid off. It hosted an estimated 38.9 million visitors in 2006 and the city added $2.3 billion in casino and hotel developments.[3] Today, the city can boast that it has 133,000 motel rooms (estimated to increase to 170,000 by 2010) and the sixth busiest U.S. airport.[4]

SUSTAINABILITY CONCERNS OF LAS VEGAS: ENVIRONMENTAL AND DEVELOPMENTAL

Las Vegas was blessed with clear and generally pure air (Elliott 1973:9). Cradled in a corner of the Mohave Desert, with a correspondingly dry climate, the city receives just 0-20 cm (0-8 inches) of rainfall annually. The local region's growing season is 239 days long. Throughout its early history, the principle environmental concern has been availability and access to water supplies. Beginning in the 1940s, nuclear fallout became an environmental hazard. It was not until 1972, that the Las Vegas basin also began to experience air pollution during parts of the year (Elliott 1973:9).

Water from the beginning was scarce everywhere in Nevada. The gushing springs located at the site of what was to become Las Vegas were described by mapmaker John Fremont in 1844 (Laxalt 1977:116). Nevada's surveyor general reported in 1865 that "Many millions of acres of land in this state now completely worthless, would be valuable if irrigated" (Elliott 1973:172). In addition to three major springs in the area, water was initially provided by wells which tapped ancient pools underneath the desert. Wells typically needed to be drilled to a depth of only 90 meters (300 feet) (Littlejohn 1999:135). The water made the site a potential relay point for railroad traffic. The Colorado River Compact in 1922 promised Nevada 3.702×10^8 cubic meters (3.0×10^5 acre feet[5]) per year of the flow of the Colorado River (Hulse 1998:214).

The development of Las Vegas accelerated after the construction of Boulder Dam (later renamed Hoover Dam) in 1931, just 48 km (30 miles) away. The dam choked the Colorado River, creating Lake Mead and providing a source for both fresh water and hydro-electric power. In the early 1950s, wells pumped continuously to provide landscaped islands in the desert with palm trees and idyllic pools for the new casinos. Yet the desert began again just beyond the fence lines that demarcated the parking areas. With water being a scarce commodity, by the mid-1950s sewage effluent was being used to keep golf courses green (Denton and Morris 2001:144). In 1956, district engineers called the developing circumstances "an impending crisis in the water supply situation" and started to pump a supplemental flow of water 304 meters (1,000 feet) up and 40 km (25 miles) out from Lake Mead (Denton and Morris 2001:145). The city, then and now, requires constant irrigation.

A major boost to the Las Vegas economy came in the mid-1960s with the Southern Nevada Water Project which tapped the balance of the Colorado River water promised in 1922. Construction of a 6.4 km (4 mile) long tunnel through the mountains from the Colorado River to the Las Vegas Valley was completed in only four months. Regardless, by the 1970s, future serious water shortages were being predicted. Solutions included developing new sources of water in addition to adding artesian wells or increasing allotments from Lake Mead. Among the ambitious schemes considered were: 1) piping water from outlying areas within 161 km (100 miles) of Las Vegas; 2) a 1,609 km (1,000 mile) aqueduct from the Columbia River in the Pacific Northwest; and 3) desalinization of water and a pipeline from the Pacific Ocean (Lexalt 1977:122). A proposal to import water from deep holes in the Amargosa Desert north of Las Vegas ran into an environmental roadblock with the discovery of one of the oldest forms of life, the pupfish, which inhabited the subterranean water (Lexalt 1977:121).

By 1982, over a quarter of a billion dollars had been invested in the water system and Clark County, Nevada boasted one of the most modern water distribution systems in the U.S. (Hulse 1998:239-240). Within a few short years, water supplies were again dwindling and by 1989, grandiose plans were once again being considered. The Las Vegas Valley Water District proposed a $1.5 billion project to drill 140 new wells in adjacent counties along with pipelines to tap new underground water sources and transport the water to the Las Vegas Valley (Hulse 1998: 239-240). Property owners and rural governments effectively opposed the proposal as it could have depleted surface and underground water supplies for their towns and farms.

Despite all efforts, by the end of the 20th century, the net result was that Las Vegas once again had a "depleted and polluted water supply" (Denton and Morris 2001:364). Areas of the basin have sunk as much as five feet, cracking and splitting foundations (Littlejohn 1999:8-9). Despite the prohibitive cost of water and the energy consumed by pumping facilities, Las Vegans currently use more water than residents of any other major city in the world, 1,230 liters (325 gallons) per capita per day. This rate of usage is a slight reduction from the 1,363 liters (360 gallons) per capita per day consumed in 1989 (Littlejohn 1999:8-9). Striding past the dancing fountains of the Bellagio Hotel, a tourist could hardly believe that water shortages could be a future problem for the city.

Considering the future availability of water for Las Vegas, the cited prediction was that "available sources of water (underground or from the Colorado River) will run up against the population wall around the year 2005" (Littlejohn 1999:35). The brash assurance from the Southern Nevada Water District is that water will always go to the highest bidder, and that Las Vegas will always be wealthy enough to cut deals with farmers, ranchers, western states and the government to get all the water it needs (Littlejohn 1999:133). Despite dire forecasts, a series of interim measures will likely provide water for continued growth through 2025, when the population is projected to reach 2.3 million (Littlejohn 1999:121).

Treated sewage discharges from the Las Vegas metropolitan area total approximately 261 million liters (69 million gallons) daily (Littlejohn 1999:137). Almost all of the effluent goes into Lake Mead at Las Vegas Bay, causing a plume of pollution consisting of potentially toxic blue-green algae, which thrives on nutrients found in the sewage. Discharges include ammonia, phosphates and nitrates (Littlejohn 1999:137). Chemical and bacterial residues cannot be filtered by conventional water treatment systems. Interestingly, the Las Vegas drinking water supply is drawn just 10.5 km (6.5 miles) downstream of the location where its treated sewage is discharged into Lake Mead (Littlejohn 1999:137). A 1996 U.S. government sponsored geological study determined that as a result of pollution, endocrine disruption of carp in Lake Mead was occurring, transforming male carps into females (Littlejohn 1999:138).

LAS VEGAS ENTERS THE NUCLEAR AGE

Nevada, a state where the federal government owns approximately 87% of the land, must have seemed ideal for nuclear testing. Las Vegas was among those cities nearest to where the nuclear age began, only 105 km (65 miles) from the primary North American testing site. As a result, Las Vegas was directly affected by the atomic age. Beginning in 1951 and over the next decade, more than 100 nuclear weapons exploded in the air or on the ground in Nye County (adjacent to and northwest of Clark County) on 404,700 hectares (one million acres) of barren desert located at the Atomic Energy Commission's (AEC) Nevada Test Site (Littlejohn 1999:121-140).

In true Las Vegas style, the explosions became yet another tourist attraction. The casinos served "Atomic Cocktails," advertised rooms for their view of mushroom clouds, and also sponsored special picnic lunches or "dawn bomb" parties outside the city to view blasts from a closer vantage point (Littlejohn 1999:121-140). The Atomic View Hotel advertised "an unobstructed sight line to the bomb blast from the comfort of one's lounge chair" (Denton and Morris 2001:140). The light from the atomic blasts caused the Las Vegas Basin to be illuminated "in a spectral false midday" (Denton and Morris 2001:141).

Despite the attraction of the testing, the reality was grim enough. After the mushroom-clouded explosions, men in white suits with Geiger counters would descend on Las Vegas. As far away as Rochester, New York, snowfall would test in the thousands of radiation counts per minute when a normal background count was only 400, causing Kodak products to be damaged while in production. It was later revealed that U.S. government scientists knew that exposure to even low levels of radiation caused cancer and leukemia. Regardless, the local director at the test site, Albert Graves, repeated to the press that "the test would not be of the slightest danger" to any citizen, not yet realizing that he was already loosing his sight from the effects of multiple radiation exposures (Denton and Morris 2001:141). A similar disingenuous public announcement would be made by authorities many years later as a result of the Three Mile Island nuclear incident in March of 1979, which eventually resulted in the evacuation of tens of thousands of residents.

Vast portions of Nevada Desert were ultimately laid to waste. At the time, the environmental impacts of atomic age activities were not widely understood. While the people, livestock, land, water and air were being poisoned with radiation, the federal government was pursuing a policy of intentionally misleading the general public (Denton and Morris 2001:141). In 1997, the National Cancer Institute confirmed that tens of thousands of cases of thyroid cancer in the U.S. had been traced to fallout from the tests of the 1950s (Hulse 1998:343). Many of those victimized never knew what had caused their disease. Michael Ventura grimly wrote of the bomb viewing outings, "...a lot of those picnickers died young" (Denton and Morris

2001:142).

The shift to underground testing occurred in 1957, evoked new protests due to contamination of water supplies and the potential of earthquakes due to the explosions, which was confirmed in the late 1960s.[6] In 1968, the Atomic Energy Commission (AEC) announced that a number of seismic aftershocks had actually followed underground nuclear tests (Elliot 1973:339).

In a National Academy of Sciences report (circa 2000), a scientific committee said that rapid growth in the Las Vegas Valley may one day cause local officials to search for water around the site—*where the extent of radioactive contamination is unknown*. A scientific committee reported that the Nevada Test Site and other federal government locations used to build nuclear bombs "will never be clean enough to allow public access to the land" (Manning 2000). At many sites, including the Nevada Test Site, radiological and non-radiological hazardous wastes will remain, posing risks to humans and the environment for the next tens of thousands of years. The report also stated, "Complete elimination of unacceptable risks to humans and the environment will not be achieved, now or in the foreseeable future" (Manning 2000).

The Las Vegas area benefited from short-term employment opportunities offered by the development of nuclear power—but has suffered from its long-term and generally misunderstood effects. Though Nevada has no nuclear-fueled electric generating capacity, one poll of Clark County voters cited nuclear waste as the most critical issue facing their community (Manning 2000). In Nevada, development of the underground nuclear waste storage facility at Yucca Mountain has become a volatile, politically charged concern among residents of the state.

OTHER SUSTAINABILITY CONCERNS

Beginning in the 1980s, local air pollution and smog, caused primarily by automobile exhaust, became an added environmental concern. Toward the end of the century, there were periods when Las Vegas had one of the highest air pollution indexes in North America, in part due to a grid of urban sprawl thrown upon a valley with scant public transportation (Denton and Morris 2001:365). Traffic movement, or lack of it, also affects the environment. Robert Parker once observed that "At peak time, it takes up to 30 minutes to travel the two blocks from Tropicana Avenue to

Flamingo Road on the strip. As a result of the traffic glut... air quality has deteriorated" (Rudd and Fainstein 1999:122). With more vehicles emitting more carbon monoxide, together with construction, high winds, and dust, air quality continues to be problematic. Las Vegas was placed on a non-compliance alert in 1997 which may threaten future federal highway funding (Littlejohn 1999:35).

Given what we know today, just how sustainable is Las Vegas? Structures in Las Vegas, like many others in the U.S., are typically designed for an economic life of forty years or less. It can be argued that the city is designed for impermanence. Las Vegas is preparing for continued and explosive growth while planning its buildings and infrastructure to solve short-term requirements. By modern standards, the casino hotels of Las Vegas are uninhabitable without mechanically induced air conditioning systems.

For Las Vegas, there always has been plenty of land but not enough water. The lack of water in the Las Vegas Valley may limit future population growth. Lack of extensive water resources creates limits to future development. In addition, substantial inflows of inexpensive energy are required for air conditioning systems, commercial lighting and water pumping, not to mention thousands of gambling machines. Such dependencies mean that disruptions to the flows of water and electricity are assuredly inconvenient and potentially catastrophic.

It might be argued that since the city was built on land that was not farmable, the selection of the site for Las Vegas might be considered environmentally appropriate. It is also arguable that the fragile desert ecosystem of the basin has been forever disrupted. Without significant inputs of resources such as energy and water, Las Vegas (to a greater extent than in most U.S. cities) could not have been built and would not have survived. While hydropower can be argued to be a sustainable energy resource, construction of Boulder Dam predated environmental advocacy. If the dam were proposed today, environmentalists might seek protection for the Colorado River and its tributaries, in hopes of preventing disruption to the local environs and habitats. The dam effectively reduced water flow to portions of northern Mexico and permanently changed the ecosystems of the lower Colorado River Basin. Maintaining Lake Mead's water level remains a difficult challenge for the region. In the spring of 2007, the water level was roughly 120 feet below normal pool levels, the lowest level since the early 1970s.[7] Low water levels also reduce the ability to generate electricity.

Meanwhile, Las Vegas suffers from air and water pollution and environmental deterioration might limit future viability. Air pollution is growing due to greater use of automotive transportation. The city experiences continuing difficulties with water pollution as well. Developmental demands for water and increases in pollution are likely exacerbated by runaway construction, once described as "an unreadable chaos of nonplanning" (Denton and Morris 2001:364). More creative and more widely deployed water conservation measures will be required in the future simply to efficiently manage the available resources.

While demonstrating certain aspects of sustainable development, the city has cumbersome long-term sustainability problems. The underlying causes include population growth, urban development, and urban energy requirements. The underlying effects of environmental impacts and changes in urban infrastructure in Las Vegas are consequences of these issues.

It was urban relocation rather than dislocation that created pressure for rapid development. People migrated to the Las Vegas basin, abandoning the cities they had previously called their homes. Key issues that limit future growth include demands for both water and energy. Without creative management of water resources, the long-term future of Las Vegas remains in question. Water from the giant reservoirs of Lake Mead and Lake Powell is required for municipal water supplies, irrigation, firefighting, and by the natural environment. Demand for water in the region is increasing with each passing year. Unless additional water resources are found and water conservation measures implemented, the rapid growth of Las Vegas can be viewed as non-sustainable since long-term habitation may unsupportable. Las Vegas is, after all, constructed on desert land. Water will always be a scarce resource.

QUALITATIVE SUSTAINABILITY INDICATORS

How does Las Vegas rate in terms of the ten selected indicators considered in Chapter 7? Which policies have been adopted and which policies have not? To answer this, each policy indicator will be explored in detail. The discussion will reveal which policies Las Vegas is using to respond to its pressing energy and sustainability issues.

Of the ten policy variables being considered for this study, Las Vegas is pursuing seven. Seven is the mode of the 25 selected Sunbelt cities. Las Vegas has slightly more of these policies than the mean of the sample (5.8).

The policies related to sustainability employed by Las Vegas also illustrate how policies and programs can be adapted to accommodate local opportunities. As will be demonstrated, values for population density, travel times to work, numbers of household vehicles, air quality index and rate of use of public transport are representative of the sampled Sunbelt cities. Like most cities, accessibility and availability of energy are not yet limiting development. From an early stage of its development, Las Vegas enjoyed inexpensive energy supplies, an important variable in sustainability. In addition, a wide range of resources have been exploited to support the development of Las Vegas. The city has unique environmental concerns.

The primary electrical utility serving Las Vegas is Nevada Power, a subsidiary of the holding company Sierra Pacific Resources. As Nevada's primary electric utility, Sierra Pacific Resources proudly proclaims that it "keeps casinos powered 24 hours a day" (Hoovers 2004:1). The utility has a rebate program for its customers to support a wide range of energy conservation efforts. The utility provides rebates to several thousand customers for improvements such as replacing air conditioning systems, improving air flow through residential ductwork, and tune-ups for existing air conditioning systems.

The city has a policy of reducing energy use in city owned and operated buildings. The following quote from the city of Las Vegas website[8] summarizes their recent efforts:

"The city has made important strides towards the conservation of energy. Last spring, the city appointed a committee with the task of identifying and implementing ways of cutting energy-related costs. As a result, about $535,000 and more than 4.8 million kilowatt hours have been saved annually. These savings have accrued as a result of numerous steps:

- *Lighting was retrofitted in city departments, 13 community centers and several senior centers. Also, the majority of the lighting at city hall has been upgraded to include motion sensors, which automatically turn the lights off when no one is present.*
- *Traffic signals have been fitted with light-emitting diodes (LED), which consume 90 percent less energy than the previously used lights. Street lighting also has been upgraded.*
- *Thermostats have been regulated throughout city buildings.*
- *The city's water pollution control facility, a major power user, has seen important energy conservation upgrades.*

The city extended its energy conservation efforts into the homes of city employees by sponsoring an internal contest to encourage staff to save energy. Employees were given a series of energy conservation tips and were asked to apply them to their everyday life. Participating employees reduced their energy consumption by approximately 104,995 kilowatt hours."

Both the city of Las Vegas and its primary local utility have policies in place to support energy conservation and energy efficiency initiatives. Starting in 2003, Nevada Power has budgeted $9.2 million to encourage energy conservation efforts (Las Vegas Review Journal 2002). The utility also funds weatherization and energy-efficient retrofit projects for local non-profit organizations. Projects include subsidies for installing solar power units, installing energy efficient equipment, time-of-use metering, air conditioner tune-ups, load management systems, and other measures that reduce electrical energy consumption (*Las Vegas Review Journal* 2002). The utility provides energy management experts to help in identifying potential projects and cost-effective energy-saving opportunities. The utility accepts and collects donations to support the Desert Research Institute, a non-profit research institute of the University and Community College System of Nevada. These funds are used to develop, promote and install GeeenPower™ systems, primarily solar, throughout the state of Nevada. Their demonstration projects include a 16 kW solar-electric system installed at the Southern Nevada Science Center in Las Vegas that was donated by Nevada Power.

There are local government policies in place to support alternative energy. Utilization of solar energy has been supported by the public school system, and a number of schools have installed solar collector arrays. The city of Las Vegas has adopted the Nevada Model Energy Code for new construction. Based on the local climate, the Model Energy Code establishes minimum standards for the energy efficiency of residential structures. The code focuses on mechanical systems and building envelope requirements.

The city of Las Vegas Strategic Plan omits sustainability from both its vision and mission statements. Despite having a number of innovative programs, sustainability is not a stated developmental goal.

As is true of most Sunbelt cities, local transportation in Las Vegas is heavily dependant on motor vehicles. There are presently no high occupancy vehicle (HOV) lanes in the Las Vegas metropolitan area. Providing HOV lanes on the interstate highways that link Las Vegas to Los Angeles has been considered. In regard to HOV lanes, Las Vegas joins the minority of the cities in this study.

Las Vegas participates in two of the three organizations used as membership indicators. While Las Vegas is not an ICLEI member, the city is an Energy Star™ Partner. Las Vegas received its Clean Cities designation in 1993. The city has since provided a system of 29 recharging stations for electric vehicles. The Clark County School District boasts one of the world's larger fleets of bio-diesel school buses with almost all of its 1,140 school buses using the alternative fuel (Las Vegas Regional Clean Cities Coalition 2002). Over 100 city vehicles have been modified to operate using compressed natural gas. A new "gas station" in Las Vegas was the first of it kind in North America to provide fuel for both hydrogen and compressed natural gas vehicles. This $10.8 million five-year demonstration project was developed by a public-private partnership in order to demonstrate the use of hydrogen as an alternative, non-polluting fuel for vehicles (City of Las Vegas 2002:1).

The city operates with a council-city manager government, with the city manager responsible for managing municipal services and city departments. According to Las Vegas Mayor Oscar Goodman, "We have an existing environmental commitment with our Clean City membership, and showcasing this technology on a daily basis is a natural and appropriate extension to our environmental commitment" (City of Las Vegas 2002:1). Such cooperative endeavors are examples of sustainability in practice. Opened in 2004, a $650 million dollar 6.5 km (4 miles) privately funded monorail system carries 20 million passengers per year and links the Las Vegas Convention Center to the Las Vegas Strip hotels (Rake 2003:1).[9] There are plans to extend the existing system to downtown by 2008 and a future phase is being considered to connect the system to McCarren International Airport.

Las Vegas not only maintains a curbside recycling program but also participates in brownfield redevelopment. While curbside recycling programs are now common among Sunbelt cities, Las Vegas, initiated residential curbside recycling in 1991, with the service provided by a vendor under contract with the city. Curbside pickup is provided on a twice monthly basis. There are also a number of locations that provide drop-off facilities for recyclable materials, and many have repurchase programs for certain types of recyclable waste.

The fact that brownfield locations exist in such a recently developed city as Las Vegas is ominous for other Sunbelt cities. Las Vegas has targeted four primary locales for study and potential redevelopment. An USEPA grant in the amount of $200,000 was awarded to Las Vegas in 1998 for a brownfield assessment demonstration project (USEPA 1998:1). Las Vegas has targeted its brownfield demonstration study on an area of the city where most of the city's disadvantaged citizens reside. The selected area has a 56% minority population with more than 30% living below the poverty level (USEPA 1998:1).

In sum, individual policies being pursued by Las Vegas support aspects of a pro-sustainability agenda, despite the fact that sustainability is not an identified goal. Las Vegas is ranked 9th among our 25 cities listed and achieves a sustainability index score of 5.4. The weighted policy score for Las Vegas is 3.5 which is higher than the average of the group of 25 Sunbelt cities. The index score for Las Vegas is 1.9, equal to the average of the 25 Sunbelt cities. Las Vegas is ranked in the lower half of the Type A cities. Total energy use in Las Vegas is increasing dramatically yet per capita energy use is declining at a faster rate than in most Type A cities. Table 10.1 identifies the policies that Las Vegas has in place.

Table 10-1. Policies Used by Las Vegas

Types of Policies	Has Policy	Lacks Policy
Local government and utility policies		
Utility rebate programs	X	
City operated energy efficiency programs	X	
Local energy conservation programs	X	
Local policy		
Sustainable development as a policy goal		X
High occupancy vehicle lanes		X
Participation in membership based organizations		
International Council for Local Environmental Initiatives	X	
Clean Cities Program	X	
Energy Star Partner	X	
Environmental policies		
Curbside recycling program	X	
Brownfield redevelopment program	X	

Sources: Las Vegas Review Journal (2002); City of Las Vegas (2002); United States Department of Transportation (2004); ICLEI membership website, http://www3.iclei.org/member.htm. 13 March 2004; U.S. Department of Transportation's website: hovpfs.ops.fhwa.dot.gov/inventory/inventory.htm; Las Vegas Regional Clean Cities Coalition (2002); United States Environmental Protection Agency (2004); United States Environmental Protection Agency (1998, May).

Creative energy policies and several broadly based programs are being implemented. Las Vegas is outpacing many other Sunbelt cities in its efforts to promote alternative energy and provide opportunities to use alternatively fueled vehicles. Policies for Las Vegas are designed to reduce energy usage and improve the environment.

Las Vegas has implemented seven of the ten policies considered in this study. Las Vegas is strong in both local government and utility policies and environmental policies. The city participates in membership based organizations yet is weak in other local policies. Las Vegas attempts to influence urban energy consumption by cooperating with its local utility, establishing energy efficiency programs, supporting curbside recycling, and promoting brownfield development. Organizational memberships are limited primarily to federally sponsored programs. Las Vegas has not established sustainable development as an urban goal. If there are limits to the growth of Las Vegas, it will be the availability of both water and energy resources that become the primary factors limiting development.

QUANTITATIVE SUSTAINABILITY ASSESSMENT

Quantitative data provide revealing information about the city of Las Vegas. Las Vegas is the largest city in Nevada with a population of 478,434 in 2000, or 23.9% of Nevada's total population. Las Vegas is a "Type A" city, yet it needs to find creative ways to accommodate rapid population growth. While not the state capital, Las Vegas is the commercial center of Nevada. Las Vegas continues to grow rapidly. As of 2006, the U.S. Census Bureau's estimate of city population was 552,539 with an MSA estimate of 1,777,500, making Las Vegas the 28th largest city in the U.S.

The climate in Las Vegas is severe. Due to its dry and arid climate, the Las Vegas basin experiences 3,214 cooling degree days, far more than is typical for most Sunbelt cities. The basin receives 259 langleys of solar radiation on a horizontal surface during a normal January, somewhat more than the average experienced by the sampled Sunbelt cities.

Total energy usage in the state of Nevada was 667.6 trillion kilojoules during 2000, an increase from the 434.2 trillion kilojoules used during 1990 (Energy Information Administration 2000). This represents an increase of 53.8% during the ten year period. Per capita energy use was 334.3 million kilojoules during 2000—a decline from 361.5 kilojoules during 1990 (Energy Information Administration 2000). The year 2000 per capita usage is substantially greater than the energy use by its Sunbelt neighbors California (265.5 million kilojoules) and Arizona (250.1 million kilojoules). Compared to the other eight non-oil exporting states represented in the sample (Arizona, California, New Mexico, North Carolina, Tennessee, Florida, Georgia, Virginia) which had an average per capita energy usage of 330.5 million kilojoules in 2000, Nevada's energy usage of 334.3 million kilojoules was barely 1.0% higher. Using this measure, energy use in Nevada is representative of the non-oil exporting states considered in this study. However, per capita energy expenditures in 2000 for Nevada were $2,413 compared to an average of $2,247 for this comparison group (7.4% greater).

When compared to the average of all selected Sunbelt states, Las Vegas continues to have below average per capita energy usage and continues to benefit from below average per capita energy expenditures. However, as indicated from the analysis in Table A5 (see Appendix), the estimated total energy use for Las Vegas has increased by 71.3% from 1990 to 2000, outpacing all other cities in the sample.

Transportation Energy Usage

Per capita transportation energy usage is 108.8 million kilojoules, 32.6% of Nevada's total energy use. Per capita transportation energy use in the state of Nevada in 2000 was greater than per capita transportation energy use in nearby California (95.4 million kilojoules) and Arizona (95.2 million kilojoules). Nevada's transportation energy usage was slightly higher than the average of the other eight non-oil exporting states represented in the sample, which had an average per capita transportation energy usage of 103.7 million kilojoules in 2000.

Residential Energy Usage

Per capita residential sector energy usage in Nevada was 66.0 million kilojoules (19.8% of total energy use) in 2000. Per capita residential sector energy use in the state of Nevada is substantially greater than per capita residential sector energy use in nearby California (41.1 million kilojoules) and somewhat greater than in Arizona (58.2 million kilojoules). Nevada's residential energy usage was slightly higher than the average of the other eight non-oil exporting states represented in the sample which had an average residential usage per capita of 64.8 million kilojoules in 2000.

Despite being located in the Las Vegas basin, a region with ample sunshine that could be exploited, Las Vegas is among those cities with the lowest percentage of residences (0.3 percent) that use alternative fuels, or no fuel, for heating.

Residential Travel

The average travel time to work (one way) for residents of Las Vegas is 25.4 minutes. This is slightly greater than the mean of the sampled Sunbelt cities. This value ranged from 18.6 minutes to 29.6 minutes with a mean value for the 25 sampled Sunbelt cities of 24.5 minutes.

The percentage of residents in Las Vegas who used public transport, walked to their places of employment, or worked from home during the year 2000, was 9.4% of the total workforce. This is slightly below the mean value of 9.9% for the 25 sampled Sunbelt cities.

The majority of Las Vegas households (51%) have no or only one vehicle and the remainder (49%) have two or more vehicles. The mean value for the 25 sampled Sunbelt cities is 51.8% indicating that Las Vegas is typical of the sampled cities.

Air Quality

The average daily air quality index (AQI) value for the city of Las Vegas during 2000 was 55, which places it in the "moderate" category. Within the group of Sunbelt cities the mean AQI value for the 25 sampled cities was 50.8 for the year 2000. The average daily air quality index for Las Vegas is above the mean of the selected Sunbelt cities. Using this measure, air quality in Las Vegas is poorer than in most Sunbelt cities.

However, there were only five days in Las Vegas during 2000 that air quality was unhealthy or unhealthy for sensitive groups. This is less than other cities in the region including Phoenix (30), Los Angeles (88), San Diego (36), and Mesa (13) but greater than Tucson (0). The mean value for the sampled Sunbelt cities during the year 2000 was 28.4 days. Only 6 of the 25 Sunbelt cities that experienced fewer days that were either unhealthy or unhealthy for sensitive groups compared to Las Vegas. Though many residents of Las Vegas may hold differing opinions, air quality is normally healthy for all groups of individuals when based on this measure. Air quality is not generally problematic when compared to the air quality of other Sunbelt cities.

Population Density

The population density of Las Vegas during 2000 was 1,197 persons per km^2 (3,100 per square mile). The mean population density for the sampled Sunbelt cities is 1,203 persons per km^2 (3,116 per square mile) indicating that the population density for Las Vegas is almost equal to the mean for the Sunbelt cities sampled.

The relative change in population compared to density for the period 1990 to 2000 provides additional evidence that Las Vegas is experiencing very rapid population growth. The population of Las Vegas has increased by 85.2% during the ten year period from 1990 to 2000. This rate of increase is less than the increase in total energy use for the period. Las Vegas experienced the largest positive change in population density, increasing by 36.2%. Las Vegas also outpaces all other sampled Sunbelt cities with a value of 49% for the highest difference between percentage of population change and percentage change in population density. The data indicate that rapid population growth is requiring both expansion of the city's land area and increases in population density in order to accommodate a growing population and an influx of new residents.

Table 10-2. Measures of Las Vegas Policies

Variable	Year 2000	Index Score
Alternative transportation	9.4%	0.24
Population density	1,197/km2	0.25
Homes heated by alternative fuels	0.3%	0.00
Days unhealthy and unhealthy for sensitive groups	5 days/year	0.96
0 to 1 household vehicles	51.0%	0.45

Sources: U.S. Census Bureau, data calculated using both U.S. *Census 1990* and U.S. *Census 2000*; U.S. Environmental Protection Agency, data from www.epa.gov/airnow/.

CONCLUSIONS

This case study delved into policies and program being pursued by the city of Las Vegas. It provides insight into its history, the course of its development, and its energy and sustainability concerns. The city grew as a result of the availability of water, access to regional transportation, and favorable taxation policies. The construction of a nearby hydroelectric plant provided the region with low cost electricity, and became the catalyst that ultimately spawned vigorous development in the Las Vegas basin. This case illustrates how layers of environmental problems evolved during the history of the basin's development and how many of these problems are associated with energy use. Local policies have been implemented to mitigate environmental and energy-related impacts.

Many of the drivers of energy and sustainability policies apply to Las Vegas. Population growth has strained the resources of the region. Urban development has created increasing demands for land. Energy use has increased substantially. These and other factors have contributed to urban environmental impacts such as increased air and water pollution, while damaging a fragile desert ecosystem. Past atomic energy research and testing has caused irreparable environmental degradation in the region. There has been extensive population migration to the city. New transportation systems, water systems, and utility system infrastructures have been required to sustain living standards for the growing population. All of these causes and effects can be linked to sustainability and energy use.

Measures of per capita energy usage indicate that Nevada in general and Las Vegas in particular use slightly more than the average aggregate energy use of the non-oil exporting Sunbelt states and more energy in the transportation and residential sectors than neighboring Sunbelt states. Las Vegas provides a number of examples of the types of identifiable energy policies and programs that cities can use.

From 1990 to 2000, total energy usage in Nevada increased dramatically while per capital measures of energy use have declined. Such a large increase in aggregate energy use is not common among Sunbelt cities. Like many other Sunbelt cities, it is clear that Las Vegas is implementing policy initiatives to address energy and sustainability issues. The declines in per capita energy use provide empirical evidence indicating that energy efficiency and energy conservation policies used by Las Vegas are having an impact.

While Las Vegas has not adopted sustainability as an urban goal, the city was found to be pursuing seven of the ten policies that were selected for study. The policies used by Las Vegas included utility rebate programs, city operated energy efficiency programs, local energy conservation programs, Clean Cities, Energy Star, brownfield redevelopment and curbside recycling.

During the 1990s, the city added and expanded policies and programs that exemplify the current direction that the city is taking in response to environmental concerns. Las Vegas is outpacing many other cities in its efforts to promote alternative energy in the transportation sector by investing in the use of alternatively fueled vehicles.

Las Vegas has a population density that is typical of the average of the other sampled Sunbelt cities. Travel time to work, and the portion of the population who use public transportation or work at home, is also about average. Atypical for Las Vegas is the low percentage of residences which use alternative fuels for heating. Las Vegas fails to fully utilize its bequest of solar energy by not designing its residences to better utilize solar resources for heating purposes. Compared to other Sunbelt cities, residents of Las Vegas spend less per capita on energy than average. They also consume more energy than people in nearby Sunbelt cities. It is possible that lower energy costs may have contributed to greater energy usage. While the air quality index averages for Las Vegas are somewhat higher than averages of other Sunbelt cities, there are far fewer days when the air is unhealthy. Rapid population growth has required expansion of the city's land area. Increases in population density continue unabated.

LESSONS LEARNED

There are lessons to be learned from Las Vegas. If not for its desert location, population growth, concerns regarding water supplies, and focus on the gaming industry, Las Vegas has developmental features not unlike many growing American metropolises. Without formally adopting the overarching goals of sustainable development, Las Vegas pursues many policies that can be categorized as sustainable initiatives. Problems, often environmental, surface and result in shifts in policy. There is a lag in time before policies begin to be effective. Environmental problems often become catalysts for policy adoption.

It is easy to wonder, given the economic resources available to Las Vegas, what the city might have been like if alternative developmental and transportation programs had been pursued. Landlocked Las Vegas had a rail line early in its history. What would the city be like today if rail and tram transportation systems had been developed as the primary means of moving people within the city? Rapid development of the city occurred after transportation by air was available. With a goal of attracting tourists and gaming enthusiasts, many of whom now fly to the city, why didn't Las Vegas choose to develop as a more pedestrian-friendly city? Taxicabs remain a primary means of moving people from the airport to hotels downtown and on the Las Vegas Strip. With water in short supply, why is it that water conservation has become a matter of concern only in the last decade?

Urban development has consumed the once spacious Las Vegas Valley. The fact of the matter is that the developmental choices made by Las Vegas, and many other U.S. cities, have locked in infrastructure that will be difficult and costly to maintain and replace. Las Vegas is the quintessential example of a city that has not purchased insurance against the possibility of an end to the Hydrocarbon Age or the depletion of its water supplies.

The policies of Las Vegas are in flux. Today, the strategy being pursued by Las Vegas is to target a set of politically palatable policies, incorporate a number of sustainability-related programs, and seek new opportunities for sustainable developments.

One example is the Las Vegas Springs Preserve. It is a new $250 million non-gaming cultural attraction that's incorporating some of the best principles of sustainable design. An article in *Earth Times* recently reported the following:

> *"The Desert Living Center (DLC) at the Springs Preserve will provide a forum where visitors can learn practical means of protecting valuable environmental resources without compromising their quality of life. A complex of five buildings, the DLC consists of the Sustainability Gallery, dialogue center, design lab and technical training center, ticketing area and general classroom, conference and office space. The DLC will offer exhibits, classroom programming, events, conferences, and activities illustrating the benefits of recycling, conservation and alternative energy sources in a fun and interactive environment for adults and children of all ages."*[10]

The idea of green buildings is being introduced. Buildings at Las Vegas Springs Preserve are being designed in accord with LEED standards. The park area will include two historic museum galleries, a children's gallery, an indoor theater, a 2,000-seat outdoor amphitheater, and a facility called the "Ori*gen Experience," which captures the essence of the land and early inhabitants who made the area their home.

Such efforts are only a beginning. Las Vegas has a long way to go in its efforts to achieve sustainability. According to Jeff McIntire-Strasberg:

> *"To say that Vegas is stretched well beyond its carrying capacity is an understatement—it's a place that has traditionally thrived on wastefulness.*

As more people are making the city and region their home, I'm guessing there's much more consideration of the relationship to the natural environment, as well as the threat of sprawl eating up some truly magnificent lands outside the city. The development of Springs Preserve shows me that a sea of change has occurred in thinking in Vegas, and I'm happy for that. Vegas itself has a long way to go, but if sustainable thinking can take hold there, then there's hope for all of us."[12]

The support of programs to increase use of alternative vehicles is a highly advertised and notable success. At a cost of roughly $100 million dollars per kilometer, the Las Vegas monorail system is itself an attraction that provided employment for the construction industry and expedites the movement of tourists to and from the hotels and casinos. However, less exotic policies, such as developing basic public transportation systems by expanding suburban bus routes, have not been widely implemented. Las Vegas serves as an example of a city that is having difficulty expanding mass transit systems to the less densely populated urban fringe.

Though embryonic, the policies being adopted by Las Vegas do not appear to be broadly successful at solving looming sustainability problems caused by rapid land development, increasing energy use, and liberal water consumption. All of these have impacted urban sustainability. Development is at the limits of controllability, energy use is greater than in neighboring states, and water consumption is higher than elsewhere in the U.S.

Las Vegas is an example of how increasing energy consumption impacts sustainability. Energy and water usage are related, as more energy is required to manufacture and deliver potable water to points of use. Increasing energy usage has contributed to higher expenditures for energy, increased urban air pollution, and variable changes in density. All of these impact urban sustainability. If the cities of Albuquerque, San Diego, and Los Angeles used energy at the same rates as Las Vegas, their energy use would be significantly greater, energy costs would be higher, and their air pollution levels would be more difficult to control.

What can Las Vegas do to reduce its energy usage and improve sustainability? The data seem to indicate that the local policies used by Las Vegas are not entirely effective. Las Vegas has a hot dry climate with ample sunshine. The winter climate creates demand for space heating, yet few buildings are heated with renewable fuels such as solar energy. Despite concerns about air pollution, air quality in Las Vegas is far healthier than in most other Sunbelt cities. Lucky Las Vegas has favorable winds and a service sector economy. Las Vegas might expand programs to improve air quality before additional air quality problems develop.

In regard to transportation, the city of Las Vegas has opened the world's first hydrogen energy station to support motor vehicles. In addition, the school district has buses that use biodiesel. Expanding the biodiesel program and obtaining more buses for the local municipal services are ideas worthy of consideration. Expanding other types of AFV programs such as promoting the use of electric and hybrid vehicles is also an idea worthy of pursuit. Supporting creative policies that will cause a reduction in the number of vehicles per household is part of the solution. Creating high occupancy vehicle lanes for alternatively fueled vehicles and carpoolers is worthy of further study. Installing bicycle lanes and offering bicycle rentals might also be viable.

The Las Vegas monorail system may help move tourists, but fails to help reduce commuting or the number of local automobile trips per year. Promoting another phase of the monorail system that would connect the system to the airport, the University of Nevada Las Vegas campus, and to the downtown center would help reduce automobile use and reduce the need for visitors to rent cars and hire taxis.

There are alternatives in development patterns and in the design of structures that can be used to a greater extent to reduce urban energy use. Designing and retrofitting homes with building envelope improvements is a reasonable solution due to the local climate conditions.

Most importantly, there is a critical need to expand incentives and programs that will result in more residences using alternative fuels for heating and installing alternative heating systems (non-electric) in existing homes. Expanding the use of solar water heating systems needs to be promoted by creating local incentives for residences and commercial enterprises. The redesign of new buildings with reduced fenestration, smaller floor plans, increased insulation and improved shading of perimeter walls are local design solutions that need to be considered by builders and developers. Expanded use of light emitting diode lighting needs to be considered for signage and display lighting applications.

More stringent water conservation policies are needed not only to reduce water consumption but also to reduce the energy consumed by water supply systems, pumping stations and sewage treatment facilities. For residences using conventional water heating systems, flow restriction devices can be installed in showers to reduce water consumption. Incentives need to be developed so that more water-saving plumbing fixtures to be installed. This will not only reduce water usage but will also reduce the energy required to pump water to where it is needed.

Other recommendations include increasing population density in the planning of new suburban developments. Reverting exterior plantings to indigenous forms of landscaping for residences and golf courses can help reduce water use as well.

Finally, Las Vegas might consider establishing sustainable development and local sustainability as an urban goal, further directing policy agendas toward these ends. Many of the recent policies that Las Vegas has chosen to pursue are in alignment with the sustainable cities agenda. Regardless, Las Vegas faces gargantuan sustainability challenges in the future that will ultimately decide its fate.

Endnotes

1. Denton and Morris (2001:7) also state that, "Upward from a dark glass pyramid beams a searchlight of 40 billion candlepower, said to be the brightest ray in the solar system, save for the sun or a nuclear blast."
2. Other U.S. examples of significant population growth from 1990 to 2000 include: Naples FL, +65.3%; Yuma AZ +49.7%; Austin, TX, 47.7%; Boise, ID, +46.1%; Phoenix, AZ, +45.3%; and Laredo, TX +44.9%. Source: U.S. Census Bureau, Census 2000 and Census 1990. See www.census.gov.
3. Yancey. K. (2007, 18 May). It's crunch time for booming Las Vegas tourism. *USA Today*. p. 1D.
4. Ibid.
5. An acre foot = 326,000 gallons, ample enough to supply two families of four people for a year.
6. Elliot (1973:314) also reports that, ."..underground testing of nuclear devices has not been completely concerned with military uses. In 1962, the AEC initiated the Plowshare Program to find peaceful uses for atomic energy. Under this program the scientists at the Nevada Test Site have conducted tests to determine the possibilities of using atomic power to dig harbors and canals, for road building, and for other types of excavation. Two other potential uses of atomic power, in mining and in developing underground water sources are of particular interest to Nevadans."
7. *Lake Mead Water Levels — Historical and Current.* www.arachnoid.com/ NaturalResources/index.html, accessed 4 July 2007.
8. City of Las Vegas. http://www.lasvegasnevada.gov/Search/search_results.asp, accessed 8 January 2004.
9. The new monorail actually replaced an earlier 1 Km monorail that connected the MGM Grand and Bally's resorts (Rake 2003:1)
10. EARTHTimes.org. (2006, 1 November). *Las Vegas' version of central park a model of sustainability.*
11. Ibid.
12. Jeff McIntire-Strasberg (2006, 2 November). *Las Vegas' new "central park"—a vision of sustainability.* http://sustainablog.blogspot.com/2006/11/las-vegas-new-central-park-model-of.html, accessed 7 January 2007.

Chapter 11

International Sustainability

"The growing interest in sustainable development as the transcendent goal for ecological planning and regulation illustrates the expanding temporal scale of environmental policy making... The idea of sustainable development, like ecosystem management, is weighted with ambiguities, so that disagreement abounds over the precise meaning... Still, both concepts resonate throughout environmental policy discourse at all governmental levels because the ideas appeal powerfully to scientific logic and planning experience."

ROSENBAUM (2005:19)

It is amazing how quickly sustainable development has captured the imaginations of policy-makers around the world. This important new paradigm of sustainability transcends national boundaries and permeates policies in all societies. One reason is that the importance of energy and environmental stewardship is now widely accepted. International efforts towards sustainability are varied, multi-dimensional and provide a new direction and a refined focus. In European Union countries and elsewhere, sustainability initiatives are a function of national policies, regional needs, and the concerns of the local political structures. Sustainable development policies in the EU countries tend to be structured by Brussels with guidelines and directives filtering down to the member countries and local governments. Individual states are responsive.

For example, Sweden has committed to phasing out its reliance on fossil fuels by the year 2020, including those used for transportation purposes. Central government funding is also available in the EU. This contrasts markedly with sustainability policies in the U.S., where a lack of central government leadership means that policies are decentralized, springing from local governments without central government influence or financing.

Sustainability is a seemingly complex and sometimes lofty goal that requires regional solutions and local initiatives. Local commitment to sustainability is required for long-term policies to be successful. It can be easier to identify sustainability problems than it is to determine their corresponding solutions. Once solutions are found, it can be even more difficult to implement them. Why is this? The solutions to regional sustainability problems require substantial capital investment, infrastructure improvements, and regional cooperation. These must also converge at the same place and point in time. This combination can be elusive. Avoiding unintended consequences that create new problems can be challenging.

This chapter discusses relevant international examples of sustainability policies and efforts.

EXAMPLES OF INTERNATIONAL SUSTAINABILITY EFFORTS AND ISSUES

Many international sustainability initiatives include the use of alternative energy systems. The city of Saarbrucken, Germany, with a population of roughly 180,000, began implementing an environmental strategy based on reducing energy consumption in the early 1980s which included using passive solar and hydropower technologies (Dincer 1999:845-854). By 1990, their program had resulted in a 15% reduction in the city's heating demand, a 45% reduction in heating energy use in municipal buildings, and a 15% overall reduction in CO_2 emissions from reduced heating and electrical requirements (Dincer 1999:845-854). Iceland is another example. Many of its cities are powered by geothermal energy.

Wind turbine generators have been appearing in Switzerland, Austria, and Hungary. Hong Kong recently installed its first wind turbine generating station and has office buildings in Kowloon that are designed to harvest solar energy. In the coastal city of Rizhao City (which translated means "City of Sunshine"), in northern China, 99% of the "households in the central districts use solar water heaters, and most traffic signals, street lights, and park lighting systems are powered by photovoltaic solar cells" (Worldwatch Institute 2007:108). As part of a strategic decision to avoid costly wiring, pole-mounted street lights in remote areas of Brazil use fixtures with LED lamps, lithium batteries, and solar photovoltaic arrays mounted on their casings. The light fixtures collect solar energy during the day, store it in concealed batteries, and use the power to provide illumination at night.

Promises to bring economic prosperity to third world countries by developing energy resources have instead yielded mixed economic results and sometimes devastating ecological damage. Shell-affiliated oil companies struck oil in Nigeria in 1956 and today employ nearly 6,000 people and own roughly 90 flow stations with pipelines throughout the

Niger delta.[1] Other companies operating in the country include Chevron-Texaco, ENI (Agip) and ExxonMobil.[2] The production of oil and natural gas has contributed to respiratory diseases, crop failures, greenhouse gas emission from hundreds of flares, and the loss of 20,200 hectares (50,000 acres) of mangrove swamps from 1986 to 2003 (O'Neal 2007:97-117). Parts of the country can no longer be cultivated. There have been almost 7,000 documented oil spills and many others that remain unrecorded. An explosion from a natural gas pipeline killed more than 700 people. While oil accounts for 80% of Nigeria's revenue, the economic benefits of oil development are unevenly distributed. The economic impact of "black gold," initially thought to bring prosperity to the country, has fomented political unrest, endemic violence and netted an annual per capita income equivalent to only $1,400 (U.S.) (O'Neal 2007:97-117).

Many villages and towns lack electricity, clean water, medicine and schools. Oil spills and acid rain in the region have stressed wildlife and decimated local fish populations. The result is unbalanced ecosystems. Amazingly, gas stations in Nigeria often are closed due to lack of refined petroleum products. After reducing its oil output by 500,000 barrels of oil

11-1. Amsterdam –Bicycles and Canals

daily for a number of years in the western regions of the Niger Delta due to military violence, attacks and bombings, Royal Dutch Shell is planning to resume full production in late 2008.[3] However, in the oil-rich Delta area, political tensions often fuel armed military movements, singling out oil companies as a means of forcing the government to invest more money in the region.[4]

Alternatively, Holland, a country of 13 million, has quietly become one of the world leaders in sustainability initiatives. The Dutch have a clearer understanding than most of the linkages between energy and sustainability. By structural adaptation, the Dutch "have developed a fighting culture to capture more land from the sea and create significant sea defenses to hold the line... The Netherlands are pre-adapted to climate change—especially sea level rise."[5] With Holland's system of dykes that protect almost half of the country, it has managed to hold back the North Sea for over 800 years. The use of wind power for water pumping has been a strategic advantage for much of its history.

Consider the Dutch city of Amsterdam. The central core of the city is pedestrian friendly. The transportation support system includes water-

11-2. High-Rise Bike Garage at Amsterdam Railway Station (photo—Mike Clust)

ways, trams, buses, and bicycle routes in addition to motorcars. While many U.S. cities struggle to find adequate parking for vehicles, Amsterdam struggles find places to park its bicycles. The city has roughly 750,000 people and 600,000 bicycles; over 85% of its residents use them. The city boasts of its multi-level parking facilities for bikes. Many residents consider bicycles the *preferred method of transportation*. By using trams and bicycles, dependence on fossil fuels for transportation purposes is reduced as are greenhouse gas emissions.

Contrasts within countries are striking. At Las Terrazes, a sustainable (albeit socialist) model community in Cuba, there is an extensive reforestation project to reclaim desolate mountain areas, terra-forming the landscape into beautiful tropical forests that have been designated as UNESCO world heritage sites (Davey 2005:26). Here the local economy is focused on ways to develop a cottage tourist industry.

Next, we consider international sustainability issues and the cities of Havana, Cuba and Venice, Italy. There are similarities and striking differences between then. These cities have a history of surviving dynamic environmental changes. They offer evidence that sustainability concerns

11-3. Residences in Los Terrazes, Cuba

have regional implications. Long term solutions must be tailored to local situations. Havana and Venice are coastal cities, and by definition are located at the edge of two ecosystems—one of water and the other of land. Energy in both cities plays a key role in achieving sustainability.

HAVANA, CUBA

Havana and its surrounds, once the capital of the Spanish Caribbean and the principal New World port, became a sugar and cigar supplier to the world (Thomas 1998:1). A place that lazes in tropical sunshine, Havana's nearby forests were once cleared to develop sugar and tobacco plantations. Despite its former wealth and long history, Havana today suffers from the burdens of its swelling population, shortages of investment capital, and decaying infrastructure. Havana's population growth has been thwarted by restrictions precluding migration from the countryside to its urban areas, restricted energy supplies, and communist economics.

Havana is the capital of Cuba. It is a city that during most of the twentieth century has an identity linked to the socialist vision of one man—Fidel Castro—whom Ernest Hemingway had once described as "brave as a badger" (Thomas 1998:918-919). Hemmingway, who supported Castro's revolution, spent ten years in Havana where he is said to have begun writing *For Whom the Bell Tolls* (McAuslan and Norman 2000:84).

Havana seems an "old world" city in a "new world" setting. Its early development resulted from its natural harbor and its advantageous location from which to control shipping lines. Havana's streets are laid out as irregular grids and bracketed by avenues that delineate neighborhoods and serve all manner of traffic. Tunnels connect the old quarter to newer development across its bay. In Havana, main streets are multi-functional. For local transportation, people drive antiquated gas-guzzling, pollution-generating, automobiles and light trucks of U.S. manufacture that date from the 1940's and 1950's, and Soviet-manufactured Ladas and Volgas sedans. Havana's history can be read in the faces of its people and facades of its buildings. The faces of Havana's people express acceptance, liveliness, pride, and struggle. Their buildings express their creativity and resourcefulness.

Havana—Background

The city of Havana has a rich and colorful history, glorious architecture, and offers exuberant hospitality. The location of present day Ha-

11-4. Street Scene in Old Havana

vana gave birth to the city in 1519. It became the capital of Cuba in 1556. The town developed as a service to the shipping industry and became known for shipbuilding. An infrastructure of warehouses, inns, brothels and gambling houses catered to the sailors who visited her harbor (McAuslan and Norman 2000:69). By the end of the 17th century, Havana had established itself as a fortified urban port. It has the largest natural harborage in the Caribbean and was convenient to shipping lanes and trade routes (McAuslan and Norman 2000:65-68). After having won Havana in 1762, the British traded their rights to Cuba in exchange for Florida, resulting in an influx of Spaniards who expanded the sugar plantations. During its history, the city has been variously called the "Key to the New World," the "Pearl of the Antilles," and the "Jewel of the Caribbean" (Baker 2003:1-15).

The city has endured the near-elimination of its native population due to famine and disease. It has experienced slavery, direct military assaults and occupation, colonialism by both the Spanish and English, U.S. "dollar diplomacy," and what in Cuba's view is economic imperialism. It was the city where Winston Churchill believed in 1895 that "anything might happen," and where Spanish poet Frederico Garcia Lorca wrote, "If I get lost look for me in Cuba" (Baker 2003:1-5). The center of thriving Cuban culture, the city and its suburbs now extend far beyond the original city limits. It is a city of contradictions—full of white coral limestone buildings, spiced with a rambunctious night-life, and alive with cultural amenities that include cinemas, historic theatres, cabarets, ancient cemeteries, breathtaking Catholic cathedrals and several Jewish synagogues.

By the mid-1920's approximately 80,000 tourists were visiting the city annually (Baker 2003:18). In the late 1950s, with Batista in power, tour-ism was providing million of dollars in foreign trade annually and creat-ing thousands of jobs. Not all benefited society. Some of these jobs resulted from gambling and prostitution. Dollars flowed to the ruling regime and mob-lead companies. However, tourist dollars allowed the construction of motels and supported public works projects including the first highway tunnel beneath Havana Harbor.

By the 1960s, the revolution brought an end to this colorful vibrancy, cleared the streets of prostitution and gambling, and shaped the basis for

11-5. The Harbor in Havana

a new communist state (McAuslan and Norman 2000:70). A failed plot in 1960 by the CIA, which conspired with Chicago gangster Johnny Roselli to poison Fidel Castro, coupled with the unsuccessful U.S.-planned and financed Bay of Pigs Invasion in 1961, caused a total breakdown in U.S.-Cuban relations.[6] Soviet attempts to place nuclear armaments on the island led to a U.S. blockade in the 1960s, economic stagnation, and an often harsh and sometimes brutal dictatorship. A U.S. trade and economic embargo was initiated in 1963—a Cold War policy which remains in effect. Afterwards, the Cuban economy was supported by the U.S.S.R. with subsidized trade, but most investment flowed to non-urban areas that had supported Castro's rise to power. After socialism failed in Eastern Europe in 1989, the flow of subsidies from the U.S.S.R. subsided and then stopped, triggering a period of severe economic hardship.

Havana is a city that survives even though it has scarce resources, sustains itself on limited tourism, and despite all of its challenges, maintains an international stature as a center for tobacco and sugar exports. Development and investment were stymied until the Castro government legalized ownership of the U.S. dollar, permitted some forms of private enterprise in 1993, and revived tourism. Prime Minister Fidel Castro has "made tourism a top economic goal, and Havana's partially restored Old City is providing a strong drawing card for visitors" (Judge 1989:293). Foreign tourists and the U.S. dollar are now dominating features of daily life in the capital. Recently, Cuba's tourist industry has been among the fastest growing in the world, increasing at an annual rate of 20% per year (Doggett and Stanley 2001:36). However, this yields only 2 million visitors annually, a number equal to about a quarter of the visitors to the U.S. territory of Puerto Rico (Doggett and Stanley 2001:36). Tourism has spawned new investment but has also helped "create two societies in Cuba—one with dollars, and the other without" (Putman 1999:14).

Fear not—the spirit of capitalist entrepreneurship is alive and well in communist Cuba. In the upside-down Cuban economy, the highest earners cater to the needs of tourists—artists, owners of small restaurants, and anyone willing to risk bootlegging cigars to foreign visitors. While a doctor might earn a furnished apartment and $20 per month, a driver of a pedicab working for tourists might earn $15 on a good day (Putman 1999:12-14). Elsewhere in Cuba, commercial agriculture is stagnating due to economic conditions while the markets for small-scale production of privately grown fruits and vegetables are improving.

Havana—Population

By 1600, Havana was a town of only 3,500 permanent residents with most of the development adjacent to the bay. It required another 200 years for the city to grow to a population of 40,000, and by 1800 Havana was the third largest city in the New World after Mexico City and Lima. It grew to 55,000 by the mid-1850s, and little land remained available for development within its fortified city walls (Baker 2003:9-12). To accommodate expansion, the city's walls were dismantled.

The pace of Havana's population growth quickened during the first half of the 20th century: to 238,981 in 1899; 302,526 in 1907; 363,506 in 1919; 728,500 in 1931; 964,000 in 1943 and 1,223,900 in 1953 (Scarpaci *et al.* 2002:120). By 1958, the population of Havana stood at 1,361,600 (Scarpaci *et al.* 2002:120). The period from 1931 to 1958 was one of phenomenal growth, increasing Havanna's population by 87%. With an influx of mixed races that included Mestizos, Spanish, Africans, Chinese, and Indians. "If ever a pot melted, it was Old Havana" (Judge 1998:297). During most periods of its history, the growth rate of Havana's population exceeded that of the rest of Cuba. However, after 1966, the growth rate of the city declined to below the national level due to restrictions on immigration from the countryside. Today, the estimated population of Cuba is 11 million and the population of Havana is 2.2 million (Doggett and Stanley 2001:36).

Rapid urban development created the need to carefully consider planning alternatives. The city was laid out in sections, mostly grid patterns intersecting at angles and differentiated by the landscape's geographical features. Avenues are either parallel or perpendicular to the sea and along with boulevards, provide definition to Havana's hierarchy of streets. Beautification efforts in the mid-1920's were led by Jean-Claude Forestier's *beaux arts* scheme which envisioned Havana as a European city rivaling Buenos Aires, Argentina. This evolved to become a city-wide master planning effort in 1926 that remained largely unfulfilled due to the fall of the Machado regime 1933. Havana implemented its first city master plan in 1963, the first ever in revolutionary Cuba (Scarpaci *et al.* 2002:139).

Havana—Environmental Issues

There has been a long history of environmental problems in Havana. Despite its age, an aqueduct constructed in 1856 still provides water to the city. As late as 1900, most houses used cesspools to collect sewage. These were occasionally drained into the harbor, causing passengers arriving by ship to be greeted with the odor of raw sewage (Baker 2003:15). By 1920, infrastructure improvements provided 426 kilometers (265 miles) of storm drains and sewers but untreated effluent was pumped directly into the sea at Cojimar, then several kilometers east of the city (Baker 2003:15).

Today, over a million people live, work or visit the shores of *Bahia de La Habana* daily. Havana's skies and harbors need improved environmental control and regulation. According to Doggett and Stanley (2001), pollution of Havana Bay remains one of Cuba's biggest environmental problems. Today, Havana's bay is dominated by heavy industry that has contributed to problems with air and water quality (McAuslan and Norman 2000:67). Wastes from chemical plants, paper mills, and inadequate sewage systems have taken a toll on the bay (Doggett and Stanley 2001:28). The impact of health problems caused by environmental pollution remains uncertain.

Recent infrastructure improvements that divert pollutants have provided some relief. Fish and birds have recently returned to *Bahia de La Habana* as pollutant levels have declined. But sources of pollution remain, such as the highly contaminated Luyano River that empties into the bay. Though pollution has moderated, much work remains, especially in the clean-up of rivers and drainage systems that feed pollutants into the bay.

By western standards, the air pollution regulations that seek to control emissions from electrical power plants are unacceptable. While Cuba now produces 80% of the oil it needs for electrical energy, power stations required costly conversion to use oil with high sulfur content.[7] Havana relies on aging electrical generation equipment that lacks rudimentary pollution-control devices. The smokestacks of electrical power plants spew untreated black smoke. Since Havana is located on the sea, winds normally disperse the pollution from these point sources. On other occasions, the winds direct pollution to downtown and suburban areas of the city.

Cuba's growing energy needs are causing concerns in the U.S. Cuba's interest in pumping oil from off-shore deposits on its northeastern coasts might cause environmental damage to its northern neighbor. Florida Senator Bill Nelson is concerned about potential losses to the Florida tourist industry and wants to find a way to block drilling in Cuba's northern waters: "Any oil spill 45 miles (72 kilometers) from Key West is going to absolutely devastate all those delicate coral reefs and the fragile Florida Keys, and endanger beaches all the way up to Fort Pierce."[8]

Despite glaring air and water quality problems, there remains a remarkable lack of an environmentally oriented culture in Havana. "This lack of concern about the environment bodes poorly for Havana" (Scarpaci *et al.* 2002:183). As you stroll along the streets of communist Havana, past the paramilitary-clad neighborhood "monitors" on busy street corners, rarely do you hear complaints about lack of infrastructure, environmental concerns, utility interruptions, or developmental issues.

Havana—Sustainability Issues

In the early history of Havana, when the city was a colonial maritime port, the surrounding woodlands were deforested to support the city's ship building industry. Nearby jungles were terra-formed into plantations and slaves were imported from Africa to work them. The local economy generally grew until the mid-20th Century, then a flight of wealth occurred in the early 1960s, and economic resources from the U.S. and elsewhere were diverted from Havana.

Havana's stifled development since the 1960s allowed much of its Spanish colonial architecture to survive to the 1990s. Afterwards, most was left to deteriorate and thousands of older homes in the city have since collapsed—a process that continues to this day. The problem has become severe. About 100 colonial-era homes are lost annually due to the lack of structural stabilization and the continuing exposure to the elements (Baker 2003:31). With minimal resources for maintenance, it is surprising that many of these homes remain habitable.

Most colonial-era buildings in Habana Vieja were designed to provide natural ventilation and control the harsh tropical sun. According to Chafin (2003), "This is achieved by deep verandas and arcades creating shade and high shuttered windows to accept the breeze." Houses integrate central courtyards in their designs, places where tropical trees and flowering plants can flourish (McBride and Black 1988:34-35). These architectural and landscape design features provide natural cooling, reducing the need for energy-intensive mechanical cooling systems.

Parts of Havana have been designated as UNESCO world heritage sites. In the Old City, there are 3,147 buildings, and more "than 900 of the area's buildings are of historical importance...144 from the 16th and 17th centuries" (Judge 1989:280-293). The last few years have seen slow and steady improvement, with many buildings being restored, especially in areas of Habana Vieja (McAuslan and Norman 2000:70). Restoration projects have included converting a number of historical buildings into restaurants, bars, and hotels that serve tourists. A special administrative zone has been created in Old Havana that allows a large portion of the income generated by tourism to be used to redevelop the historic sections of the city. Building-by-building, block-by-block, areas of Old Havana are being reclaimed and restored.

The experience of Cuba demonstrates the importance of energy to sustainability. When the Cold War ended, Cuba's economy crashed due to the loss of 85% of its markets and over half of its fuel supply. Cuba had relied on shipments of 255,000 barrels of oil per day from the Soviets under "preferential terms" that were traded for products such as sugar cane and tobacco.[9] In June 1992, the government began a policy of periodic electrical blackouts. There were periods of no air-conditioning, fans, refrigeration or lights in sections of the city (Baker 2003:32). Fuel for transportation was unavailable. Human and animal labor replaced oil-operated equipment (Baker 2003:32). Factories closed down, distribution systems failed, all manner of goods were rationed, and the economic output of Cuba fell 34% from 1990 to 1994 (Baker 2003:31). The situation was so grim that a report from an international relief agency noted, "In the cities, buses stopped running, generators stopped producing electricity, and factories became silent as graveyards."[10]

Energy remains in short supply in Havana, and most of it is used

11-6. Havana Taxi

to support manufacturing, business needs, tourism and personal use. Almost all private homes have electricity but their occupants are subject to electrical and natural gas supply disruptions (both scheduled and unscheduled), in order to enforce mandatory energy conservation policies (McAuslan and Norman 2000:58). Increasing tourist-sector energy use and inadequate maintenance of transmission lines are often cited as the causes of the island's electrical blackouts.[11] Youth brigades have been sent into homes to install energy-saving fluorescent lamps to help reduce loads on the aging electrical infrastructure and help prevent electrical blackouts. Commercial establishments, such as hotels, utilize roof-top solar hot water heating systems.

Years after oil exports from the communist bloc were suspended; Cuba established a program to send several thousand doctors along with hundreds of teachers and engineers to Venezuela, trading their labor for 53,000 barrels of oil daily from Venezuela.[12] To conserve transportation fuels, Havana relies heavily on buses and manufactures durable, two cylinder open-cab vehicles that are used as taxis. These three-wheeled, three-seat vehicles are both economical and fuel efficient.

After 1990, agricultural exports to the U.S.S.R subsided and Cuban crop production has since declined. However, Cuba has the technology to extract biomass energy from sugar-cane production and intends to exploit this energy source in the future to produce fuel-grade ethanol.[13]

Domestic oil and gas production in Cuba has increased from 20,000 barrels per day in the Havana and Matanzas provinces a decade ago to 80,000 barrels today.[14] The country's greatest hope is its 112,000 km^2 (43,000 square mile) exclusive economic zone in waters northeast of its costs that may contain 4.5 to 9 billion barrels of medium grade crude—a reserve almost as large as Alaska's ANWR (and much closer to U.S. markets).[15] A commercially viable discovery could make Cuba an oil-exporting county. Exploratory drilling in this reserve has been undertaken by China's Sinopec and Repsol, YPF, an integrated oil and gas company with assets in Spain and Argentina.[16] Companies from Britain, Brazil, and Venezuela are also considering obtaining rights for future exploration.

After agricultural production in Cuba stumbled, obtaining enough food for the day became the primary activity for many Cubans. An urban agriculture movement began in the 1990s. Lacking land for large-scale projects, residents began growing gardens in yards, parking lots and urban rooftops.[17] Using permaculture, a process of growing vegetables that minimizes energy use, Cubans grew "local organic produce out of necessity," developing "bio-pesticides and bio-fertilizers as petrochemical substitutes," and changing diets to include more fruits and vegetables.[18]

Cuba's commercial food supply system is not self-sustaining and food production continues to decline. Two-thirds of Cuba's fruits and vegetables are produced by its small farmers and its 3,500 cooperatives.[19] After food production dropped 7% in 2006, Vice President Carlos Lage admitted to municipal leaders in Cuba that "production is insufficient and commercialization is deficient."[20] Prices for milk and meats have been increasing and wages are not keeping pace. People complain that fruits and vegetables sold privately are too expensive and that government salaries are adequate to purchase only a few items each month at local markets.[21] Today, 82% of Cuba's subsidized food (distributed by ration), is imported at an annual cost of $1.6 billion—one-third of it coming from the U.S.[22]

Food shortages have led to over-fishing along Cuba's shores and in many reef areas, fish large enough to eat are now rare (Doggett and Stanley 2001:31). Conch and tortoise shells were sold in Havana markets while black coral reef formations were being destroyed for their commercial value (Doggett and Stanley 2001:28). Attempts made by the Cuban government in 2000 to lobby the UN to allow the sale of rare hawksbill turtle shells to Japan were thwarted by a UN vote against the proposal (Doggett and Stanley 2001:28). Having lived in the oceans for 400 million years, there is concern that these turtles may not survive more than another two decades (Doggett and Stanley 2001:28).

Piped water and sewage treatment are not always available in Havana. A significant percentage of homes lack running water and areas of the city often have their water supplies shut off for much of the day due to the inability to meet the demand (Baker 2003:65). Only about 50% of Havana's inhabitants have functional sewage treatment systems (Baker 2003:65).

Despite these challenges, there have been noteworthy successes. For example, health and education are areas of great improvement in Cuba. The literacy rate in Cuba is roughly 95%. Cuba has one teacher for every 42 inhabitants (Putman 1999:27). The World Health Organization in 2000 ranked Cuba's public health system 39th of 191 countries worldwide (Putman 1999:37-38). 98% of the children under the age of two have been immunized against ten common diseases (Putman 1999:4). These are significant achievements for a country with such limited resources. However, Havana has the highest rate of asthma in the world—likely caused by dust and mold due to decaying infrastructure and shortages of paint and sealants (Baker 2003:66).

Like other countries in the Americas, Cuba had a history of slavery in the 17th century and endemic racism that endured through the 1940s. Noteworthy are the successful efforts to improve race relations and sexual equality in a country that is 66% white, 12% black, and 22% mixed ethnicity (Baker 2003:46). Cuba's revolutionary government outlawed institutional discrimination in the 1950s. Today, Cuban society is "more intermixed than any other on Earth and racial harmony is evident everywhere on the streets of Havana" (Baker 2003:46). Cuba is also among the top 20 nations in which women have the highest rates of participation in business and politics.

VENICE, ITALY

Italy is a country struggling to uphold its historical and architectural heritage in a time of declining birthrates. Venice, is a European city like no other—truly the essence of enchantment. Goethe wrote of Venice, "water was at once street, square and promenade. The Venetian was forced to become a new creature; and Venice can only be compared to itself" (Muraru and Graber 1963:20). And what will become of this "Jewel of the Adriatic"—a city lovingly referred to by its people as "La Serenissima" (the most serene)? Will the seas rise and flood her stone streets permanently? Will the world simply pay admission for day trips to see a flooded urban museum? Or will a sea wall be constructed to protect her?

Venice is over 1,000 years old and appears on the horizon as a city of walls floating on water. Venice, a port city, arose from 117 islands spread across a brackish estuary. She has known many forms of government including her own independence as a city-state controlling the region of Veneto. Today, Venice fights to hold back rising tides, and at times sections of the city are literally inundated with water.

Venice's people have a history of independent thinking, international awareness, and democratic values. Their history also includes the widespread cultural approval of risk-taking, gambling, and prostitution. Being an open-minded city with a tendency toward liberal behavior, residents seem to welcome everyone. To accommodate visitors, Venice is filled with quaint hotels and quiet pensiones. Venice accepted Armenians and Jews at times when most of Europe chose to despise them. Once the port-of-entry for goods headed to the heart of Europe, now Venice accepts tourists by car, train, and cruise ships.

11-7. View of Venice from the Campanile in Piazza San Marco

Venice is the capital of its region. Venice is an original—it has no counterpart. She is a world asset, yet few cities seem so delicate and fragile. It is among the quintessential pedestrian cities. The French architect La Corbusier once commented, "a functional city, Venice, extraordinarily functional."

Venice—Background

Venice was born on a collection of islands in a protected lagoon fed by several fresh water rivers, including the Benda, the Sile and the Piave. The early Venetians were a community of boatmen. The origins of Venice date to 452 A.D., when Atilla the Hun invaded Italy and people fled to the protection of the islands of the lagoon.[23] Laguna Veneta[24] shielded its residents from Pepin, son of Charlemagne, who seized land-based towns during his rampage through northern Italy in 810 A.D.[25] After 1,000 A.D., Venice became a seagoing nation, sailing and trading throughout the entire Mediterranean (Lane 1978:1).

Apart from slaves, the primary trade good was lumber which provided raw material for ship building. Mediterranean Europe suffered from deforestation and the Venetians, at the head of the Adriatic, tapped one of the few remaining areas in Europe where timber remained plentiful. Since quality timber was considered a strategic defense material, popes and emperors often forbade its sale (Lane 1978:8). With Rome a fair distance away, the Venetians placed their business interests before either ecclesiastical or imperial edict and become a regional center for shipbuilding. The superior supplies of lumber formed the basis for a division of labor between the inhabitants of the lagoons and distant Mediterranean peoples from whom they purchased wine, wheat and oil (Lane 1978:8). Other natural resources in the immediate area were important to the early growth of Venice. Salt and fish, both plentiful in the lagoon, were also key trade goods. After the first books were produced in Venice in 1469, the city became the European center of publishing.

The shallow lagoon provided islands for construction. When built out, growth was accommodated by creating more land, either by filling in shallow lagoon areas or by creating settlements on the nearby is-

11-8. Canal in Venice

lands. The city rests on millions of wooden stilts pounded into marshy ground or on stone carried in as ballast for shipping (Broad 2000). Over the ages, people living in the lagoons have had to haul in countless tons of dirt, silt, wood and stone to fortify their swampy islands and keep their homes and cities from disappearing into the sea (Broad 2000).

The city plan of Venice is a maze of narrow pedestrian streets and canals.[26] Transportation in Venice is mainly by walkway or waterway. The main "street" of Venice is an "S" shaped canal (3 kilometers long) named the "Canale Grande" whose banks are surrounded by beautiful residences, palaces, boarding houses and wharfs. It is crossed in only three locations by pedestrian bridges that serve as walkways over the waterways, the most renowned being the Rialto Bridge. Crossing the canal elsewhere requires a gondola, boat, or water-taxi.

Renowned for artwork, surrealistic masks, and Murano glass, tourism is the city's primary industry. Venice faces a costly struggle to preserve its architecture and upgrade its outdated infrastructure. With a proud history and uncertain future, Venice draws down her metal shop doors in the afternoon for peaceful siestas. Evenings come early and quietly to Venice. With the exception of bars that entertain tourists, there is little active night life. Families prefer to head home and indulge themselves in the world's most tantalizing pasta dishes and the fine locally produced table wine.

Venice—Population

In 1200, the population of Venice was 80,000, growing to 120,000 a century later with 160,000 in the lagoon area as a whole (Lane 1978:18). In 1330, the population of Paris approached only 100,000 and more populous Venice was among Europe's largest cities (Lane 1978:19). The population dropped after a series of plagues, then rebounded to 120,000 in 1500 and to 190,000 by 1570 (Lane 1978:19). Development of the city was limited by the high cost of multistory buildings constructed of stone and the expense of building land using pilings, dirt, ballast and debris.

During the 17th and 18th centuries, population fluctuated between 100,000 and 160,000, dropping again to 120,000 in the historical center by 1969 (Lane 1978:21). It is remarkable that during a 750 year period, the population of the original islands of Venice fluctuated within a range from 80,000 to 160,000 inhabitants with growth accommodated by mainland construction. Today, the total population of the newly defined Venice metro area has grown to 367,800. This includes the mainland suburbs of Mestre and Porto Marghera (Lane 1978:454). Yet the population of its historic city-center has declined to 60,000. Many residents are gone, having grown weary of wailing sirens that signal the advance of *Acqua Alta* (high water) usually caused by tidal flooding.

Venice — Environmental Issues

During the Middle Ages, the domestic water supply for Venice was provided by rain water collected from rooftop drains and from courtyards that were sloped to catch runoff and filter it through sand into a cisterns capped by wells (Lane 1978:206). Raw sewage was dumped directly into the lagoon. Such actions were not uncommon for a medieval city, and in the case of Venice, were practiced well into modern times. Most of the rivers feeding the lagoon were diverted away from the estuary over 500 years ago, minimizing flooding from these sources. The rivers carried silt, a threat which could clog the lagoon. Eventually, dikes were constructed on the landward side to keep out fresh water. This changed the ecology of the estuary and its wildlife habitats. It also reduced the amount of water flowing through the lagoon and on to the Adriatic Sea, making the lagoon more stagnant. Charles Dickens commented about "the prevailing Venetian odor of bilge water and an ebb tide on a weedy shore" (Pemble 1995:20). Still later, travel writer Anne Buckland advised ladies that a pocket handkerchief was indispensable in passing through the smaller canals, since these were used as open sewers (Pemble 1995:20).

To accommodate growth, islands were connected and shallow areas were filled in, creating new land. To facilitate movement and trade, canals, bridges, stone walkways, and docking facilities were constructed. Venice has a long history of accommodating specialized activities by moving them away from the central islands. The glassworks was ordered to the island of Murano in 1292 to lessen the danger of fire (Lane 1978:16). In 1493, the island of Lazzaretto was made a lodging place for the ill to reduce the spread of communicable diseases (Lane 1978:18). More recently, industries have developed on the mainland in the city of Mestre (now part of metropolitan Venice) to accommodate growth.

The lagoon, the biological basis for the city's life, was a constant concern. Maintaining shipping lanes was vital to the economic success of the city. To protect the three primary seaward entrances to the lagoon facing the Adriatic, Napoleon ordered that jetties be constructed. The barrier beaches, important for holding back storms on the Adriatic, were damaged by the burning of pine trees to clear land and by the removal of sand used for ballast by ships. These actions were eventually prohibited. Some shorelines were strengthened by stone sea walls finished in 1783. These lasted until 1966 when the sea walls were severely damaged by storms which submerged Venice under three to six feet of water (Lane 1978:452).

Venice and its lagoon have been assaulted by man-made environmental damage. One cause of pollution in the lagoon is runoff from agricultural wastes and chemicals. Agricultural crop fields were a source of nitrogen pollution affecting the lagoon's ecosystem (Franco *et a.l* 1996:1). Until the 1960s, runoff was controlled by large hedgerows and vegetated strips. Most of these were dismantled in the last decades of the 20th century to create a "modern" agricultural cropping system. A "new" landscape planning model, designed for the needs of local agriculture, has proven effective. The Venetian Municipality is presently instituting several low-tech actions to control lagoon pollution, including the construction of windbreaks and replanting hedgerows. This control process (with an efficiency of about 50%) has proven effective for nitrates, the most dangerous pollutant in local water quality assessments (Franco *et al.* 1996:4). In addition, the hedgerows stopped more than 85% of solids, and consequently reduced phosphorus levels (Franco *et al* 1996:4).

The Porto Marghera industrial complex, on the northwestern shore of the lagoon, is one of the larger concentrations of Italian industry with chemical and petrochemical plants, metallurgical and engineering works, and thermoelectric stations. By the 1970s, the influx of population and industries at Porto Marghera were causing pollution, overcrowding and urban sprawl. The lagoon suffered from lead and mercury contamination from industrial sources. Venice and its surrounds were also affected by the Chernobyl Nuclear Power Plant accident in the Ukraine, causing significant local exposure to radioactive fallout.

That Venetians live a few precarious inches above sea level is evident to today's visitors. Its canals teem not only with gondolas and motorboats but also with fish, algae and seaweed (Broad 2000). On hot summer days, the canals at low tide give off a pungent odor, said to be based partly on decomposing sea life and partly on the outflows of the city's ancient and often poorly maintained plumbing systems (Broad 2000).

Approximately 20,000 workers commute to the historic center of Venice daily (Lane 1978:457). Most commuters arrive by train, bus, or automobile via an umbilical cord-like causeway connecting Venice to the mainland or by boat. Automobiles are housed at central parking facilities near the maritime center. Others commute by ferry. This mix of commuting technologies minimizes air pollution and reduces traffic congestion.

The city plan of Venice, with its winding and sometimes narrow walkways and canals, successfully disperses and facilitates pedestrian movement by commuters, residents, and tourists. However, Venice's reliance on motorboats creates wave action that deteriorates the foundations of the buildings along the canals. Regulations which slow motorboat speeds have not eliminated their impact.

Damage from wave action, tidal surges, and flooding is common. The brick foundation walls of many buildings located on the canals are coated with corrosive sea salt that has been absorbed by the masonry, weakening them and causing them to crumble (Broad 2000). Low wooden doors that provide access to the canals are stained or rotted away at the bottom because of past flooding. Commercial shop doors often have metal guides that allow the quick installation of waterproof gates to prevent water from flowing into buildings (Broad 2000). The first floor areas of many residences have been abandoned and made uninhabitable due to repeated exposure to water. To allow people to walk about the city, raised walkways are set up during winter floods.

Venice—Sustainability Issues

Venice is the ultimate example of urban sustainability. Venice has survived for centuries, a tribute to the determination and resiliency of her people. For Venice, there was always too much water and not enough land. The site location avoided the use of prime agricultural land available on the mainland. The city's stone buildings were constructed to last hundreds of years. The lack of land mass in the central part of the lagoon and the costs of creating new land, limited growth and increased population density. After all of the land was developed more had to be created. In order to survive, Venetians had to import substantial quantities of food and harvest timber from the mainland for trade. Venice was forced to mold the lagoon into a usable form to facilitate construction, develop water-based transportation systems, and construct pedestrian bridges. The city continually redesigned itself. Its growth evolved by enlarging islands in its lagoon and by forcing industrial development away from the central city. This has allowed Venice to protect its waterways and architectural heritage.

Venice maximized the use of transportation technologies available in the middle ages. The city exploited the more efficient means of transporting people and cargo—over water rather than land. Unlike most other cities in Europe, the water was literally at the doorway of everyone's home. In addition, urban densification occurred due to the high costs of constructing new land mass.

Since there were no railroads, no automobiles, nor airplanes at the time Venice was developed, these central features of our modern cities were not included in its original design. However, as these transportation technologies became available, they were accommodated in a manner that failed to significantly alter Venice's urban form. A causeway was constructed to bring auto and rail transportation to the edge of the original city. The seaport was moved and expanded. The enlargement of land area in the northern portion of the lagoon adjacent to Mestre provided space for the Marco Polo International Airport. In keeping with the Venetian tradition of arriving by water, visitors arriving at the airport can use water-taxis to get from the airport to the city.

Venice has a history of using legislation to stem environmental problems. Examples include regulations to limit sand and tree removal from barrier islands. The primary sustainability problem for Venice is high water in the lagoon due to tides and floods, often augmented by the stormy winds of the Adriatic.

Does Venice demonstrate sustainable development? Dumping sewage into the lagoons was a misuse that has caused permanent damage to local aquatic life. Today, Venice is constructing a central sewage system to serve its tourists (20 million in 2007) and its remaining residents. Canals are being temporarily drained, piping systems installed, and connections made to sewage treatment systems. As each section is completed, the dikes are removed and the canals are refilled.

Since the water is rising and the land is sinking, the encroaching waters of the Adriatic Sea could devastate this celebrated city of art and architecture (Broad 2000). One cause of the sinking was the introduction of electric water pumps at Marghera in the 1930's to draw water for irrigation, industrial, and potable uses (Lane 1978:456). This is thought to have contributed to the land's sinking, which varies from place to place. Earlier, the rate of sinking averaged about ½ inch every ten years and increased to about 2 inches every ten years (Lane 1978:456). Today, Venice is sinking more slowly since pumping water from the aquifer underneath the city has been reduced.

The future of Venice is in doubt—the combination of sinking land and rising Adriatic water remains unresolved. The islands are now being undermined by the extraction of inexpensive methane gas deposits locat-

11-9. Installation of Sewage System in Venice

ed beneath the Adriatic near Venice. The situation may worsen if global warming creates more intense storm surges and causes seas to rise. After a major flood in 1966, a private consortium of engineers and architects, the Consorzio Venezia Nuova (CVN) was tasked to develop a plan to keep Venice from becoming a modern-day Atlantis. A project is underway to raise sidewalks and shores in low-lying areas by one meter. To this end, CVN has raised 960 hectares (2,370 acres) of land, 80% of their target.[27] Surging waters may well overwhelm their other major initiative to fight the impending danger—the construction of 79 mobile floodgates designed to cut off the Venetian lagoon and its islands from the highest of the high tides (Broad 2000).[28] The plan called MOSE—the Italian name for Moses—is estimated to cost €4.3 billion, and has won backing from Italian companies eager for the work.[29] Critics of the plan claim that rising sea waters caused by global warming will make the project ineffective and that less expensive options are available. The costs of the project may make the debate moot. By 2006, only €1.46 billion had been allocated for the project, leaving a funding shortfall of almost €3 billion.[30]

    Regardless, Venice has passed the test of time. The most important long-term sustainability issue for Venice concerns environmental changes which are beyond its control—the rising water in the lagoon. If global warming increases water levels further, the problem may be even more difficult to resolve. The answer for Venice lies in the ability to engineer and construct systems that can effectively control the waves, tides and storms of the Adriatic and reduce their effects on the lagoons. This will be a difficult task. The sinking process induced by man's intervention such as water pumping and the extraction of methane gas, must be halted.[31] Finally, creative ways of restoring buildings and foundations must be found. Unless Venice finds a solution that manages the impacts of the rising seas, the city's future sustainability is indeed questionable.

SUMMARY

The idea of sustainable development has gained a foothold in cities around the world. In Holland, Brazil, the U.S., the UK and elsewhere, the concept has been accepted as an overarching principle in the planning of communities. The concept has been used to format local responses to energy and planning decisions. Many examples of sustainable development problems are linked to energy, environmental impacts, urban development, and population growth. While localities worldwide have looming, challenging, often burdensome, and seemingly insurmountable sustainability problems, most remain unaddressed.

Havana and Venice have created distinctive urban forms and international urban images. Venice is a city of the Middle-Ages and the Renaissance; Havana is arguably the first major city of European origin in the New World. Both were constructed on seacoasts in locations which avoided the use of valuable farmland. Venice, the distinctive "Jewel of the Adriatic" flourished from the 9th to the 16th centuries and was constructed on islands in a shallow lagoon. Havana, the "Jewel of the Caribbean" and capital of an island nation, was constructed in lowland areas next to a magnificent harbor. Both cities were once centers for the shipbuilding industry. Both have experienced continuing difficulties with water pollution. Both have developed into major tourist centers and have social structures that support hospitality industries. The populations of both cities have suffered from the nightmares of diseases and plagues, often causing population losses, yet both have found ways to persevere. Each city evolved through periods of rapid and dynamic population growth.

Each developed due to its ability to trade and offer services other cities shunned.

Venice is sinking due a combination of causes. The rising of the Adriatic is exacerbating local flooding. The primary rivers supplying Venice's lagoon with fresh water have been permanently diverted. Havana had the benefit of two freshwater rivers to supply water. While both cities are struggling toward sustainability, they also have unresolved long-term sustainability problems. Environmental issues limit their future viability. Venice suffers from an over-abundance of water. Both Venice and Havana suffer from infrastructure problems that hinder water quality. Havana relies on older vehicles and buses, while Venice's automobiles are parked at its rear gates. Venice is the quintessential pedestrian city. These factors are key issues in assessing their sustainability.

Venice and Havana hold cultural amenities from the past. Venice invites a quiet stroll along its canals. Sundials in Venice are inscribed with the words: *Horas non numero nisi serenas* (I count only the happy hours). Havana offers its Malecon, a seaside walkway, for strolling residents and

11-10. Pedestrian Bridge in Venice

tourists who linger there in the evenings.

The growth of Havana in the later part of the 20th century was controlled by government mandates, which reduced the flow of immigrants from the countryside. A shortage of land limited the growth of Venice. Like Venice, Havana has a history that can be seen in the facades of its buildings—ones that are prohibitively expensive to maintain. Both Venice and Havana suffer from lack of investment in urban infrastructure, Havana perhaps to a greater extent. Havana's sustainability is strongly linked to communist politics and the lack of economic resources. Venice has a democratic tradition and history.

For Venice, the expense of filling the lagoon to create land for construction limited development, yet provided an incentive to jettison its industries to nearby islands and to the mainland. Today, as in the past, the health of its lagoons relate directly to the viability of the city. Unless the root causes of the sinking of Venice are mitigated by engineered solutions, Venice will one day be uninhabitable.

Havana and Venice share an economic need to attract income from tourism. Havana is surrounded by suburbs and has aging infrastructure, especially in Old Havana, that is worthy of preservation. Both cities are creatively trying to find ways to maintain and improve their centuries-old architectural treasures. The costs of maintaining buildings and structures in Old Havana and in Venice are burdens they both bear. As buildings crumble in the Havana's tropical conditions and the foundations of Venice continue to sink, time for remedial action is diminishing.

Venice is connected to its newer suburbs on the mainland by a causeway and a fleet of boats. The lower levels of buildings, abandoned due to rising water levels, remind the Italians that action must be taken quickly in order for Venice to survive and preserve its irreplaceable cultural and architectural heritage.

Havana struggles to support its teeming population with limited resources. As it does it must find hard currency with which to purchase imported food and energy supplies. Despite a trade embargo, about one-third of its food is imported from the U.S. After the end of the Cold War, Cuba lost access to oil from the U.S.S.R. It has subsequently increased domestic oil production and today produces most of the oil it needs for electrical energy. However, domestically produced oil is high in sulfur content. Since electrical generating stations lack pollution-control equipment, air pollution results from its use.

Cuba is placing its hopes on developing offshore oil resources in economically recoverable quantities in the near future. If this were to happen, Cuba could become an oil-exporting country. However, we have learned from lessons in Nigeria and elsewhere that developing oil and natural gas resources often fails to bring regional prosperity. It often damages eco-systems and reduces natural capital. There are concerns in Florida that development of Cuba's off-shore oil deposits might result in oil spills that may damage reefs, coasts, beaches and waterways.

Venice offers a glimpse of how coastal cities might cope with environmental sustainability in the future. Developing and implementing adaptation strategies for human settlements can be challenging. There are three ways to cope with environmental stress and change: 1) flee—migrate away from the impact; 2) attempt to fight the impact by making structural changes, or 3) accommodate the impact. Venice has pursued all three. The population of Venice's central islands is declining. Many former residents have fled the area due to economic conditions and the relentless *Acqua Alta*. Efforts to fight the flooding include raising walkways and barriers, increasing the height of land in low-lying areas, instituting legislation to minimize damage to barrier islands, and proposing a floodgate system. Accommodation has been Venice's primary response during most of its history, abandoning the lower levels of buildings and homes. Other efforts to control flooding include reducing the practice of pumping water from beneath the lagoon and slowing the speeds of motor boats. For the last couple of centuries, as Venice has continued to slowly sink, residents have resigned themselves to accept the slowly rising waters of Laguna Veneto as part of life in their city.

Despite the difficulties involved, cities around the world are identifying issues related to local sustainability and are developing potential solutions. Cities must be responsive and programs must be customized to deal directly with local sustainability concerns. Local actions involve mitigation, intervention and cooperation. Comprehensive programs and policies are needed. An intergenerational perspective is necessary—most quick-fix responses fail to resolve the longer-term consequences of sustainability issues. Permanent solutions typically require substantial capital investment, infrastructure improvements, and regional cooperation if cities are to achieve their goals and become more sustainable.

Endnotes

1. Amnesty International. (2004, 1 August). *Human rights and oil in Nigeria*. web.amnesty.org/library/Index/ENGAFR440232004?open&of=ENG-398, accessed 4 July 2004.

2. Ibid. Oil production in April 2007 was roughly 2.2 million barrels of oil per day.
3. Mouawad, J. (2007, 4 April). *Nigeria: Shell to raise Nigerian oil production.* www.corp-watch.org/article.php?id=14444, accessed 4 July 2004.
4. Ibid.
5. Thomas, E. *Venice or Amsterdam? — Adapting to climate change for coastal communities.* http://www.lgat.tas.gov.au/webdata/resources/files/Venice_or_Amsterdam_adapting_to_Climate_ Change_for_Coast_.pdf, accessed 25 June 2007.
6. Willing, R. (2007, 27 June). CIA discloses past abuses. *USA Today.* p. 1A.
7. Fawthrop, T. (2003, 13 November). Cuba's new oil industry. *BBC News.* http://news.bbc.co.uk/2/hi/business/3266509.stm, accessed 2 July 2007.
8. Martin, S. T. (2006, 8 May). Cuba seeks oil near Keys. *St. Petersburg Times.* http://www.sptimes.com/2006/05/08/Worldandnation/Cuba_seeks_oil_near_K.shtml, accessed 2 July 2007.
9. Frank, M. and Boadle, A. (2004, 24 July). Exploratory oil drilling done off Cuba. *Energy bulletin.* http://www.energybulletin.net/1212.html, accessed 2 July 2007.
10. Quinn, M. (2005, 30 November). *The power of community: How Cuba survived peak oil.* http://globalpublicmedia.com/articles/657, accessed 22 June 2007.
11. Frank, M. and Boadle, A. (2004, 24 July). Exploratory oil drilling done off Cuba. *Energy bulletin.* http://www.energybulletin.net/1212.html, accessed 2 July 2007.
12. Ibid.
13. Fawthrop, T. (2003, 13 November). Cuba's new oil industry. *BBC News.* http://news.bbc.co.uk/2/hi/business/3266509.stm, accessed 2 July 2007.
14. Frank, M. and Boadle, A. (2004, 24 July). Exploratory oil drilling done off Cuba. *Energy bulletin.* http://www.energybulletin.net/1212.html, accessed 2 July 2007.
15. Martin, S. T. (2006, 8 May). Cuba seeks oil near Keys. *St. Petersburg Times.* http://www.sptimes.com/2006/05/08/Worldandnation/Cuba_seeks_oil_near_K.shtml, accessed 2 July 2007.
16. Frank, M. and Boadle, A. (2004, 24 July). Exploratory oil drilling done off Cuba. *Energy bulletin.* http://www.energybulletin.net/1212.html, accessed 2 July 2007.
17. Quinn, M. (2005, 30 November). *The power of community: How Cuba survived peak oil.* http://globalpublicmedia.com/articles/657, accessed 22 June 2007.
18. Ibid.
19. Snow, A. (2007, 1 July). Looming crises prompts Cuba to repay farmers. *The Courier-Journal.* p. A13.
20. Ibid.
21. Ibid.
22. Ibid.
23. Merali, Z. (2002, 19 August). An ambitious plan seeks to prevent a modern Atlantis. *Scientific American.* www.sciam.com/article.cfm?articleID=00088E1F-D709-1D5B-90FB809EC5880000, accessed 23 June 2007.
24. The lagoon is also called La Laguna di Venezia, or the Lagoon of Venice.
25. Ibid.
26. *Vaporetto* is derived from the Italian word for "steamer." However, the boats today are propelled by diesel and gasoline engines.
27. Merali, Z. (2002, 19 August). An ambitious plan seeks to prevent a modern Atlantis. *Scientific American.* www.sciam.com/article.cfm?articleID=00088E1F-D709-1D5B-90FB809EC5880000, accessed 23 June 2007.
28. Ibid.
29. Ficara, C. (2005, 29 September). *Venice flood protection project to begin in Italy.* www.allheadlinenews.com/articles/7000319920, accessed 23 June 2007.
30. McMahon, B. (2006, 7 August). Flood barrier systems for Venice faces axe due

to budget shortfall. *The Guardian.* http://www.guardian.co.uk/italy/story/0,,1838678,00.htm, accessed 24 June 2007.

31. Given the compact nature of the historical center, I wonder if there might be a way to pump massive volumes of water to recharge the aquifer under the city and stabilize the sinking, offering Venice a reprieve until better solutions can be found. Perhaps the melt-water from the receding Alpine glaciers could be piped to Veneto and used for this purpose.

32. Thomas, E. *Venice or Amsterdam? — Adapting to climate change for coastal communities.* http://www.lgat.tas.gov.au/webdata/resources/files/Venice_or_Amsterdam_adapting_to_Climate_ Change_for_Coast_.pdf, accessed 25 June 2007.

Chapter 12

What the Future Holds: Creating a Sustainable World

"Make the wrong choices now and future generations will live with a changed climate, depleted resources and without green space and biodiversity that contribute both to our standard of living and our quality of life. Each of us needs to make the right choices to secure a future that is fairer, where we can all live within our environmental limits. That means sustainable development. This is an agenda for the long term. There is no magic wand that government or any one else can wave to make sustainable behavior and activity the norm overnight. We will only succeed if we go with the grain of what individuals and businesses want, and channel their creativity to confront the environmental challenges we face. Development, growth and prosperity need not and should not be in conflict with sustainability."

Blair, T. (2005, March). Securing the future—
delivering UK sustainable development strategy.
Presented to Parliament by the Secretary of State for the Environment,
Food and Rural Affairs. Norwich, England: HM Government.

Be prepared—the world is on the verge of revolutionary changes that will affect every human being. Imagine a world whose climate is in flux, temperature is increasing (albeit very gradually): a world in which energy and fresh water resources are not only scarce, but increasingly expensive.

By 2045, there will be over 1,000 cities with populations of over 1,000,000, more than 2½ times as many as there are today. Glaciers are melting, sea levels are rising, storms are more intense, wildfires more frequent, deserts are expanding, flooding more commonplace, atmospheric carbon levels are unmanageable, and species are in decline. Solutions to these and other sustainability problems will be costly. The world looks for solutions and discovers that there are no easy ones, no "silver bullets," no "killer-applications" that solve all of the problems we face—and there may be no salvation. There is growing awareness that past technologies have been imperfect and have created many of the problems we face today. The technologies of the future will also be imperfect, at times solving problems, at other times creating more. Regardless, we are moving forward. We must be cautious—the potential for unintended consequences is always present.

The economies of the world are using more energy. While we are certainly witnessing the peaking of the Hydrocarbon Age, it is doubtful that the age of the Hydrocarbon People is truly over. At long last there is acceptance that the time has arrived for keepers of the earth—the third rock from the sun—to be placed on a hydrocarbon diet.

The planet's future is being formatted by changes in policies, driven by economic necessity, and mitigated by the hope offered by new technologies. The sustainability of the world hangs in the balance. Ecosystems are in flux. The underlying causes of sustainability problems—population growth, urban development and increasing energy use—will continue and their forces will strengthen. The underlying impacts that stress sustainability—environmental deterioration, urban dislocation, and changes in urban infrastructure—remain unresolved. There is a nagging reality that we face daunting challenges.

In order to assess the ways to a sustainable future, policies and programs need to be continually monitored and assessed. While research is important, researching sustainability problems is not the solution—especially if the research investigates problems for which workable solutions have already been identified. Implementing workable mitigation strategies that result from successful research can pave the path to *resolving* problems. Research needs to be an on-going effort and remediation efforts need to be pursued in tandem with it. It is time for pilot projects to prove-out technologies, and time to commercialize those that have been tested and are known to be effective. There is no alternative but to take concerted action in an effort to achieve global sustainability.

The process of urbanization offers both opportunity and peril. Latin America has already transitioned structurally, with 77% of its population living in urban areas (Worldwatch Institute 2007:8). Currently, 700 to 800 million of the 1.3 billion Chinese live in the countryside and half of these people will migrate to cities over the next 20 years (Friedman 2005:479). Development is taking place—but with little or no planning. Human settlements are also being created (without planning guidelines) in third world countries in Africa and parts of Asia. While Africa is the least urban continent, this is rapidly changing. People are on the move, escaping from drought, flooding, local conflicts and political turmoil, and the processes of deleterious urbanization are beginning.

This type of urban development is marked by a lack of resources and basic infrastructure, including running water, sewage treatment, and electricity. What is happening is that developments are being constructed on a large scale in a form so base that slums by western standards would be an improvement. Squalor, disease and a lack of ability to fulfill basic human needs are the results. This process of rapid urbanization will continue during the 21st century and dealing with sustainability will be both unavoidable and challenging.

Sustainable development policies are constantly evolving. They are being defined and programs to carry them out are being implemented. The idea that it is man's manifest destiny to dominate and exploit nature without affecting ecosystems is passé. People are no longer viewing clean air, water, and natural environmental amenities as free goods but rather as depreciable natural assets. New technologies offer hope that sustainability concerns can be addressed and mitigated. To this end, there is a consistent trend toward the integration of interrelated sustainability policies.

THE ECONOMICS OF OIL AND SUSTAINABILITY

Depending on your point of view, oil can either be "black gold" or the *"excremento del Diablo."*[1] Oil is a commodity that drives our economies... and our tank is getting empty. Oil companies faced with competing alternatives, such as bio-fuels and more efficient vehicles, will control pricing not only by managing the availability of crude oil, but by also opting not to invest in refineries. In this way, they can limit capacity and the amounts of refined products they bring to market. This tactic provides them with flexibility in the ways they support higher price levels... and profit margins.

This began to happen in 2007 as oil companies scaled back plans to expand U.S. domestic refining capacity by nearly 40%.[2] According to Mike Cooper of the Consumer Federation of America, "By creating a situation of extremely tight supply, the oil companies gain control over the wholesale price level" and the refining industry "has no interest in creating spare capacity."[3] The result is that the supply of crude oil will diminish as the primary factor driving increases in the costs of refined oil-based products.

The U.S. within 20 years or less might find itself in the unenviable situation of importing over 80% of its oil supplies while being plagued by seemingly perpetual trade deficits. China, once able to produce all the oil it needed from domestic sources, is now the world's second largest importer of oil. Energy consumption in China increased 65% from 2002 to 2005 (Friedman 2005:497). If current trends continue, within 10 years China and the U.S. will combine to consume half of the world's oil supplies. Will the rest of the world stand idly by and allow this to continue? After all, 25 countries in the world hold 97% of all known oil reserves.

The idea that oil production will create sustainable economies in the producing countries that are dependent on oil revenues is a myth. Natural resource abundance is negatively correlated to per capita incomes. Per capita income in OPEC nations actually declined by 35% from 1965 to 1998, while developing nations increased their per capita GNP by 105% during the same period.[4] These results are due to the ways many developing nations tend to diversify their economies. During the same period, of the 65 countries in the world that are rich in natural resources, only four managed to increase their GDP over 25%, roughly equal to most industrial countries that lack them.[5] While this seems counter-intuitive, the question remains—why does this happen? According to Gylfason, "natural resource abundance may blunt private and public incentives to save and invest and thereby reduce economic growth."[6]

Oil production is not only unlikely to bring wealth to oil producing states, but it helps keep non-democratic regimes in power. Oil exploration and extraction pollutes air and water resources, devastates farmlands and wetlands, and reduces biodiversity. Oil-rich countries have very high rates of infant mortality, malnutrition, low life expectancies and poor health care. They are also more likely to have corrupt governments and experience more frequent secessionist civil wars. Since 1967, civil wars have occurred in the oil-rich countries of Algeria, Angola, Chad, Columbia, the Congo Republic, Indonesia and the Sudan. The countries of Iraq, Nigeria and Yemen have each experienced two civil wars in the last 40 years. The world's dependence on oil imports from these countries creates opportunities for new wars in the future.

New oil discoveries that are economically viable will occur in decreasing frequency. Venezuela has the potential to supersede Saudi Arabia as the world's No. 1 oil producer if new finds in the Orinoco region, roughly the size of West Virginia, prove out and if economic forces allow development. Known oil reserves in the region may increase from 80 million barrels to over 300 million barrels.[7] However, like many new discoveries, these reserves are "extra heavy oil" that must be refined into light, synthetic crude before they can be processed into gasoline.[8] As oil increases in value, deposits such as these will be developed.

There remains hope of finding new ways to use traditional fuels. Coal will continue to be part of the problem—and also part of the solution. The U.S., China and other countries will rely on their coal reserves to supplement declines in oil production. Coal gasification and carbon sequestration offer hope for coal—but their environmental costs may be greater than their benefits. The use of these technologies may not help us fashion a sustainable future.

If current trends continue, China will soon have the dubious honor of being the No. 1 emitter of atmospheric greenhouse gases, overtaking the U.S. This is due to the China's increasing use of coal and the proliferation of automobiles. In 2006, China increased production of automobiles by 27% from 2005 levels and is now the world's second largest automobile market.[9] More disconcerting, sales of luxury models increased by 37%, and now account for half of the vehicles sold in China.[10] 30,000 new vehicles are being purchased monthly in Beijing alone (Friedman 2005:479).

Today, 16 of the world's 20 cities with the poorest air quality are located in China and environmental degradation, made worse by the increasing use of automobiles, is costing the country an estimated $170 billion annually (Friedman 2005:479). Finding ways to acquire more oil is important to China's agenda.

As the linkages between sustainability and energy become more widely understood, sustainable solutions that directly address these linkages will be the most successful. Developing solutions to sustainability problems is challenging. The critical path of problem-to-solution cycles is blurring.

Despite the ecological damage that has been done in the past, we live on a planet that still has large undisturbed areas. Energy is one key to sustainability. The battles for sustainability will be fought not about whether or not there is a need for concerted action, but about exactly what changes to make, when and how we must intervene, and which technologies we should employ. As it becomes more widely accepted that action must be taken to assure the sustainability of the planet, alternative solutions will be developed and implemented. Many will be successful; others will yield unintended consequences; and some will fail. Sustainable solutions that focus on ways to use energy more effectively have the greatest potential for success.

POLICIES

Many of the most successful sustainable policy developments have been locally initiated. This may signal a drift away from centralized policy-making. Why? Centralized planning and decision-making created many of the important problems we now face, and has proven ineffective at identifying, funding and implementing effective solutions. In the age of the internet, access to answers is readily available—*and the ability to control information has already been decentralized.*

Eric Bonabeau, a complexity theorist and scientist at Icosystem Corporation in Cambridge, MA, asserts that "We are not used to solving decentralized problems in a decentralized way."[11] Not everyone is comfortable with the decentralization of policy decisions. The sustainability revolution will change how resources are allocated, and in the process change how constituents view the decisions of their governments. Political and economic forces will cause shifts in how money is invested.

As a result, political power structures will wrestle with their agendas and shift direction as they discover that their constituents view inaction as a non-solution. Those feeling the effects and paying the price for unsustainable policies will see the transparency of stalling tactics and "greenwashing" attempts. While governments will support investments in both traditional and alternative fuel sources, the focus is shifting from managing surpluses to managing demand. Decentralized production and energy efficiency solutions will be seen as the preferred energy options in the future. Why is this so? *These solutions are much less carbon-intensive than using energy from traditional sources, and simply make economic sense.*

Some governments are leading, taking individual actions, hoping to not be considered anachronistic. Electrical utility producers in Australia are required to obtain 2% of their energy from renewable resources or else pay a penalty to the government of $40 (A) per megawatt hour for any shortfall.[12] This adjustment created a new market for renewable energy and provided incentives for utilities to generate power using landfill gas, bio-fuels, wind and solar energy.[13] Elsewhere in Australia, local governments are taking action by establishing new goals and making new policies. The city of Newcastle stabilized its electrical demand, reducing it by 50%, and established a "no-regrets" policy for regulations regarding climate change.[14]

In the U.S., policy activity is moving from the local to the regional, from the city councils to the state capitals. Legislation passed in 2007 in

New Jersey requires the state to reduce greenhouse gas emissions to 1990 levels by 2020, and to further reduce emissions by 80% below 2006 levels by 2050. In Minnesota, Governor Tim Pawlenty signed the nation's strongest renewable energy mandate in February 2007. The legislation requires energy companies to provide 25% of their power from renewable sources by 2025. Minnesota's largest electricity provider, Xcel Energy, is required to provide 30% of its electrical power from renewable sources by 2020.[15] An additional 5,000 megawatts of energy from renewable sources will be added, roughly eight times more than is presently generated from renewables. In Minnesota, the power sources proposed for future electrical generation include wind turbines, hydrogen, biomass, and solar power. Such changes are only a beginning. Soon, many more states will pass similar legislation.

States are working together to address climate change. "In the absence of meaningful federal action, it is up to the states to take action to address climate change," says Arizona Governor Janet Napolitano.[16] A recent agreement, named the Western Regional Climate Action Initiative, is an unprecedented step toward collaboration. The states of Arizona, California, New Mexico, Oregon, and Washington have agreed to establish regional targets to lower greenhouse gases by implementing a "cap-and-trade" system. This program allows companies that are unable to meet their emission reduction targets to purchase credits from companies that have reduced carbon dioxide emissions.[17] These programs attempt to cap total emissions at a given level, and reduce allowable emissions gradually over a period of time.

According to California Governor Arnold Schwarzenegger, the Western Regional Climate Action Initiative provides a framework to develop a national cap-and-trade program and shows that states can lead the U.S. by addressing climate change.[18] In September 2006, California's Global Warming Solutions Act was approved by the Governor. It calls for reducing California's greenhouse gas emissions by 25% by 2020 and by 80% by 2050. Also in 2006, the California Solar Initiative, the country's largest solar energy incentive program, allocated $3.2 billion for solar energy rebates through 2018. The initiative will subsidize 3,000 megawatts of solar power development.

Despite all attempts to use new resources and implement energy efficiency initiatives, utility costs will continue to increase. Efforts to keep energy costs low have failed. The idea that such attempts might have worked hindered the replacement of obsolete equipment, inefficient plants, and infrastructure. It has increased our dependence on carbon-based fuels. Most energy resources are undervalued in a social sense, as there are subsidies in place. Many of the subsidies are environmental externalities that are known to deplete natural capital. As these costs become integrated into energy prices and commoditized, energy costs will increase.

Recently, some religious organizations have become interested in sustainability. Taking care of Earth's endowments certainly has religious implications. According to Gardner (2003), "Worldwide, the major faiths are issuing declarations, advocating new national policies, and designing educational activities in support of a sustainable world—sometimes in partnership with the secular environmental community." Surprisingly, the world's largest 10 or so religious organizations own or control 7-8% of the world's land area. In July of 2007, Vatican City announced that it would become the world's first sovereign state to become carbon neutral, offsetting emissions by planting trees in Hungary.[19] In addition, the papal audience hall next to St. Peter's Basilica will have solar photovoltaic panels installed on its roof to provide heating, cooling and lighting for the building.[20] If a trend toward environmental management continues to evolve within organized religions, the impact on the drive toward sustainability could be enormous.

CORPORATIONS

There is a past record among U.S. corporations of resistance to considering and implementing solutions for sustainability. In their view, lack of leadership at the federal governmental level has contributed to a regulatory quagmire. Implementing sustainability seems so inherently complex that individual actions by corporations are thought to be counterproductive in highly competitive markets. As local and regional sustainability policies are adopted, the same corporations that resisted and opposed federal intervention will face a complicated patchwork of local and regional rules and regulations. This will contribute to future financial and regulatory uncertainties, with subsequent impacts on corporate investments and profits.

Many companies have developed successful operations across jurisdictions that help achieve the goals of sustainability. Waste Management, a company that operates 281 landfills across the U.S., is turning the production of electricity from renewable landfill gases into a major business opportunity. Landfill gases are typically composed of 40% to 50% carbon dioxide and 50% to 60% methane.[21] Landfill gas projects operate about 95% of the time, providing base-load electrical power and reducing greenhouse gases that would normally be flared.

Waste Management currently operates 103 Landfill Gas-to-Energy (LFGTE) plants. It has plans to create new revenue streams by developing 10 additional sites in 2007, with another 60 more to come over the next five years, adding 230 megawatts its generating capacity.[22] These projects will be located in Texas, Virginia, New York, Colorado, Massachusetts, Illinois and Wisconsin. When completed, the company will be a mini-utility. According to Paul Pabor, the company's Vice-President of Renewable Energy, "This initiative is a major step in Waste Management's ongoing efforts to implement sustainable business practices."[23] Combined with their existing landfill gas generation facilities, the total electrical energy produced when all projects are completed will be 700 megawatts—roughly the equivalent of a fossil-fuel power plant.[24]

Utilities are taking notice—and taking action. They are diversifying their portfolios of investments and upgrading the environmental performance of their generating facilities. State regulations are changing electrical rate structures that were designed to penalize utilities that implemented demand reduction strategies.

Vectren, a regulated Midwestern gas and electrical supplier, serves as an example. Since 2002, the utility has upgraded its coal-fired plants with over $300 million in emissions control equipment, with another $70 million worth of improvements underway.[25] According to Neil Ellerbrook, Vectren's Chairman, the improvements will mean that their fleet of generating plants will be "100% scrubbed for sulfur dioxide, 90% controlled for nitrogen oxide, and will further reduce mercury emissions."[26] Vectren is also developing a renewable portfolio for electrical generation. In 1992, the company formed a non-regulated subsidiary, Energy Systems Group (ESG), to implement energy savings performance contracts. To date over 750 energy-related projects have been completed by ESG. One recent project involved the development of a landfill-gas-fueled, electrical co-generation facility in Johnson City, Tennessee. Another project at Union College in Barbourville, KY, provided energy-saving improvements for the campus including new windows, lighting systems, and geothermal heating and cooling systems.

According to Vectren's 2006 Annual Report, ESG generated $114 million in sales, contributing $3.1 million in income.[27] This type of diversification by utilities will continue to play an important role in reducing energy use. Investments such as these will expand in scale and scope in the foreseeable future.

More companies are considering renewable energy solutions. In a recent survey of energy engineers and managers, a surprising 21.6% of respondents indicated that photovoltaic technologies were likely to be installed at their facilities over the next three years.[28] (Thumann *et al.* 2007).

The global market for renewable energy technology is expected to grow to $167 billion by 2015.[29] As carbon trading becomes firmly established, investments in alternative energy systems will increase. Carbon trading introduces market reforms that reduce carbon emissions. Purchased carbon credits are used to offset the carbon emissions of producers. Market theorists believe that if the price of emitting carbon dioxide increases enough, at some point company executives will consider it less expensive to reduce their own industrial emissions than to continue purchasing emissions credits.[30]

Since 2005 in the EU, 12,000 industrial plants have been able to buy and sell rights to release carbon into the atmosphere.[31] The EU Emissions Trading Scheme is a way that companies exceeding individual CO_2 emissions targets can purchase allowances from companies that have successfully met their targets.[32] The EU program includes industries such as power generation, iron and steel production, glass, cement, pottery and masonry product manufacturing.[33] Trading has been surprisingly active with emissions being factored into the operations of installations throughout the EU; over 780 million European allowances having been traded.[34] In the U.S., the Chicago Climate Exchange is the only current carbon trading market. According to investor Sunil Paul, "There are huge forces at work right now. With the subsidies that are already in place in California, and markets for carbon credits emerging, you have the perfect conditions for innovative companies to capture a piece of the $1 trillion U.S. electricity market."[35]

TECHNOLOGIES

In the future, sustainability will be achieved by means of proven— and also exciting new technologies. Many technologies that improve energy efficiencies remain largely untapped. There is a backlog of aging energy-intensive infrastructure in use today that has exceeded its economic life. Over the next decade, much of it will need to be replaced. This includes automobiles, lighting systems, mechanical systems, and power production equipment. Millions of individual decisions as to what to do and how to them will be made. The technologies we choose to use today will become tomorrow's infrastructure. Care must be taken to select the best solutions available.

There is great interest in breakthrough technologies that may provide new solutions to achieve sustainability. Researching the problems we face will help us to enhance our options. Despite our best efforts, we will find no single cure-all. Research will expand into areas of developing technologies that seemed like science fiction only a few decades ago. Once the world's economies become focused, we will discover a mother-load of options, some more feasible than others. Research can result in possible solutions, but it can also divert our attention and resources from workable "on-the-shelf" technologies that have been tested and are easily implemented.

There are many available technological solutions that have yet to be fully implemented. Atomic power (a centralized electrical generation technology) will have a place in the future of electrical generation as more countries, including the U.S., turn to it in the future. Atomic energy produces no greenhouse gases (except in the production of uranium). France gets 80% of it electrical power from atomic energy by using a standardized reactor design. However, in the U.S. and elsewhere, localities will fight construction of atomic plants and many projects will be delayed or aborted. Decentralized atomic power applications are on the horizon. For example, Russia is constructing sea-going atomic power plants to serve remote localities, despite environmental concerns about installing the plants over water. Eventually they will sell the technology to other countries.

Many technological advances will lead us to a more sustainable future. High pressure steam injection is already being used to recover oil from fields which were previously deemed exhausted. While likely to be more widely used in the future, this process consumes water, and more energy is required to extract the oil. Fuel cells have been available for over 100 years, yet are only now economically viable for selective applications. Using Connecticut's Clean Energy Fund, six fuel cell energy projects—generating 68 megawatts of electricity—were started in 2007.[36]

The best technologies are those that reduce the need for energy. Yet many simple solutions that can be easily commercialized have yet to become standard. Take for example the compact fluorescent lamp (CFL). The

CFL lamp emits light evenly across its surface, uses only 25%-30% as much electricity as an incandescent lamp. It reduces air conditioning loads and saves money. While more expensive than the incandescent lamp, the compact fluorescent lamps last 5-10 times longer. CFL lamps were invented at Lawrence-Livermore Laboratories in the U.S., but the technology was initially commercialized by the Japanese. Yet due to their low initial costs, incandescent lamps are the standard in U.S. homes. In 2007, laws banning the use of incandescent lamps were proposed in California, Connecticut and New Jersey. Australia has announced plans to become the first nation to phase out the use of incandescent lamps and replace them with more efficient fluorescent lamps. Their legislation will reduce greenhouse gas emissions by 4 million tons annually by 2012, according to Environment Minister Malcolm Turnbull.[37] Bans or phase-outs of incandescent lamps are also being considered in Brazil, Canada, New Zealand, and elsewhere. Reductions in energy use and carbon emissions are the primary reason for these bans.

ALTERNATIVE ENERGY

Alternative energy technologies promise that sustainability goals can be accomplished. Electrical power plants contribute over 25% of the world's CO_2 emissions. In 2006, Airtricty, a Dublin-based developer of wind power, announced plans for a 10,000 megawatt European supergrid that will use 2,000 offshore wind turbine generators (Kaihla 2007:69). The electrical production would be equivalent to the electrical output of several conventional power generating plants. When compared to energy from traditional sources, CO_2 emissions will be reduced by the equivalent of 60 million tons per year.

Solving the problem of greenhouse gas emissions from transportation sources involves finding new fuel sources. Hydrogen vehicles will be manufactured and used in the future but other less exotic fuels are more readily commercialized. In Brazil, gas stations now offer automotive fuels derived from ethanol made from sugar cane, which is a more efficient source of fermentable carbohydrates than corn. Using flex-fuel vehicles, their ubiquitous dial-a-blend fuel pumps allow motorists to select ethanol-and-gasoline ratios for their vehicles. Brazilian cars using ethanol have displaced roughly 40% of the fuel that would be required if their vehicles were operated on gasoline alone. Regardless, Brazil continues to use far more oil than ethanol.

The transition to ethanol has added new environmental concerns. Since sugar cane fields are burned prior to harvesting, the skies in sugar cane producing areas are blackened during harvest season. In some parts of Brazil, the expansion of sugar cane fields has decreased biodiversity. Regardless, the success of ethanol production in Brazil illustrates how countries can diversify their energy resources. There are plans to export Brazilian bio-fuel products to Japan and elsewhere. In the U.S. there are now 120 ethanol production facilities that can produce 6 billion gallons annually with another 76 plants either under construction or being expanded.[38] More and more farmland is being used to grow corn.

Gasohol fuels can be made from varieties of willow, soybeans, wood byproducts and sunflowers. But some of these processes use more energy than they create.[39] Regardless, biodiesel fuels, made from processed corn and soybean oil, are becoming commonplace in the U.S. Their use has environmental benefits—unburned hydrocarbons are reduced by 20%, carbon monoxide and particulate matter emissions are both reduced by 12%.[40] The use of biodiesel creates a "growing" market for corn in the U.S. and for soybeans in states such as Iowa, Minnesota, Illinois and Indiana. Biodiesel is most often used in 2% to 20% blends. There are also 100% bio-fuel products, directly replacing petroleum. It has the benefit of improving the lubricity of fuel, potentially extending equipment life. Also used in farm equipment, biodiesel makes engines operate more smoothly and with less objectionable exhaust. With these benefits, expect the price of tortillas to increase as more corn production is converted to ethanol.

Like most fuels, the costs for biodiesel vary depending on the proximity of a fuel supplier to a biodiesel manufacturer. School districts in Las Vegas and Louisville now use blended forms of biodiesel for their school buses. Alternatively, the city of Berkeley, CA, tried using 100% biodiesel fuels in 200 trucks but suspended its use—the city was unable to obtain a consistently clean product, or to manage bacterial contamination.[41] Fuel additives may ultimately resolve biological contamination problems.

Fuels of the future will be developed from surprising sources. There will be new ways to convert waste products into energy sources. ConocoPhillips and Tyson Foods formed a strategic alliance in 2007 to manufacture a synthetic diesel fuel from waste fat generated by processing beef, pork and poultry. Production ultimately will yield 175 million gallons annually of renewable diesel.[42] Unlike ethanol or blended biodiesel fuels, this unique fuel will have the same molecular and chemical characteristics as normal diesel fuel. In California, several dairies have installed methane gas digesters. Cow dung is a significant source of destructive greenhouse gases. The digesters extract the methane gas and use it to create electricity by means of a turbine generator system (Datta & Woody 2007:88).

Westport Innovations is a company that offers spark-ignited natural gas engines for buses and a high-pressure, direct-injection system that allows semi-trucks to use liquefied natural gas (Kaihla 2007:70). Within the next 10 years, direct-injection engines will be available for vehicles that run on hydrogen fuels.

A number of companies are developing technologies that use plasma-arc processes to eliminate wastes and create interesting by-products. Prototype systems are already in operation. These plants are a means of converting post-consumer municipal solid wastes into usable products. Plasma-arc gasification uses a high temperature (1,650° C or 3,000° F), lightning-like process that breaks down shredded solid waste into molecular building blocks which provide marketable by-products: 1) a combustible synthesis gas (syngas); 2) metal ingots; and 3) a glass-like solid that can be used for floor tiles or gravel (Durst 2007:78). Materials that can be processed using plasma furnaces include municipal and household wastes and hazardous wastes such as oil sludge and those generated by hospitals. Plasma-arc incineration plants have been constructed in Yoshii, Japan (2000), Utashinai City, Japan (2002), Mihama, Japan (2002), and Ottawa, Canada (2007).[43] However, dioxin can be emitted and chorine is often found in the wastes.[44] Plants using this technology in Australia and Germany were shut down because they were unable to meet local emission standards.[45] Carnival Cruise Lines uses a similar system developed by PyroGenesis to process five tons of waste daily on one of its ships, reducing the waste to a few kilograms of sand (Durst 2007:78).

A cabinet-sized prototype hydrogen "fueling station" capable of being located in a corner of a residential garage, and using current from a solar collector, has been developed in Australia, potentially eliminating the need to create a hydrogen supply infrastructure. According to Datta and Woody (2007:83), "the home fueling station is expected to produce enough hydrogen to give your runabout about 100 miles without emitting a molecule of planet-warming gas." This newly developed process also eliminates the need to create hydrogen gas by burning a fossil fuel.

One common problem with electrical generation has to do with having adequate supplies available to meet periods when electrical demand loads are highest. When demand moderates, generators must still remain online, thus producing more power than is needed. Electricity has to be used the instant it is generated. Right? Well... not really. Think of all the greenhouse gas emissions that could be reduced if additional generating capacity (new power plants) were not required for peak periods. Wouldn't it be nice if there was a way to store large amounts of electrical power from generating plants and save it for peak load conditions? Well, actually... there is.

NaS (Sodium and Sulfur) battery banks that can store large amounts of electricity are available today. In the past, lead-acid batteries were widely used and they required a warehouse full of interconnected batteries to store electricity. Battery life was only about five years. The lead in these batteries also contaminates the environment if disposal is not handled properly. By bridging electrodes with a porcelain-like material, a room-sized bank of durable NaS batteries will last 15 years and may transform the way we think about delivering peak power.[46]

According to Stow Walker of Cambridge Energy Research Associates, by "using NaS batteries, utilities could defer for years, and perhaps avoid, construction of new transmission lines, substations, and power plants."[47] This is possible since the NaS batteries can be charged during periods of low electrical demand, such as nights or weekends when electrical costs are lowest, and discharge power during peak demand conditions. Green utilities benefit since the battery banks can provide back-up power during power outages and be used in combination with solar and wind power production.[48] In these applications, electricity is stored during peak production periods and released during the night (when there is no sun) or on windless days.

Costs for NaS batteries are approximately $2,500 per kilowatt or about 10% more than the cost of building a new coal-fired power plant to produce electricity—but battery systems are much smaller, do not require additional fuel, and do not emit additional greenhouse gasses.[49] NaS battery systems are a proven technology that is already being used by electric utilities in the U.S. and Japan.

REDUCING CARBON EMISSIONS

Technologies are widely available today to reduce energy consumption and cut our dependency on fossil fuels. Many offer economic benefits. In most cases, trade-offs are necessary. U.S. anthropogenic emissions of greenhouse gases have increased 16.9% from 6,113 metric tons in 1990 to 7,147 metric tons in 2005 and are not subsiding.[50]

What can be done regarding unaddressed environmental problems such as CO_2 gas emissions? Carbon capture and sequestering technologies are becoming available and offer hope. CO_2 currently being released into the atmosphere can be stored in oceans, in caves, abandoned oil and natural gas fields, and elsewhere. Not all of these sequestration techniques are environmentally friendly. When stored in water, carbonic acid can be formed, affecting marine life. Though costly, the use of liquid sodium hydroxide (lye) holds promise as it has the ability to absorb and capture CO_2 from airstreams. It is most feasible for locations where emissions are concentrated and are being exhausted, such as the stacks of coal-fired power plants. Planting trees and other forms of vegetation is perhaps one of the most cost-effective solutions—they are natural systems that not only store carbon but also generate oxygen.

Renewable energy will be an important part of our portfolio of sustainable energy solutions. Sustainable energy development has its own "upside-down" economics. While solar and wind power often have higher initial costs, they also have negligible carbon impact, and there is no need to purchase fuel. Conventional fuel solutions have lower initial costs, burdensome carbon emissions, and must use ever-more costly fuels. As the right to emit carbon into the atmosphere becomes more costly and non-carbon energy sources become less expensive, alternative energy solutions will become more viable.

CITY AND COUNTY GOVERNMENTS

Urban growth in the past century depended on by the availability of inexpensive and abundant carbon-based energy supplies. Urban growth in the future will depend on ways creative cities can become sustainable by using costly and limited energy resources more effectively. Cities are continuing to "go green." Energy use in their buildings accounts for more that half of their greenhouse gas emissions. These emissions are greater in cities that use energy inefficiently.

As we have seen, large cities have common sustainability issues. The "C40 Large Cities Climate Summit" provided an opportunity for local governments to consider strategies to reverse the trends of climate change.[51] A total of 16 cities including New York City, Chicago, Houston, Toronto,

Mexico City, London, Berlin, Tokyo, and Rome will become involved in multi-billion dollar initiatives to cut their emissions.[52] This arrangement involves several global banking institutions and a number of energy service companies. The "green makeovers" will include projects to replacing heating, cooling and lighting systems with energy-efficient systems; make roofs more reflective to deflect more of the sun's heat; seal windows and install new ones that let more usable light into occupied spaces; and provide sensors that control lighting and air conditioning systems.[53] The ability to reduce energy costs provides a key incentive to make these improvements. According to Sadhu Johnston, Commissioner of the Chicago Department of the Environment, lighting systems improvements in the city's buildings over the past six years have already resulted in about $4 million in annual savings.[54]

Cities are customizing their programs to meet their specific circumstances. The city of Cambridge, MA, is launching a $100 million energy efficiency initiative called the "Cambridge Energy Alliance (CEA)." According to *City View*, this non-profit organization plans to "design, market, finance, manage and document unprecedented efficiency improvements in the use of energy, water and transportation in Cambridge."[55] City Manager Robert Healy says that the program's goal is to make strides to "reduce greenhouse gas emissions and at the same time make our households; businesses and institutions more resilient against rising energy prices."[56]

The objectives of the CEA include: 1) reduce peak electricity demand by 15%; 2) reduce the city's electricity and water usage by 10%; 3) achieve a 50% participation rate in the program; and 4) reduce city-generated greenhouse emissions by 10% by 2011.[57]

The program applies to all neighborhoods in the city and to all municipal, university, commercial and residential facilities. Financing for these initiatives will be obtained primarily from private sources, with 20% coming from utility incentive programs.[58] Cambridge has also been recognized for its sustainability efforts. It received an Environmental Merit Award for climate protection, an award for excellence in affordable green housing, and the Sustainable Community Forestry Award.

St. Lucia County in Florida is on the verge of having the first plasma arc incineration facility in the U.S. Landfill wastes will be gasified to yield synthetic gas, slag for road construction and steam for a nearby factory.[59] In addition, the $425 million project will provide 120 megawatts of electrical power using turbines, one-third of which will be used to operate the facility.[60] Tropicana Products, Inc. will purchase the 80,000 pounds of steam generated daily for its juice plant's turbines.[61] Faced with an accumulation of 4.3 million tons of trash collected over 30 years, increasing waste disposal costs, and a shortage of landfill space, the county plans to resolve all of these problems by eliminating the need for the landfill. When completed, this plant will vaporize 3,000 tons of garbage per day, enough to consume the wastes in the county's landfill within 18 years.[62] While the process emits CO_2, it is much less than the amount produced from traditional carbon-based fuel sources. According to Louis Circeo, director of Georgia Tech's plasma research department, this technology "could not only solve the garbage and landfill problems in the U.S. and elsewhere, but it could significantly alleviate the current energy crises."[63]

BUILDINGS

According to former U.S. President Bill Clinton, "If all buildings were as efficient as they could be, we'd be saving an enormous amount of energy and significantly reducing carbon emissions. Also, we'd be saving a ton of money."[64] The energy used by buildings in New York City accounts for 79% of the city's total output of heat-trapping gases. The combination of energy prices for source fuels such as oil, natural gas and electricity, along with changes in building codes and improved state and local incentives, will motivate building owners and managers to investigate opportunities for energy efficiency.

There is great potential to reduce the energy used for buildings and they are becoming more "energy smart." Buildings of the future will have systems that can change the architectural features of their exteriors to control the impact of environmental conditions. Exterior shell materials will be available that reflect sunlight during cooling seasons and absorb solar radiation during the winter. Many of these technologies are already available.

The next generation of "super windows" will be both highly insulated and have dynamic features. Substantial amounts of heat losses and gains are due to heat transfer through building fenestration. Windows that are much more efficient at managing heat transfer are now being used. Glazing is already available that can dynamically modulate transmittance, morphing characteristics such as color and reflectivity. Using highly insulated windows that have dynamic solar heat gain control features has the potential to provide a "net zero" heating consumption impact and to reduce cooling loads by as much as 80%—eliminating the need for 4.3

quadrillion kilojoules (4.1 quadrillion Btus) of energy in the U.S annual-ly.[65] Windows will also be configured to harvest solar energy and generate electricity.

Building computer systems designed to reduce energy use will do more that just turn systems on and off. They will monitor building oc-cupancy by counting the number of people in heating and cooling zones, and adjusting interior temperatures and the amount of fresh outside air supplied to the occupied spaces accordingly.

The planned San Francisco Public Utility Commission Headquar-ters, a 12-story office building, will use a "thermal chimney" design to provide natural cooling when conditions permit. Solar panels embedded in its outer walls and wind turbines on its roof will generate electricity. The structure will produce 40% of its energy requirements and will drop off from the electrical grid altogether on windy, sunny days.[66] The General Manager of the Public Utility Commission, Anthony Irons, had an ambi-tious goal: "I wanted us to design a building completely unconnected to the utility grid."[67]

Electrical utilities will reach inside buildings with new technologies that will control refrigerators, washing machines, ovens, and water heat-ers, cycling them off during periods of high electrical demand. Smart con-trols for central air conditioning systems that cycle electric compressors off during periods of peak demand are already on the verge of becoming ubiquitous.

Water consumption in buildings is being reduced by changes in codes and the greater use of appliances, such as dish and clothes washers, that incorporate water-saving devices and technologies. High-efficiency toilets and waterless urinals are minimizing the amount of water used in buildings. Rainwater collection systems will become commonplace in the future and the water will be directed to non-potable uses. During dry con-ditions in the U.S., water consumption for landscaping purposes increas-es and can account for half of the water consumed. In many parts of the country, residences and small commercial operations are charged sewer fees even though the water is being used for landscaping. To reduce util-ity expenses, sophisticated irrigation control systems are becoming com-monplace. These systems can monitor the soil moisture and the presence of sunlight, and then provide irrigation at times when it offers the maxi-mum benefits. High-tech irrigation systems that use data from satellite-monitored ground weather stations are already being used in California to manage the amounts of water that is supplied for landscaping.

GREEN PRODUCTS

Get ready for the green product revolution! Silberberg (2003:245) believes that although "engineers and chemists will be in the forefront in exploring new energy directions, a more hopeful energy future ultimately depends on our wisdom in obtaining and conserving planetary resources." Future products will more resource stingy. In a world striving to be green, manufacturers are redesigning consumer products in an effort to minimize the resources required for production and operation. This is already happening for refrigerators, electronic devices and garden equipment. "Green consumerism" may not be an oxymoronic phrase after all.

Many types of outdoor equipment produce more pollution than automobiles. New "green machines" are already entering the marketplace: propane lawn mowers, bio-diesel all-terrain vehicles, hydrogen-powered lawn carts, hybrid lawn mowers, and snow blowers.[68] Solar-powered vehicles that can serve four passengers are available. Fuel-cell-powered nailing tools are being used in the construction industry. Entire lines of products, components and materials are being brought to market to meet

12-1. Four-Passenger Solar-Powered Vehicle

the growing demand for green construction materials.

There is a quiet revolution underway to make consumer electronic products greener. Changes will be incremental but sweeping and the opportunities are enormous. The costs of leaving personal computers operating when unused total almost $2 billion annually in the U.S. About 19.8 billion kilowatt hours of electricity is wasted each year—enough to supply 1.9 million residences.[69] Dell has developed a new line of computer servers that use 42% less energy when coupled with air conditioning systems.[70] They are also manufacturing laptop computers that don't contain lead or polyvinyl chloride, and have expanded their computer recycling programs.[71] Panasonic has developed a lead-free solder for use in its electronic products. Motorola has a program to encourage its customers to recycle telephones and has worked to make cell phone chargers more energy-efficient.

Despite the fight for shelf-space, packaging will be designed with a "less is more" philosophy. There is an industry trend to be more efficient and environmentally friendly by using smaller containers and less packaging in products.[72] Concentrated products, not just "Simple Green," will become the norm at retail outlets. These changes are in direct response to efforts to reduce wastes and lower the amounts of petrochemicals used for product containers. P&G and Unilever will be selling concentrated cleaning solutions. A broader selection of detergents will be available that are designed to use less water and work effectively in cold water.

On the other hand, there are consumer products on the market today which will no longer be acceptable due to their environmental impact. Cypress mulch, used for landscaping, is responsible for the loss of cypress in areas of the southeastern U.S.—harvesting has caused irreparable damage to fragile ecosystems. Plastic grocery bags and plastic water bottles are also examples. Americans annually use 30 billion single-serving plastic water bottles (derived from crude oil and delivered by trucks using diesel fuel) annually and only 14% are recycled.[73] According to Jennifer Gitlitz of the Container Recycling Institute, "It is the most environmentally egregious way to distribute water."[74]

PARTNERSHIPS TO ACHIEVE SUSTAINABILITY

Perhaps the most exciting development in the move towards sustainability is the ability of cities to integrate policies and programs. The future of cities can be found in how creatively they will determine ways to achieve sustainability goals, and how they work with their county governments and local businesses to turn policies into programs that make sustainable living a reality.

The city of Chico, California, an urban area of 105,000 in Butte County, has become one of the national leaders in implementing sustainability. Chico has been engaged in sustainability measures for over 20 years. The "North State Renewable Energy (NSRE)" partnership was formed with private entities, city and county governments, and a local university to encourage the use of renewable energy.[75] The city values its environmental stewardship. It focuses its programs on maximizing the environmental benefits of sustainability initiatives and reducing operating costs. Energy projects are a key component of their agenda. The city is a U.S. EPA green power producer. It is heavily involved in energy conservation, alternative energy production, green construction practices, water conservation, improving transportation systems, and implementing sustainable land use planning programs. The comprehensive set of policies and programs being pursued by Chico provide an integrative example of how future cities will achieve sustainability goals. Here is what Chico's has done: [76]

- Installed a passive solar power installation for a new city parking structure.

- Installed a 1.0 Megawatt solar tracking facility at a wastewater treatment plant designed to maximize output during peak demand periods.

- Used solar powered emergency telephones at a local park.

- Installed LED traffic signals and a computer-controlled traffic signal system to reduce vehicles fuel use.

- Upgraded city buildings with energy-efficient lighting systems, occupancy controls, and new energy management and control systems.

- Installed a cogeneration system at a pollution control plant that uses methane gas to produce electricity.

- Purchased eight hybrid-electric vehicles.

- Implemented new landscaping strategies that utilize drought-tolerant plants and installed irrigation systems that minimize water consumption.

- Exceeded the "California Integrated Waste Management Act (AB939)" requirements and has achieved a waste diversion rate for residents and businesses of 61% by offering low-interest loans to convert or utilize recycled materials in products and operations.

- Instituted a program that recycles asphalt from road construction and uses repaving emulsions that contain recycled waste tire products.

- Adopted a municipal code that requires residential property owners to meet energy conservation standards prior to re-sale of their properties.

- Adopted urban planning standards to maintain its historic downtown, encourage narrow streets, institute traffic-calming design principles, and expand bikeways.

- Adopted an urban area redevelopment plan that allows for mixed-use development, requires underground utilities, and provides for the rehabilitation and retention of historic properties by authorizing the redevelopment agency to purchase structures of a historic nature.[77]

Chico's sustainability activities were coupled with others being taken within its county. Butte County installed a 1.18 megawatt solar generation facility for the County Government Center, receiving a $4.2 million renewable energy rebate from its electric utility (PG&E) that paid for half of the project's cost.[78] The city of Oroville (population 13,000) is 32 kilometers (20 miles) from Chico. Oroville was proclaimed Solar City USA, generating an average of 200 watts of solar energy annually per resident. In Oroville, solar generating facilities are located at Police and Fire Headquarters, the Public Works Yard, the State Theater, the Pioneer Museum, and Oroville City Hall. Other alternative energy initiatives in Butte County include:[79]

- The Butte County Rice Growers Association installed a 200 kW, $1.5 million solar array for a rice dryer system.

- Chico-based Sierra Nevada Brewing installed a fuel cell system that generate one megawatt of electrical power.

- Heritage Partners, a land development firm, installed a photovoltaic system generating 96 kW.

- The Sewer Commission, which serves 15,000 families in the Oroville area, installed a 622 kW solar generating system and received a $2.4 million rebate from its utility.

- Chico-based FAFCO, a producer of solar pool heating panels, installed a thermal energy storage system that reduced peak electrical demand by 410 kW.

- The Kiwanis Chico Community Observatory installed a 1.5 kW solar-electric system to power computers, robotic telescopes and night vision lighting.

California State University, Chico (CSU Chico) has a long commitment to issues relating to the environment and the wise use of scarce resources. CSU Chico is one of the first campuses in the nation to sign a long-range commitment to reduce greenhouse gas emissions and become "climate neutral" in its effects on the environment. This commitment was codified in a new strategic plan for the University that contained the following proclamation:

> *"Believing that each generation owes something to those which follow, we will create environmentally literate citizens, who embrace sustainability as a way of living. We will be wise stewards of scarce resources and, in seeking to develop the whole person, be aware that our individual and collective actions have economic, social, and environmental consequences locally, regionally, and globally."*[80]

In 2006, CSU Chico President Paul Zingg joined six other campus executives in signing the American College & University Presidents Climate Commitment (ACUPCC). The use of energy efficient technologies, energy-saving retrofits, and educational programs have helped CSU Chico control and reduce its energy use. By cutting energy consumption, the campus will save money and reduce its impact on the environment. Here are examples of the strategies that the campus is using to meet its newly adopted sustainability goals:

1. *Renewable Energy Resolution*—This campus policy recommends the use of green building practices for all new buildings and requires that sustainable building practices (using LEED guidelines) be followed.

2.   *Thermal Energy Storage System*—The campus has a program that involves the use of its thermal energy storage system. The strategy allows chillers to operate at night, storing chilled water in a tank for use the following day. This avoids running the chiller during peak load conditions, reducing electrical costs and electric demand by 1.5 megawatts.

CONCLUSION

This journey of discovery concerning the new revolution in sustainable development has brought us full circle—from its definition, though its history, then to problems, then to possibilities—and finally to solutions. We are at the beginning of a new phase of human history. The wave of sustainable development will wash over us, redefining our future. Our journey toward a sustainable future offers hope that solutions can be found—and many are ready now. Achieving sustainability requires us to consider new policies, devise new programs, and consider new technologies. We must have the willpower to change the unsustainable practices to which we have grown accustomed. Solutions will be found to the challenging problems that we face.

The time for inaction has passed. It is no longer feasible to maintain the status quo and do business in ways that worked in the past. Without proactive changes we will not resolve the problems that we now face—and will only exacerbate them. Mistakes have occurred in the past. Our past irresponsibility does not absolve us from the potential of a future replete with environmental degradation, resource scarcity, and economic ruin.

The time for motivation and action is now upon us. Sustainability is not only about those of us living today. It is about the conditions of the world that our descendents will face. It concerns how we manage our dowry of natural capital. Failure to take action will certainly lead to a breakdown in the faith we have in changing our future, and in our democratic institutions which must identify sustainable policies to meet these challenges. The worst-case scenarios are almost too terrible to consider—ecological failures, failed institutions, declining resources, and regional anarchy.

To achieve sustainability we must minister to the needs of our growing populations, have the fortitude to implement incremental changes, and monitor and adjust our approaches. We must find the means and resources to resolve our environmental problems. All this is possible. However, we will require every tool in our sustainable development toolkit—and others that have not yet been invented. Becoming increasingly reliant on carbon intensive buildings and processes will not lead to a sustainable future. Infrastructure decisions must be based on long-term resource allocation implications and their impact on natural capital. Focusing on known solutions that decrease dependence on carbon-based energy sources which rely solely on inefficient processes, is both sensible and necessary. Energy conservation, energy efficiency and alternative energy play an important role in our efforts to become sustainable. Technology offers hope that we can develop and implement solutions to the problems we face. We must be open to accept the potential—and benefits—that they offer.

Endnotes

1. Translated from Spanish as "the devil's excrement."
2. Herbert, H. J. (2007, 13 June). Oil industry trims refinery plans. *The Courier-Journal*.
3. Ibid.
4. Gylfason, T. (2002). *Paradox of the plenty: the management of oil wealth*. Report for ECON, Center for economic analysis. Oslo, Norway.
5. Gylfason, T. (2001, August). *Lesson's from the Dutch disease: causes, treatment and cure.* http://www.ioes.hi.is/publications/wp/w0106.pdf, accessed 10 July 2007. p. 9.
6. Ibid., p. 8.
7. Lynch, D. (2007, 4 April). Oil—Venezuela's lifeblood—is also a political flashpoint. *USA Today*. pgs. 1B-2B.
8. Ibid.
9. Pong. P. (2007, 4 February). Cheaper fuel to spur car industry. *Sunday Morning Post.* p. 1.
10. Ibid.
11. Miller, P. (2007, July). Swarm theory. *National Geographic*. p. 146.
12. Thomas, E. *Venice or Amsterdam? —adapting to climate change for coastal communities.* http://www.lgat.tas.gov.au/webdata/resources/files/Venice_or_Amsterdam___adapting_to_Climate_Change_for_Coast_.pdf, accessed 25 June 2007.
13. Ibid.
14. Ibid.
15. Minnesota Northstar. (2007, 22 February). *Governor Pawlenty signs strongest renewable energy requirement in the nation*. www.governor.state.mn.us/mediacenter/pressreleases, accessed 1 March 2007.
16. Five states line up against global warming. *USA Today*. 27 February 2007. p. 3-4
17. Associated Press (2007, 27 February). *Five governors agree to work together on climate change*. www.gainsville.com/apps/pbcs.dll, accessed 4 March 2007.
18. Ibid.
19. Rocca, F. (2007, 26 July). Vatican takes role in keeping God's earth green. *USA Today*. p. 9D.
20. Ibid.
21. Renewable Energy Access. (2007, 27 June). *700 Mw from electricity to come from landfill gas*. http://www.renewableenergyaccess.com/rea/news/story?id=49123, accessed 7 July 2007.

22. Ibid.
23. Ibid.
24. Ibid.
25. Vectren. (2006, 16 February). *2006 Summary Annual Report*. Evansville, Indiana.
26. Ibid.
27. Ibid.
28. In the same survey, 10.1% of respondents believed that their companies would install windpower systems in the next 10 years.
29. Copeland, M. and McNichol, T. (2006, November). Here comes the sun. *Business 2.0*. p. 100.
30. Hotz, R. (2007, 11 February). Wanted: carbon emissions. *The Courier-Journal*. p. A11.
31. EurActiv (2007, 29 June). *EU emissions trading scheme*. http://www.euractiv.com/en/sustainability/eu-emissions-trading-scheme/article-133629, accessed 10 July 2007.
32. Ibid.
33. Ibid.
34. International Emissions Trading Association. (2006, 13 October). *IETA position paper on EU ETS market functioning*. http://www.ieta.org/ieta/www/pages/getfile.php?docID=1926, accessed 9 July 2006.
35. Copeland, M. and McNichol, T. (2006, November). Here comes the sun. *Business 2.0*. p. 100.
36. Connecticut clean energy fund selects FuelCell Energy's power plants (2007, May). *Energy and power management*. 32 (5).
37. BBC News (2007, 20 February). *Australia pulls plug on old bulbs*. www.bbc.co.uk/2/hi/asia-pacific/6378161.stm.
38. Kirchhoff, S. (2007, 25 July). Ranchers, farmers battle over corn. *USA Today*. p. 3B.
39. Cornell University. (2005, 6 July). Ethanol and biodiesel from crops not worth the energy. *Science Daily*. http://www.sciencedaily.com/releases/2005/07/050705231841.htm, accessed 7 July 2007. According to this study, "Ethanol production requires large fossil fuel input" with corn requiring "29% more fossil fuel than energy produced"; switchgrass 45% more; soybean 27% more; and sunflower plants 118% more.
40. U.S. DOE. (2007, 6 March). *Using biodiesel in vehicles*. http://www.eere.energy.gov/afdc/afv/bio_vehicles.htm, accessed 7 July 2007.
41. Artz, M. (2005, 18 March). *City halts use of pure biodiesel fuel, citing build-up of bacteria mold*. http://www.mindfully.org/Energy/2005/Biodiesel-Fuel-Berkeley18mar05.htm, accessed 8 July 2007.
42. Souza, K (2007, 17 April). Tyson, ConocoPhillips turning fat into fuel. *Times Record*.
43. Wikipedia. *Plasma arc waste disposal*. http://en.wikipedia.org/wiki/Plasma_arc_gasification, accessed 7 July 2007.
44. Muller, P. *Syngas in Johnson County*. http://mypc.press-citizen.com/story.php?id_stories=304, accessed 7 July 2007.
45. Ibid.
46. Davidson, P. (2007, 5 July). New battery packs power punch. *USA Today*. p. 3B.
47. Ibid.
48. Ibid.
49. Ibid.
50. Energy Information Administration (2005). Emissions of greenhouse gases in the United States, executive summary. www.eia.doe.gov/oiaf/1605/ggrpt/pdf/executive_summary.pdf, accessed 30 July 2006.
51. 16 cities to receive "go-green" makeovers. *The Courier-Journal*. 17 May 2007. p. A3.

52. Additional cities involved in the building plan are Mumbai, India; Karachi, Pakistan; Seoul, South Korea; Bangkok, Thailand; Melbourne, Australia; Sao Paulo, Brazil; and Johannesburg, South Africa.

53. 16 cities to receive "go-green" makeovers. *The Courier-Journal*. 17 May 2007. p. A3.

54. Kugler. S. (2007, 16 May). 16 cities go green under Clinton plan. www.thruthout. org/docs_2006/051707.shtml, accessed 8 July 2007.

55. City launches innovative new energy conservation initiative. (2007, Spring/Summer). *City View*. http://www.cambridgeenergyalliance.org/press/Cityview%20Ne wsletter%20Spring-Summer07.pdf, accessed 4 July 2007. p. 1.

56. Ibid., p. 3.

57. Ibid.

58. Ibid.

59. Zaks, D. (2006, 12 September). *Geoplasma—plasma arc incineration*. http://www. worldchanging.com/archives/004926.html, accessed 7 July 2007.

60 Ibid.

61. Florida county plans to vaporize landfill trash. (2006, 9 September). *USA Today*. http://www.usatoday.com/news/nation/2006-09-09-fla-county-trash_x.htm, accessed 7 July 2007.

62. Ibid.

63. Ibid.

64. Byers, J. (2007, 17 May).Toronto buildings get energy retrofit. *Toronto Star*. http:// www.thestar.com/article/214889, accessed 1 July 2007.

65. Arasteth, D. *et al.* (2006, 13-18 August). *Zero energy windows*. U.S. Department of Energy LBNL-60049. http://gaia.lbl.gov/btech/papers/60049.pdf, accessed 22 July 2007.

66. King, J. (2007, 13 April). *San Francisco hopes to set example with new green tower*. www.sfgate.com/cgi-bin/article/cgi?file=/c/a/2007/04/13/MNGFP852Q1.DTL, accessed 15 June 2007.

67. Ibid.

68. Woodyard, C. (2007, 31 May). "Green machines" generate weak sales. *USA Today*. p. 5B.

69. Kessler, M. (2007, 17 July). Tech's green problems. *USA Today*. pgs. 1B-2B.

70. Ibid.

71. Ibid.

72. Sewell, D. (2007, 24 May). Getting green on top of clean. *USA Today*. p. 5B.

73. Breslau, K. (2007, 9 July). *A good drink at the sink*. Newsweek. p. 14.

74. Ibid.

75. Tan, D. and McNall, S. (2005, 27 January). *North state renewable energy collaborative to host solar conference. www.greatvalley.org/solar/pdf/chico_pr.pdf*, accessed 8 July 2007.

76. *City of Chico's effort toward sustainability*. (2006, 26 October).

77. Chico Revelopment Agency (2004, 14 June). *Greater Chico urban area redevelopment project*.

78. Butte County (2004, 27 October). *Soaking up the sun*. www.buttecounty.net/solar/ dedication_oct_27_04.html.

79. Tan, D. and McNall, S. (2005, 27 January). *North State Renewable Energy Collaborative to host solar conference. www.greatvalley.org/solar/pdf/chico_pr.pdf*, accessed 15 July 2007.

80. Chico State University. (2006, May). *Updating CSU, Chico strategic plan for the future*. http://www.csuchico.edu/prs/documents/pdf/strategicPlan5_06.pdf, accessed 7 July 2007. p.15.

Bibliography

Abbott C. (1981). *The new urban America*. Chapel Hill, North Carolina: The University of North Carolina Press.

Air Conditioning & Refrigeration Institute (2002). *Cooling industry produces 130 millionth air conditioner –Achievement celebrated at New York stock exchange*. Arlington, Virginia: Ed Dooley Communications.

Al-Homound, M.S. (2000). Total productive energy management. *Energy engineering journal*. 98 (1).

Andrews, R.L. (1999). *Managing the environment, managing ourselves*. New Haven, Connecticut: Yale University Press.

Aston, A. (2007, 22 January). Who will run the plants? *Business Week*.

Association of Energy Engineers Hungarian Chapter (2001). May it be saved? *Energy efficiency, energy markets and environmental protection in the new millennium*. Sopron, Hungary: AEE Hungarian Chapter.

Association of Energy Engineers (2004). *Membership*. www.aeecenter.org/membership.

Ayres, R.U. (1998). *Eco-restructuring: Implications for sustainable development*. Tokyo: United Nations Press.

Babbie, E. (2001). *The practice of social research*. Stanford: Wadsworth/Thompson Learning.

Baker, C.P. (2003). *Moon handook's Havana*. Eneryville, California: Avalon Travel Publishing.

Ballard, S.C. and James, T.E. (1983). *The future of the Sunbelt—Managing growth and change*. New York: Praeger Publishers.

Banham, R. (1969). *The architecture of the well-tempered environment*. London: Architectural Press.

Bardarch, E. (2000). *A practical guide to policy analysis*. New York: Seven Bridges Press.

Barnett, L. (1957). *The universe and Dr. Einstein*. New York: Bantam Books.

Bartik, T.J. and Smith, V.K. (1987). Urban amenities and public policy. *Handbook of regional and urban economics*. 2.

Beatley, T. (2000). *Green urbanism: Learning from European cities*. Washington, D.C.: Island Press.

Begley, S. (2007, 16 April). Curbing emissions won't be enough. *Newsweek*.

Bell, S. and Morse, S. (1999). *Sustainability indicators, measuring the immea-*

surable. London: Earthscan.

Benfield, F.K., Terris, J. and Vorsanger, N. (2001). *Solving sprawl—Models of smart growth in communities across America*. New York: National Resources Defense Fund.

Bennett, M. and Newborough, M. (2001). Auditing energy use in cities. *Energy Policy*. Elsevier Science Ltd. 29. p. 125-134.

Berke, P.R. (2002). Does sustainable development offer a new direction for planning? Challenges for the Twenty-first century. *Journal of planning literature*. 17 (1).

Berke, P.R. and Manta-Conroy, M. (2002). Are we planning for sustainable development? An evaluation of thirty comprehensive plans. *Journal of the American planning association*. 66 (1).

Bernard, R.M. and Rice, B.R. (1983). *Sunbelt cities: Politics and growth since World War II*. Austin, Texas: University of Texas Press.

Berry, M.A. and Rondinelli, D.A. (1999). Greening of industry network conference best practice proceedings. *Environmental citizenship in multinational corporations: Social responsibility and sustainable development*. Chapel Hill: University of North Carolina.

Bertalanffy, L.V. (1933). *Modern theories of development*. Oxford: Oxford University Press.

Bertalanffy, L.V. (1968). *General system theory, foundations, development, applications*. New York: Braziller.

Bezdek, R.H., Wendling, R.M. and Jones, J.D. (1989). The economic and employment effects of investments in pollution abatement and control technologies. *Ambio* 18 (5).

Bishop, I.D., Ksemsan, S. and Yaqub, H.W. (2000). Spatial data infrastructures for cities in developing countries. *Cities*. 17 (2).

Blair, J.P. (1995). Local economic development. *Market areas and economic development strategies*.

Blair, T. (2005, March). *Securing the future—delivering UK sustainable development strategy*. Presented to Parliament by the Secretary of State for the Environment, Food and Rural Affairs. Norwich, England: HM Government.

Bluestone, B. and Harrison, B. (1982). *The deindustrialization of America*. New York: Basic Books, Inc.

Borenstein, S. (2001). The trouble with electricity markets: Understanding California's electricity restructuring disaster. *Journal of economic perspectives*. 16 (1).

Boucher, M. (2004). Resource efficient buildings—Balancing the bottom line.

Proceedings of the 2004 world energy engineering congress, Austin, Texas, September 22-24 2004; Atlanta: Association of Energy Engineers.

Bowman, A.O. (2003, August). *Green politics in the city.* Presented at the annual meeting of the American Political Science Association, Philadelphia, Pennsylvania.

Brainard, J., Jones, A.P., Bateman, I.J., Lovett, A.A. and Fallon, P.J. (2002). Modeling environmental equity: Access to air quality in Birmingham, England. *Environment and planning.* 34.

Breslau, K. (2007, 16 April). The green giant. *Newsweek.*

Broad, W.J. (2000, 29 August). That sinking feeling again, as Venice's past haunts city's future. *New York Times.*

Brown, L.R. (2001). *Eco-economy - Building an economy for the earth.* New York: W.W. Norton & Company.

Brown, L.R. (2006, 4 August). *U.S. population reaches 300 million heading for 400 million.* Earth Policy Institute. www.earch-policy.org/Updates/2006/Update59.htm.

Brown, M.A. (2001). Market failures and barriers as a basis for clean energy policies. *Energy policy.* 29 (14).

Brown, R.E. and Kooney, J.G. (2003). Electricity use in California: past trends and present usage patterns. *Energy policy.* 31.

Burgess, R., Carmona, M. and Kolstee, T. (1997). *The challenge of sustainable cities.* London: Zed Books.

Burgess, R. and Jenks, M. (2000). *Compact cities—sustainable urban forms for developing countries.* London: Spon Press.

Caldwell, L.K. (1963). Environment: A new focus for public policy? *Public administration review.* 22.

Capek, S.M. and Gilderbloom, J.I. (1992). *Community versus commodity-Tenants and the American City.* Albany, New York: State University of New York Press.

Carson, E. (1996). The legacy of Habitat II. *The urban age.* 4(2).

Cassedy, E.S. (2000). *Prospects for sustainable energy: A critical assessment.* Cambridge: Cambridge University Press.

Castells, M. (2000). Urban sustainability in the information age. *City.* 4(1).

Casten, S. (2003). Five aces and a winking dealer—The costs and reasons for the U.S.'s failure to invest in energy efficiency. *Strategic planning for energy and the environment.* 23 (2). Liburn, Georgia: Association of Energy Engineers.

Chafin, L. (2003). *Cuba: the history, the economy, the architecture...the experience.*

Chasek, P.S. (2000). *The global environment in the twenty-first century: Pros-*

pects for international cooperation. New York: The United Nations University.

Chertow, M.R. and Esty, D.C. (ed.) (1997). *Thinking ecologically—The next generation of environmental policy.* New Haven, Connecticut: Yale University Press.

Chouri, N.I. (ed.) (1993). *Global accord.* Cambridge, Massachusetts: The Massachusetts Institute of Technology Press.

Cineros, H.G. (1995, March). *Regionalism: the new geography of opportunity.* Washington, D.C.: U.S. Department of Housing and Urban Development.

City of Atlanta Online (2003). *Energy conservation program history.* www.ci.atlanta.ga.us.mayor.energyconservation.

City of Houston (2004). Solid Waste Management Department. *Recycling Services.* www.ci.houston.tx.us/swd/recycling.htm.

City of Las Vegas (2002). *City of Las Vegas dedicates the world's first energy station featuring hydrogen and electricity.* Public Works Department Media Release. 14 November. http://www.ci.las-vegas.nv.us/108_4563.htm.

City of San Diego (2002, 18 October). *City of San Diego Mayor Dick Murphy commissions the city's first "energy independent" municipal building utilizing solar power.* Environmental Services News Release.

City of Santa Monica (1977). *Santa Monica sustainable city program, sustainability indicators.* www.ci.santa-monica.ca.us/environment/policy/indicators.htm.

City of Tucson (1998, 26 June). *Civano impact system memorandum of understanding in implementation and monitoring process.*

Clayton, M.H. and Radcliffe, N.J. (1996). *Sustainability—A systems approach.* Westview Press. Edinburgh.

Clements, J. (2003, 21 December). Big house can crack budget. Your money, a guide to spending, saving and investing. *The Courier-Journal*: Louisville, Kentucky.

Colonbo, U. (1992). Development and the global environment. In Hollander J.M. (ed.), *The energy-environment connection.* Washington D.C.: Island Press.

Columbia Encyclopedia (2001). Sixth Edition.

Columbia Gazetteer of North America (2000). New York: Columbia University Press.

Columbia University (2001). See http://www.ciesin.columbia.edu/idicators/ESI.

Commoner, B. (1971). *The closing circle. Nature, man and technology*. New York: Knopf.

Couch, C. and Dennemann, A. (2000). Urban regeneration and sustainable development in Britain. *Cities*. 17 (2).

The Courier-Journal (2002, 2 September). *Russia agrees to sign Kyoto Protocol but won't state timetable*.

Clayton, K.L. (2003) The energy manager's role in sustainable design. *Solutions for energy security and facility management challenges*. Atlanta: Association of Energy Engineers.

Clayton, M.H. and Radcliffe, N.J. (1996). *Sustainability—A systems approach*. Edinburgh: Westview Press.

Crenson, M. (1971). *The un-politics of air pollution: a study of non-decision making in the cities*. Baltimore, Maryland: John Hopkins University Press.

Cresswell, J.W. (2003). *Research design—Qualitative, quantitative, and mixed methods approaches*. Thousand Oaks, California: Sage Publications, Inc.

Cusimano, M.K. (2000). *Beyond Sovereignty Issues for a Global Agenda*.

Daeschner, S.W. (2005). Color us green. *Sustain*. 12.

Dahl, R.A. (1967). *Pluralist democracy in the United States: Conflict and Consent*. Chicago: Rand McNally and Company.

Daly, H., Droste, B. von, Goodland, R. and Serafy, S.E. (ed.) (1991). *Environmentally sustainable economic development: Building on Brundtland*. Paris: United Nations Educational Scientific and Cultural Organization.

Datta, S. and Woody, T. (2007, January). Eight technologies for a green future. *Business 2.0*. 8 (1). p. 81-88.

Davey, J. (2005). Sustainable development after the revolution: Las Terrazas, Cuba. *Sustain*. 11.

Davidson, P. (2007). TXU takes $32B deal, cuts coal plans. *USA Today*. p. B1.

Davis, A. (2007, 29 April). Shopping for energy savings. *The Courier-Journal*.

DeCanio S. (1993). Barriers within firms to energy-efficient investments. *Energy policy*. 29 (9).

DeCanio S. (1998). The energy-efficiency paradox: Bureaucratic and organizational barriers to profitable energy-saving investments. *Energy policy*. 26 (5).

Denton, S. and Morris, R. (2001). *The money and the power, The making of Las Vegas and its hold on America, 1947-2000*. New York: Alfred A.

Knopf.

Desai, U. (2002). *Environmental politics and policy in industrialized countries.*
Cambridge, Massachusetts: MIT Press.

DeSimone, L. and Popoff, F. (1997). *Eco-efficiency—The business of sustainable development.* Cambridge MA: The Massachusetts Institute of
Technology Press.

Devuyst, D., L. Hens and Lannoy, W.D. (ed.) (2001*). How green is the city?
Sustainability assessment and the management of urban environments.*
New York: Columbia University Press.

DiGaetano, A. and Klemanski, J.S. (1999). *Power and city governance, comparative perspectives on urban development.* Minneapolis: University of
Minnesota Press.

Dincer, I. (1999). Environmental impacts of energy. *Energy policy.* Elsevier
Science.

Dincer, I. and Rosen, M. (2004). Exergy as a driver for achieving sustainability. *International Journal of green energy.* Marcel Dekker.

Doggett, S. and Stanley, D. (2001). *Havana—revolution, rumba and rum.*
Victoria, Australia: Lonely Planet Publications.

Drakakis-Smith, D. (2000). *Third world cities.* London and New York:
Routledge.

Duany, A., Plater-Zyberk, E. and Speck, J. (2000). *Suburban nation: The
rise of sprawl and the decline of the American dream.* New York: North
Point Press.

Durst, S. (2007, January). Overfishing. *Business 2.0.* 8 (1).

Dyllick T. and Hockerts, K. (2002). *Beyond the business case for corporate
sustainability. Business strategy and the environment.* 11. p. 130-141.

Eckstein, B. and Throgmorton, J.A. (2003). *Story and sustainability.* Cambridge Massachusetts: The MIT Press.

Edwards, A.R. (2005). *The sustainability revolution, portrait of a paradigm
shift.* Gariola Island, Bitish Columbia: New Society Publications.

Ehrlich, P. (1971). *The population bomb.* London: Ballantine.

Eley, C. (2006, May). High performance school characteristics. *ASHRAE
Journal.* 48.

Elliott, L. (1998). *The Global Politics of the Environment.* New York: New
York University Press.

Elliott, R.R. (1973). *History of Nevada.* Lincoln, Nebraska: University of
Nebraska Press.

Energy Information Administration (2000). State energy data. www.doe.
eia.gov. Washington, D.C.

Bibliography 427

Energy Information Administration (2001, February). Monthly energy review. DOE/EIA-0035. U.S. Government Printing Office: Washington, D.C.

Engardio, P. (2007, 29 January). Beyond the green corporation. *Business Week.*

EurActiv.com (2006). *Corporate America gets hot on climate change.* www.euractiv.com/en/sustainability, accessed 1.6.07.

EuroCouncil (2003, April). *Council Report V.* Belgium.

European Commission (1996, March). *European sustainable cities.* Directorate General XI, Brussels.

European Commission (1997, March). *Agenda 21- The first five years.* European Community progress on the implementation of Agenda 21, 1992-97. Brussels: European Commission.

Feagin, J.R., Anthony, M.A. and Sjoberg, G. (1991). *A case for the case study.* Chapel Hill, North Carolina: University of North Carolina Press.

Fetterman, M. (2006, 25 September) Wal-Mart grows "green" strategies. *USA Today.*

Fisher, A.C. and Rothkoph, M.H. (1989). Market failure and energy policy: A rationale for selective conservation. *Energy policy.* 17 (4.)

Forrest, J. (1995,1996). *Exploring for a better world: Corporations and Sustainable Development.* www.bwz.com/BWZ/9610/explore.htm, accessed 12/23/06.

Fortis (2004). *Corporate sustainability statement.* www.fortis.com/sustainabilty/index.asp, accessed 12.27.06.

Fowler, J. (2003, 17 December). U.N. agency says 2003 third-hottest on record. *Courier-journal.* Louisville, Kentucky.

Franco, D., Perelli, M. and Scattolin, M. (1996). Buffer strips to protect the Venice lagoon from non point source pollution. *Proceedings of the international conference on buffer zones, their processes and potential in water protection.* Heythrop: UK.

Friedman, T.L. (2005). *The world is flat.* New York: Farrar, Straus and Giroux.

Gallagher, P. (2000). *Local sustainability frameworks — Santa Monica, CA.* www.ci.santa-monica.ca.us/environment.

Gardner, G. (2007, 7 February). Engaging religion in the quest for sustainability. *World Watch Institute.* www.worldwatch.org/node/1552. Accessed 18 April 2007.

Gilderbloom, J.I. and Capek, S.M.(1992). *Community vs. commodity — ten-*

ants and the American city. Albany, New York: State University of New York Press.

Gladwin T., Kennelly, J. and Krause T. (1995). Shifting paradigms for sustainable development: implications for management theory and research. *Academy of management review* 20(4): 874–907.

Gonchar, J. (2003, January) Green building industry grows by leaps and bounds. *Engineering News-Record*.

Goodman, O.B. (2002a). *City of Las Vegas budget in brief for fiscal year 2002*. Las Vegas, Nevada.

Goodman, O.B. (2002b). *Mayor's state of the city address, Las Vegas, Nevada (7 January 2002)*. Las Vegas: City of Las Vegas Communications Director.

Gordon, K. and Ode, S. (2005). *High performance cities*. ICLEI.

Gordon, P. and Richardson, H.W. (1997). Are compact cities a desirable planning goal? *Journal of the American planning association*. 63 (1).

Gore, A. (2007) *The Assault on Reason*. New York: Penguin Press.

Gottdiener, M., Collins, C.C. and Dickens, D.R. (1999). *Las Vegas—The social production of an all-American city*. Malden, Massachusetts: Blackwell Publishers, Inc.

Grumman, D.L. (ed.) (2003). *ASHRAE green guide*. Atlanta: ASHRAE.

Guber, D.L. (2003). *The grassroots of the green revolution: Polling America on the environment*. Cambidge, MA: MIT Press.

Guralnik, D.B. (1972). *Webster's new world dictionary of the American language*. New York: World Publishing.

Habitat (1996). *An urbanizing world: Global report on human settlements, 1996*. Oxford: Oxford University Press.

Hall, P. (1988). *Cities of tomorrow*. Oxford UK: Blackwell Publishers.

Hansen, S.J. and Weisman, J.C. (1998). *Performance contracting: Expanding horizons*. Lilburn, Georgia: Fairmont Press.

Hamel, J., Dufour. S., and Fortin, D. (1993). *Case study methods*. London: Sage Publications.

Harrington, J.J. and Vogel, R.K. (2000). *Political change in the metropolis*. New York: Addison-Wesley Education Publishers, Inc.

Haughton, G. and Hunter, C. (1994). *Sustainable cities*. London: Jessica Kingsley.

Hawken, P. (1993). *The Ecology of Commerce*. New York: HarperBusiness.

Hawkin, P., Lovins, A. and Lovins, L.H. (1999). *Natural Capitalism*. Boston: Little, Brown and Co.

Healy, M. (2007, 23 Janaury). Scientists predict Alpine glacier loss. *USA*

Today.

Hebert, H.J. (2004, 4 February). Tougher air conditioner efficiency standards will take effect in 2006. *Courier-journal.*

Heonen, K. (2006, 1 September). *New standard approved for high performance schools.* Press release, the Collaborative for High Performance Schools.

Hempel, L.C. (1996). *Environmental governance — The global challenge.* Washington, D.C.: Island Press.

Hens, L. (1996). The Rio conference and thereafter, in B. Nath, L. Hens and D. Devuyst, (eds.). *Sustainable development.* Brussels: Vrije Universiteit Brussel Press.

Heywood, P. (1997). The Emerging Social Metropolis: Successful Planning Initiatives in Five New World Metropolitan regions. *Progress in planning.* 47 (3).

Hewings, H. and Moshe, Y. (2002). *Job jolt — the economic impacts of repowering the midwest.* Environmental Law and Policy Center: Chicago, Illinois.

Holdren, J.P. (1991). Population and the energy problem. *Population and environment.* 12 (3).

Holland, L., Holland, A. and McNeil, D. (2000). *Global sustainable development in the 21st century.* Edinburgh: Edinburgh University Press.

Hoover's, Inc. (2004, 22 February). *Sierra Pacific Resources.* http://www.hoovers.com/free/co/factsheet.xhtml?COID=11353.

Huber, H. (2005). Sustainable buildings—The systems perspective. *Climate Change — Energy Awareness — Energy Efficiency.* Fourth international conference, June 8-10, 2005. Visegrad, Hungary.

Hulse, J.W. (1998). *The silver state.* Reno, Nevada: University of Nevada Press.

ICLEI (2003). International council for local environmental initiatives. www.iclei.org.

Inoquchi, T., Newman, E. and Paoletto, G. (ed.) (1999). *Cities and the environment — New approaches for eco-societies.* United Nations University Press: New York.

ICMA (2002). *Getting to smart growth.* Washington, D.C.: International City/County Management Association.

iiSBE (2005). *International initiative for a sustainable built environment.* http://iisbe.org/iisbe/start/iisbe.htm, accessed February 2005.

IPMVP New Construction Subcommittee (2003). International Performance Measurement & Verification Protocol, Vols. 1, 2 & 3. April

2003.

IPPC (2001). International panel on climate change. *Third assessment report, climate change 2001: The scientific basis.* www.ippc.ch.

Iwata, E. (2007, 14 February). Businesses grow more socially conscious—More think strategy can also be profitable. *USA Today.* p. 3B.

Jackson, K.T. (1985). *Crabgrass frontier.* New York: Oxford University Press.

Jaffee A.B.and Stavins, R.N. (1994) Energy-efficiency investments and public policy. *The energy journal.* 15 (2).

Johnson and Johnson (1997). *Environmental Report 1997.* New Brunswick, NJ: Johnson and Johnson.

Johnson Controls (2006). *2006 Business and Sustainability Report.* Creating Smart Environments. Milwaukee: Johnson Controls.

Judd, D.R. and Fainstein, S.S. (1999). *The tourist city.* New Haven, Connecticut: Yale University Press.

Judge, J. (1989, August). The many lives of Old Havana. *National Geographic Magazine.* 176 (2).

Kablan, M.M. (2003). Decision support for energy conservation promotion: an analytic hierarchy process approach. *Energy policy.* Elsevier Sciences Ltd.

Kahn, M.E. (2000). The environmental impact of suburbanization. *Journal of policy analysis and management.* 19 (4).

Kaihla, P. (2007, January). Global warming. *Business 2.0.* 8 (1).

Karem, D. (2005). Louisville waterfront park: the greening of a city. *Sustain.* 12.

Katz, P. (1994). The New Urbanism—Toward an architecture of community. *Environmental science,* Vol. 45. New York: McGraw-Hill, Inc.

Kelbaugh, D.S. (2002). *Repairing the American metropolis, commonplace revisited.* Seattle: University of Washington Press.

Klevas, V. and Minkstimas, R. (2004). The guidelines for state policy of energy efficiency in Lithuania. *Energy policy.* Elsevier. 32.

Kneese, A.V. and Brown, F.L. (1981). *The Southwest under stress.* Baltimore: Johns Hopkins University Press.

Ko, J. (2000). *An integrated assessment of energy and resource efficiency trends at regional.* State University of New York.

Koeha, T. (2004) What is sustainability and why should I care? *Proceedings of the 2004 world energy engineering congress,* Austin, Texas, 9.22-24. Atlanta: Association of Energy Engineers.

Kolankiewics, L. and Beck, R. (2000). *Sprawl city.* Arlington, Virginia:

www.sprawlcity.org.

Koven, S.G. and Lyons, T.S. (2003). *Economic development—Strategies for state and local practice.* Washington, D.C.: International City/County Management Association.

Kraft, M. (2004). *Environmental Policy and Politics,* Third Edition. New York: Pearson/Longman.

Kuhn, T.S. (1996). *The structure of scientific revolutions.* Chicago, Illinois: The University of Chicago Press.

Kunster, J.H. (1994). *The geography of nowhere.* New York: Simon and Schuster, Inc.

Kurokawa, K. Architects and Associates (1998). *Public space conceptual design along central axis in Shenzhenu City Center, Stage II.*

Kutzhanova, N. and Roosa, S.A. (2003). Sustainable business incubation using public-private partnerships. *Shaping our urban future utilizing creativity, vision, policy and advocacy to make change.* Cleveland, Ohio: Urban Affairs Association.

Lafferty M.L. and Meadowcroft, J. (2000). *Implementing sustainable development: Strategies and initiatives in high consumption societies.* Oxford University Press.

Lane, F.C. (1973). *Venice - A maritime republic.* Baltimore and London: Hopkins University Press.

Las Vegas Regional Clean Cities Coalition (2002). Clark County School District operates largest fleet of bio-diesel school buses. *Valley skylines.* 2(4).

Las Vegas Review Journal (2002, 1 October). *Nevada Power wants to spend $9.2 million on energy conservation.* Las Vegas.

Laszlo, E. (1972). *The relevance of general systems theory.* New York: Braziller.

Laxalt, R. (1977). *Nevada, a bicentennial history.* New York: W.W. Norton & Company, Inc.

Layzer (2006). *The Environmental Case.* Washington D.C.: CQ Press.

LeGates, R.T. and Stout, F. (ed.) (2000). *The city reader.* Bury St. Edmunds, Suffolk: St. Edmundsbury Press.

Levine, R.S. (1990). Sustainable development. *Conference report of the first international ecocity conference.* Berkeley, California.

Lindblom, C.E. (1958). Policy analysis. *American economic review.* 48.

Lindblom, C.E. (1965). *The intelligence of democracy—Decision making through mutual adjustment.* London: Collier-Macmillan Limited.

Ling J.A. (2006). High performance buildings can teach sustainability

24/7. *Proceedings of the 2006 world energy engineering congress.* Atlanta: Association of Energy Engineers.

Lipman, B. (2006, October). A heavy load: the combined housing and transportation burden of working families. *Center for housing policy.* Washington D.C. www.cnt.org/repository/heavy_load_10_06.pdf, accessed 26 April 2006.

Littlejohn, D. (1999). *The real Las Vegas.* Oxford and London: Oxford University Press.

Low, N., Gleeson, B., Elander, I. and Lidskog, R. (2002) *Consuming cities—The urban environment in the global economy after the Rio Declaration.* London; Rutledge.

Low, N. (2002). Eco-socialization and environmental planning: A Pollyannian approach. *Environment and planning.* 34.

McAuslan, F. and Norman, M. (2000). *The rough guide to Cuba.* Rough Guides: New York.

McBride, S and Black, A. (1988). *Living in Cuba.* St. Martin's Press.

Maclaren, V.W. (1996a). *Developing indicators of urban sustainability: A focus on the Canadian experience.* Toronto: ICURR Press.

Maclaren, V.W. (1996b). Urban sustainability reporting. *Journal of the American planning association.* 62 (2).

McGowan, J. (2004, February). Selling real time metering up the management chain. *Energy user news.*

Maffatt, I. (1996). *Sustainable development: principles, analysis, and policies.* New York: Parthenon.

Malcolm, D.R. (2006). *An energy diet for electric motors.* www.plantservices.org/articles/2006/313.html, accessed 12.21.06.

Manning, M. (2000). Report says cleanup of test site impossible. *Las Vegas sun.* August 8.

Masoner, M. (1988). *An audit of the case study method.* New York: Praeger.

Mazria, E. (1979). *The passive solar energy book.* Emmaus Pennsylvania: Rodale Press.

Meadows, D.H. (1989, 15 October). Clouds of dispute over global warming. *Los Angeles times.*

Meadows, N. and Gleeson, B. (1998). *Justice, society and nature—An exploration of political ecology.* London: Routledge Press.

Meckler, M. (ed.) (1994). *Retrofitting buildings for energy conservation.* Lilburn GA: The Fairmont Press.

Merkel A. (1997). *Der preir des uberlebens: Gedanken und Gessprache uber zu-*

kunftige aufgaben der umweltpolitik. Stuttgart: Dt. Verlags Anst.

Meyerson, M. and Banfield, E. (1955). *Politics, planning and the public interest*. Glencoe, OL: The Free Press.

Miami-Dade County (2004). Division of Solid Waste Management. *Curbside Recycling*. www.co.miami-dade.fl.us/dswm/curb_rec.asp

Miller, J.G. (1978). *Living systems*. McGraw-Hill: New York.

Miller, W. (1998). Citizenship: a competitive asset. *Industry week*. 227 (15).

Mitchel, J. (ed.) (1983). *Random house encyclopedia*. New York: Random House.

Moehring, E.P. (1989). *Resort city in the Sun Belt: Las Vegas 1930-1970*. Reno and Las Vegas: University of Nevada Press.

Moen, L.K. (2007). Resource Management in Local Government. *Strategic Planning for Energy and the Environment*. 26 (4).

Mohl, R.A. (ed.) (1990). *Searching for the sunbelt—Historical perspectives on a region*. Knoxville: The University of Tennessee Press.

Morgan, R. (2005). Sustainable Austin. *Sustain*. 12.

Morris J. (ed.) (2002). *Sustainable development—Promoting progress or perpetuating poverty*. London: Profile Books.

Morris, W. (ed.) (1969). *The American heritage dictionary of the English language*. New York: The American Heritage Publishing Company.

Muraru, M. and Graber, A. (1963). *Treasures of Venice*. World Publishing Co.

Naisbitt, J. (1994). *Megatrends*. New York: Avon Books.

Nathan, R. and Adams, C. (1976). Understanding central city hardship. *Political science quarterly*. 91 (1).

National Pollution Prevention Center (1989, 9 March). Regulatory barriers to conservation. *NPPC*. Issue Paper 89-10.

Neff, E. (2002, 2 February). Poll cites nuclear waste as top concern, voters rate economy as second most critical issue. *Las Vegas sun*.

Nemtzow, D.M. (2003). The way it should be: Clean, cheap and practical. *Strategic planning for energy and the environment*. 23 (3).

Newbold, P. (1988). *Statistics for business and economics*. Englewood Cliffs, New Jersey: Prentice-Hall, Inc.

Newman, P. and Kenworthy, J. (1989). Gasoline consumption and cities: A comparison of U.S. cities with a global survey. *Journal of the American planners association*, 55.

Newman, P. and Kentworthy, J. (1997). More cars in cities—not! *Journal of the conservation law foundation*. 2.

Newman, P. and Kenworthy, J. (1999). *Sustainability and cities: Overcoming automobile dependence*. Washington, DC: Island Press.

Newman, P. (2003). *Sustainability—Cities, transport, land use, the environment and renewable energy*. Perth, Australia: Murdock University Press.

Nijkamp P. and Pepping, G. (1998). A meta-analytical evaluation of sustainable city initiatives. *Urban studies*. 35 (9).

North Carolina Department of Transportation (2004). *HOV lane experience around the U.S.* www.ncdot.org/hov/us.

Ohlemacher, S. (2007, 5 April). Without immigrants, nation's big metro areas would shrink. *Courier-Journal*.

O'Meara, M. (1999). Exploring a new vision for cities, in L. Brown, C. Flavin and H.F. French (eds.), *State of the world 1999*. New York: W.W. Norton and Company.

O'Neal, T. (2007, February). Curse of the black gold, hope and betrayal in the Niger Delta. *National Geographic*. p. 97-117.

Organization for Economic Cooperation and Development (1990). *Environmental policies for cities in the 1990's*. Paris, France: OECD.

Organization for Economic Cooperation and Development (1998). *Towards Sustainable Development—Environmental indicators*. Paris, France: OECD.

Owen, D. (2005). Green Manhattan. *Sustain*. 12.

Palmeri, C. (2006, 11 September). The green stamp of approval. *Business Week*. p. 96.

Panjabi, R.K. (1997). *The Earth Summit at Rio*. Boston: Northeastern University Press.

Pattee, H.H. (1973). *Hierarchy theory—the challenge of complex systems*. New York: Braziller.

Pemble, J. (1995). *Venice rediscovered*. Oxford England: Clarendon Press.

Perhac, R.M. (1989, September). A critical look at global climate and Greenhouse gasses. *Power engineering*.

Perloff, H.S. (1980). *Planning the post-industrial city*. Chicago: Planners Press.

Peterson, P. (1981). *City limits*. Chicago: University of Chicago Press.

Phillips, K. (1969) *The emerging republican majority*. New Rochelle, NY: Arlington House.

Pitzer, B. (2007, 26 March). Why are bluefin tuna on the decline? *The Courier-Journal*. New York Times Syndicate.

Platt, R.H. (2000, 10 November). *Ecological cities: The state of science and practice in the United States*. Ecological Cities Symposium, Boston

College Law School. Boston, Massachusetts.

Popenoe, D. (1986). *Sociology*. Englewood Cliffs, New Jersey: Prentice-Hall, Inc.

Porter, G. and Brown, J.W. (1996). *Global environmental politics*. Oxford: Westview Press.

Portney, K.E. (2003). *Taking sustainable cities seriously*. Cambridge, Massachusetts: The MIT Press.

Puget Sound Energy (2007, February) *Greenhouse gas policy statement*. www.pse.energyEnvironmentl/policyGreenhouseGas.aspx, accessed 3.38.07.

Putman, J.J. (1999, June). Evolution in the revolution—Cuba. *National Geographic Magazine*. 195 (6).

Prescott-Allen, R. (2003). *The well-being of nations at a glace*. Island Press.

President's Council on Sustainable Development (1996, February). *Sustainable America—A new consensus*. U.S. Government Printing Office: Washington, D.C.

Rake, L. (2004, 21 March). Las Vegas monorail test runs to begin. *Las Vegas times*.

Ramsey, J.R. (2005). U of L parnership to create "green" city. *Sustain*. 12.

Reba, E. (2005, March). *NSWMA's 2005 Tip Fee Survey*. NSWMA Research Bulletin 05-3.

Reiber, B. (2005). *Hong Kong*. Hoboken, NJ: Wiley Publishing.

Register, R. (2002). *Ecocities—Building cities in balance with nature*. Berkeley: Berkeley Hills Books.

Rennings, K. and Wiggering, H. (1997). Steps toward indicators of sustainable development: linking economic and ecological concepts. *Ecological economics*. 20.

Rifkin, J. (2007, 17 December). The risks of too much city. *Washington Post*. p.8.

Rocks, L.R. and Runyon, R.P. (1972). *The energy crisis*. New York: Crown Publishers.

Rogers, R. (1997). *Cities for a small planet*. London: Westview Press.

Roosa, S.A. (1988). The economic effects of oil imports: A case for energy management. *Strategic planning for energy and the environment*. 8 (1).

Roosa, S.A. (2002a). A discussion of the economic aspects of implementing energy conservation opportunities. *Solutions for energy security and facility management challenges - Proceedings of the 25th world energy engineering congress*. Atlanta: Association of Energy Engineers.

Roosa, S.A. (2002b). Measurement and verification applications: Option

"A" case studies. *Energy engineering*. 99(4).

Roosa, S.A. (2004) *Energy and sustainable development in North American Sunbelt cities*. Louisville, KY: RPM Publishing .

Rosenbaum, W. (2005). *Environmental politics and policy*. Washington, D.C.: CQ Press.

Rudd, D.R. and Fainstein, S.S. (ed.) (1999). *The Tourist City*. New Haven and London: Yale Press.

Rusk, D. (1993). *Cities without suburbs*. Baltimore, Maryland: The Johns Hopkins University Press.

Russell, J.S. (2003). When suburbs become mega-suburbs. *Architectural record*. (8). New York: McGraw-Hill.

Saha, P.C. (2003). Sustainable energy development: a challenge for Asia and the Pacific region in the 21st century. *Energy policy*. 31.

Sale, K. (1975) *Power shift, the rise of the southern rim and its challenges to the eastern establishment*. New York: Random House.

San Antonio Community Portal (2004a). *Curbside recycling*. www.sanantonio.gov/enviro/solidwaste/recycling.asp.

San Antonio Community Portal (2004b). *What is CRAG?* http://www.sanantonio.gov/CRAG/crag.asp?res=800&ver=true.

San Diego Association of Governments (September, 2000). *Report card of the San Diego region's livability*. San Diego, California.

Sanghi, A.K. (1992, July). *Economic impacts of electrical supply options*. New York State energy office.

Savitch, H.V. and Kantor, P. (2002). *Cities in the international marketplace*. Princeton, New Jersey: Princeton University Press.

Savitch, H. and Vogel, R. (2000). Metropolitan consolidation versus metropolitan governance in Louisville. In Campbell, R.W. (ed.). *State and local government review*. Carl Vinson Institute of Government: Athens, Georgia.32 (3).

Scarpaci, J.L., Segre, R. and Coyula, M. (2002). *Havana—Two faces of the Antillean Metropolis*. Chapel Hill, North Carolina: The University of North Carolina Press.

Schamess, L. (2006). Everything Old is New Again. *Planning*. Chicago: American Planning Association. Vol. 72 (1).

Schell, O. (2007, 16 April). Where China's rivers run dry. *Newsweek*.

Self, P. (2000). *Rolling back the market—Economic dogma and political choice*. New York: St. Martin's Press.

Selman, P. (1996). *Local sustainability*. London: Paul Chapman.

Shell, A. and Krantz, M. (2007, 27 February). Global warming a hot spot for investors. USA Today. pgs. B1-B2.

Shelton, B.A., Rodriguez, N.P., Faegin, J.R., Bullard, R.D. and Thomas, R.D. (1989). *Houston—Growth and decline in a Sunbelt boomtown*. Philadelphia: Temple University Press.

Silberberg, M. (2003). *Chemistry* 3rd edition. New York: McGraw-Hill Companies.

Sioshansi, F.P. (2001). Will the new economy use more energy or less energy? *Strategic planning and energy management*. 21 (1).

Smith, P. and Warr, K. (1991). *Global warming issues*. London: Holder and Stoughton.

Soleri, P. (1971). *The sketchbooks of Paolo Soleri*. Cambridge, Massachusetts: The Massachusetts Institute of Technology Press.

Southern California Edison (2000). *SCE and your community—commitment to energy efficiency*. www.sce.com/sc3/004.

Southern California Edison (2004). *Edison's commitment to green power*. www.sce.com/sc3/004_sce_comm/04e_envcomm/Green_Power. htm.

Speth, J. (2004). *Red sky at morning: America and the crises of the global environment*. New Haven: Yale University Press.

Spirn A.W. (1984). *The granite garden: Urban nature and human design*. New York: Basic Books.

Sports Illustrated (2004). Men's college basketball. CNN/Sports Illustrated. http://sportsillustrated.cnn.com/basketball/college/men/conferences/belt/.

Stecky, N. (2004, 22-24 September). Introduction to the ASHRAE Greenguide for LEED, *Proceedings of the 2004 world energy engineering congress*, Austin, Texas. Atlanta: Association of Energy Engineers.

Steffes, D.W. (2002). United States energy dependency and terrorism, *Strategic planning and energy management*. 23 (3).

Stein, J.M. (Ed.) (1995). *Classic readings in urban planning*. New York: McGraw-Hill.

Stephens, G.R. and N. Wilkstrom (2000). *Metropolitan government and governance*. New York: Oxford University Press.

Stone, C.N. and Sanders, H.T. (1987). *The politics of urban development*. *Lawrence, Kansas*: University of Kansas Press.

Stoten, P. (2000). Strategic planning, sustainability and urban regeneration, in Carmona M. (ed.) *Globalization, urban form and governance*. Delft: Delft University Press.

Sun, J.W. (2001:1). Is GNP-energy model logical? *Energy Policy*. 29. Elsevier.

Sustainable Seattle (1993). *Sustainable Seattle indicators of sustainable community: A report to citizens on long terms trends in their community.* Seattle, Washington: Sustainable Seattle.

Swanstrom, T. (1988). Semisovereign cities: The politics of urban development. *Polity—The journal of the northeastern political science association.* 21.

Synergy (1997). *A "tale of two cities" -- a Dickens of a dilemma.* 1(1).

Tabachnick, B.G. and L.S. Fidel (2001). *Using multivariate statistics.* New York: Allyn and Bacon.

Tauzin, B. (2003). A 21st century energy policy for America. *Strategic planning and energy management.* 8 (1).

Thomas, H. (1998). *Cuba or the pursuit of freedom.* Da Capo Press.

Thompson, S.P. (1910). *Life of William Thomas Baron Kelvin of Lars.* London.

Thumann, A. and Marie, R. (2007). *The market survey of the energy industry.* www.aeecenter.org, accessed 23 May 2007.

Travers, M. (2001). *Qualitative research through case studies.* London: Sage Publications.

Underwood, A. (2007, 16 April). Mayors take the lead. *Newsweek.*

Union of Concerned Scientists (1999). Clean energy—Public benefits of renewable energy use. *Powerful solutions: Seven ways to switch America to renewable electricity.*

United Nations (1992). *Global partnership for environment and development: A guide to agenda 21.* Geneva: United Nations.

United Nations (1993). United Nations conference on trade and development. *Environmental management in transnational corporations: Report on the benchmark corporate environmental survey.* ST/CTC/149. New York: United Nations.

United Nations (1999). *Report of the United Nations conference on environment and development.* New York: United Nations.

United Nations (2001). *Indicators of sustainable development: Guidelines and methodologies.* New York: United Nations.

United Nations (2003). *Activities: Urban energy for cities.* New York: United Nations Environment Programme, International Environmental Technology Centre: www.uneo.or.jp/activities/urban/energy_city. asp

United States Census Bureau (1999). *American factfinder for Clark County Nevada.*

United States Census Bureau (2000). *Statistical abstract of the United States.*

U.S. Department of Energy (2003). *Weatherization and intergovernmental program –rebuild America.* http://www.rebuild.org/index.asp.

U.S. Department of Energy (2004). *What are clean cities?* www.ccities.doe.gov/what_is.html.

United States Department of Energy, Energy Information Administration (2000). *State energy data.*

United States Department of Energy, Energy Information Administration (2006), *Annual energy outlook.* Washington, D.C.

United States Department of Transportation (2004). *HOV facility inventory.* Federal Highway Administration. hovpfs.ops.fhwa.dot.gov/inventory/inventory.htm.

United States Environmental Protection Agency (1998, May). *Brownfields assessment pilot fact sheet —Las Vega, Nevada.* EPA 500-F-98-161. Washington D.C.

United States Environmental Protection Agency (1999, June). *Brownfields assessment demonstration project - Albuquerque, New Mexico.* EPA 500-F-99-148. Washington D.C.

United States Environmental Protection Agency (2000). http://yosemite.epa.gov/oar/globalwarming.nsf/content/Climate.html. 1.7.

U.S. Environmental Protection Agency (2003). *Understanding the AQI.* Epa.gov/air/data/gosel.html.

United States Environmental Protection Agency (2004). *Join Energy Star— Improve your energy efficiency.* www.energystar.gov. accessed February 2004.

U.S. Green Building Council (2002, November). *Green building rating system for new construction and major renovations Version 2.1.*

U.S. Green Building Council (2003) *Green building rating system for new construction and major renovations Version 2.1 reference guide.*

U.S. Green Building Council (2004, October). *LEED—certification process.* www.usgbc.org/LEED/Project/certprocess.asp, accessed October 2004.

U.S. Green Building Council (2004). *LEED green building rating system.*

U.S. Green Building Council (2006). *About USGBC.* www.usgbc.org/DisplayPage.aspx?CategoryID=1, accessed 1.3.07.

U.S. Green Building Council (2006, October). *Green building facts.* www.usgbc.org/DisplayPage.aspx?CategoryID= 2027. Accessed 12.26.06.

URS Europe (2005). http://www.urseurope.com/services/engineering/engineering-breeam.htm, accessed February 2005.

Valley of the Sun Clean Cities Coalition (2004). *Compressed natural gas public fueling stations in the metropolitan Phoenix area.* http://www.cleanairaz.org/HTM/LocalStations.htm.

Vergano, D. and O'Driscoll, P (2007, 4 April). Is Earth nearing its 'tipping points'? *USA Today.*

Verrengia, J.B. (2002a, 14 August). U.S. not embracing expanded renewable energy development, *The Courier-Journal.*

Verrengia, J.B. (2002b, 30 August). U.S. fights back at world summit, cites "new approach" to problems, *The Courier-Journal.*

Verrengia, J.B. (2002c, 15 September). Global warming taking over Alaskan island, *The Courier-Journal.*

Vig, N.J. and Kraft, M.E. (2003). *Environmental policy—New directions for the twenty-first century.* Washington, D.C.: CQ Press.

Walters (ed.) (2002, 15 December). Last gasp. *Fresno bee.* Fresno, California.

Watson, J. (2007, 11 April). U.N. report: climate change to have global impact. *The Courier-Journal.*

Weiss, C.H. (1972). *Evaluation research.* Englewood Cliffs, New Jersey: Prentice-Hall, Inc.

Wheeler, S. (2000). Planning sustainable and livable cities, in LaGates R.T. and Stout F. (ed.) *The City Reader.* New York: Routledge, Taylor and Francis Group.

White, J.C. (1992). *Global climate change.* New York: Plenum Publishing Corporation.

Wholey, J.S., Hatry, H.P. and Newcomer, P.E. (ed.) (1994). *Handbook of practical program evaluation.* San Francisco: Jossey-Bass Publishers.

Wolf, P. (1999). *Hot towns.* New Brunswick: Rutgers University Press.

Worldwatch Institute (2007). *State of the world, our urban future.* New York: W.W. Norton.

Yergin, D. (1991). *The prize—The epic quest for oil, money and power.* New York: Simon and Schuster.

Zarnikou, J. (2003). Consumer demand for 'green power' and energy efficiency. *Energy policy.* 31.

Zhang, Z. (2004). The challenging economic and social issues of climate change: introduction. *Energy policy.* 32.

Appendix

ABBREVIATIONS

ACUPCC	American College & University Presidents Climate Commitment
AEC	Atomic Energy Commission
AEE	Association of Energy Engineers
AFV	Alternatively Fueled Vehicle
AIA	American Institute of Architects
ANOVA	Analysis of Variance Analysis
ANSI	American National Standards Institute
ANWR	Arctic National Wildlife Refuge
AQI	Air Quality Index
ASHRAE	American Society of Heating, Refrigerating & Air-Conditioning Engineers
BART	Bay Area Rapid Transit
BREEAM	Building Research Establishment Environmental Assessment Method
BRT	Bus Rapid Transit
BTL	Biomass-To-Liquid
Btu	British Thermal Units, equal to 1.055 kilojoules
C	Celsius
CDD	Cooling Degree Days
CDM	Clean Development Mechanism
CEA	Cambridge Energy Alliance
CEO	Chief Executive Officer
CFC	Chlorofluorocarbons
CFL	Compact Fluorescent Lamps
CH_4	Methane
CHP	Combined Heat and Power
CHPS	Collaborative for High Performance Schools
CIA	Central Intelligence Agency

CO_2	Carbon Dioxide
COP	Coefficient of Performance
CSR	Corporate Social Responsibility
CTL	Coal-To-Liquid
CVN	Consorzio Venezia Nuova
DCL	Desert Learning Center
DfE	Design for the Environment
EC	European Community
EIA	Energy Information Administration
EMCS	Energy Monitoring and Control System
EMS	Environmental Management System
EPA	Environmental Protection Agency
ER	Energy Rating
EROEI	Energy Return on Energy Investment
ESG	Energy Systems Group
ESI	Environmental Sustainability Index
ESPC	Energy Savings Performance Contract
ETS	Emissions Trading Scheme
EU	European Union
EV	Electric Vehicle
EVO	Efficiency Valuation Organization
F	Fahrenheit
ft	Foot
GDP	Gross Domestic Product
GHG	Greenhouse Gas
GHP	Ground-source Heat Pumps
GW	Gigawatts
HID	High Intensity Discharge
HFC	Hydrofluorocarbons
HOV	High Occupancy Vehicle
HVAC	Heating, Ventilating and Air Conditioning
IAQ	Indoor Air Quality
ICLEI	International Council for Local Environmental Initiative

ICMA	International City/County Management Association
IECC	International Energy Conservation Code
IESNA	Illuminating Engineering Society of North America
IISBE	International Initiative for a Sustainable Built Environment
IPMVP	International Performance Measurement and Verification Protocol

K12	Kindergarten through 12th Grade
kBtu	Thousand British Thermal Units
Kcal	Kilocalorie
Kj	Kilojoule, equal to .9478 Btus
Km	Kilometer
Kph	Kilometers per hour
kW	Kilowatt
kWh	Kilowatt hour

lbs.	Pounds
LA21	Local Agenda 21
LCD	Liquid Crystal Display
LED	Light-Emitting Diodes
LEED	Leadership in Energy and Environmental Design
LEED-EB	Leadership in Energy and Environmental Design—Existing Buildings
LEED-NC	Leadership in Energy and Environmental Design—New Construction
LFG	Landfill Gas
LFGTE	Landfill Gas to Energy

M	Meter
MARTA	Metropolitan Area Rapid Transit Authority
M&V	Measurement and Verification
MERV	Minimum Efficiency Reporting Value
MIT	Massachusetts Institute of Technology
Mpg	Miles per gallon
Mph	Miles per hour
MSA	Metropolitan Statistical Area
MTR	Mass Transit Railway
MUV	Multiuse Vehicle
MW	Megawatt

NaS   Sodium and Sulfur
NASA   National Aeronautics and Space Administration
NAFTA   North American Free Trade Agreement
NEPA   National Environmental Policy Act (U.S.)
NIMBY   Not In My Backyard
N_2O Carbon Dioxide
NOTE   Not Over There Either
NPPC   National Pollution Prevention Center

OAPEC   Organization of Arab Petroleum Exporting Countries
ODGI   Ozone Depleting Gas Index
OECD   Organization for Economic Cooperation and Development
OPEC   Organization of Petroleum Exporting Countries
ppm   Parts per million
PTC   Production Tax Credits
PSE   Puget Sound Energy
PVC   Polyvinylchloride

R^2 Coefficient of Determination
RPS   Renewable Portfolio Standards

SBIC   Sustainable Buildings Industry Council
SF   Square foot
PSE   Puget Sound Energy
SUV   Sport Utility Vehicle

TOD   Transit-Oriented Development
TVA   Tennessee Valley Authority
TGV   Train à Grande Vitesse

UAV   Urban Assault Vehicle
UK   United Kingdom
UN   United Nations
UNICEF   United Nations Children's Fund
US   United States
USDOE   United States Department if Energy
USEPA   United States Environmental Protection Agency
USGBC   United States Green Building Council
USSR   Union of Socialist Soviet Republics

Table A-1. Population, Land Area and Estimates of Population Density

Sunbelt City	State	2000 Population (1)	2000 Land Area in Square Miles	Population per Square Mile	1990 Population (2)	1990 Land Area in Square Miles	Population per Square Mile	Change in Population Density	2000 Land Area in Km²	Population per Km²	1990 Land Area Km²	Population per Km²
Los Angeles	California	3,694,820	469.1	7,876	3,485,398	469.3	7,427	6.1%	1,215	3,041	1,215	2,868
Houston	Texas	1,953,631	579.5	3,371	1,630,553	539.9	3,020	11.6%	1,501	1,302	1,398	1,166
Phoenix city	Arizona	1,321,045	474.9	2,782	983,403	419.9	2,342	18.8%	1,230	1,074	1,088	904
San Diego city	California	1,223,400	324.4	3,771	1,110,549	324.0	3,428	10.0%	840	1,456	839	1,323
Dallas city	Texas	1,188,580	342.6	3,469	1,006,877	342.4	2,941	18.0%	887	1,340	887	1,135
San Antonio	Texas	1,144,646	407.6	2,808	935,933	333.0	2,811	-0.1%	1,056	1,084	862	1,085
Jacksonville	Florida	735,617	757.7	971	635,230	758.7	837	16.0%	1,962	375	1,965	323
Austin	Texas	656,562	251.5	2,611	465,622	217.8	2,138	22.1%	651	1,008	564	825
Memphis	Tennessee	650,100	279.3	2,328	610,337	256.0	2,384	-2.4%	723	899	663	921
Nashville-Davidson	Tennessee	545,524	473.3	1,153	488,374	473.3	1,032	11.7%	1,226	445	1,226	398
El Paso	Texas	563,662	249.1	2,263	515,342	245.4	2,100	7.8%	645	874	636	811
Charlotte	North Carolina	540,828	242.3	2,232	395,934	174.3	2,272	-1.7%	628	862	451	877
Fort Worth	Texas	534,694	292.6	1,827	447,619	281.1	1,592	14.8%	758	706	728	615
Oklahoma City city	Oklahoma	506,132	607.0	834	444,719	608.2	731	14.0%	1,572	322	1,575	282
Tucson city	Arizona	486,699	194.7	2,500	405,390	156.3	2,594	-3.6%	504	965	405	1,001
New Orleans	Louisiana	484,674	180.6	2,684	496,938	180.6	2,752	-2.5%	468	1,036	468	1,062
Las Vegas	Nevada	478,434	113.3	4,223	258,295	83.3	3,101	36.2%	293	1,630	216	1,197
Long Beach	California	461,522	50.4	9,157	429,433	50.0	8,589	6.6%	131	3,536	129	3,316
Albuquerque	New Mexico	448,607	180.7	2,483	384,736	132.2	2,910	-14.7%	468	959	342	1,124
Fresno	California	427,652	104.4	4,096	354,202	99.1	3,574	14.6%	270	1,582	257	1,380
Virginia Beach	Virginia	425,257	248.3	1,713	393,069	248.3	1,583	8.2%	643	661	643	611
Atlanta	Georgia	416,474	131.8	3,160	394,017	131.8	2,990	5.7%	341	1,220	341	1,154
Mesa	Arizona	396,375	125.0	3,171	288,091	108.6	2,653	19.5%	324	1,224	281	1,024
Tulsa	Oklahoma	393,049	182.7	2,151	367,302	183.5	2,002	7.5%	473	831	475	773
Miami city	Florida	362,470	35.7	10,153	358,548	35.6	10,072	0.8%	92	3,920	92	3,889

(1) U.S. Census Bureau, U.S. Census 2000.
(2) U.S. Census Bureau, U.S. Census 1990.

TableA-2. Raw Values for Selected Indicators or Measures of Energy Use

Sunbelt City	Energy Use Per Capita Million KiloJoule/Yr. (1)	Energy Costs Per Capita US $s (1)	Alternate or no Fuels for Heating % (2)	Single Occupant Households % (2)	Mean Travel Time to Work Min. (2)	Alternative Transport % (2)	0 to 1 Vehicles per Household % (2)	AQI Index (3)	Days Unhealthy & Unhealthy for sensitive groups (3)	Population Density Per Sq KM (2)	Population Change - Population Density % (2)
Los Angeles, California	265.5	2,098	4.4	28.5	29.6	17.9	56.8	72	88	2,868	-0.1
Houston Texas	586.6	3,551	0.9	29.6	27.4	10.5	44.5	53	53	1,166	8.2
Phoenix, Arizona	250.1	2,059	0.9	25.4	26.1	8.8	48.4	68	30	904	15.5
San Diego, California	265.5	2,098	0.5	28.0	23.2	11.8	47.2	59	36	1,323	0.2
Dallas, Texas	586.6	3,551	0.6	32.9	26.9	8.2	56.9	49	35	1,135	0.0
San Antonio, Texas	586.6	3,551	0.6	25.1	23.8	8.2	49.5	38	3	1,085	22.4
Jacksonville, Florida	260.5	1,951	1.2	26.2	25.2	6.8	47.5	0	2	323	-0.2
Austin, Texas	586.6	3,551	0.5	32.8	22.4	10.4	50.3	41	12	825	18.9
Memphis, Tennessee	375.9	2,419	0.5	30.5	23.0	6.6	58.3	54	28	921	8.9
Nashville/Davidson, Tennessee	375.9	2,419	0.7	33.8	23.3	7.2	41.8	54	22	396	-0.1
El Paso, Texas	586.6	3,551	0.4	19.2	22.4	6.5	46.1	54	11	811	1.6
Charlotte, North Carolina	464.2	2,404	0.3	29.5	25.1	7.9	47.0	55	31	877	38.3
Fort Worth, Texas	586.6	3,551	0.5	28.6	24.6	5.3	49.1	46	20	615	4.7
Oklahoma City, Oklahoma	428.2	2,706	0.6	30.7	20.8	5.3	48.3	42	7	282	-0.2
Tucson, Arizona	250.1	2,059	0.9	32.3	21.6	9.8	56.0	45	0	1,001	23.7
New Orleans, Louisiana	936.6	4,638	0.7	33.2	25.7	21.6	69.6	45	20	1,062	0.0
Las Vegas, Nevada	334.3	2,413	0.3	25.0	25.4	9.4	51.0	55	5	1,197	49.0
Long Beach, California	265.5	2,098	3.3	29.6	28.7	10.0	57.6	72	88	3,316	0.9
Albuquerque, New Mexico	327.9	2,259	0.6	30.5	20.4	8.0	46.5	53	3	1,124	31.3
Fresno, California	265.5	2,098	1.1	23.3	21.7	7.5	53.0	80	132	1,380	6.1
Virginia Beach, Virginia	343.5	2,372	0.8	20.4	23.9	5.5	35.7	40	0	611	0.0
Atlanta, Georgia	357.2	2,416	0.6	38.5	28.3	22.3	66.0	61	60	1,154	0.0
Mesa, Arizona	250.1	2,059	0.5	24.2	25.9	6.6	47.0	53	13	1,024	18.1
Tulsa, Oklahoma	428.4	2,706	0.3	33.9	18.6	7.5	51.6	45	10	773	-0.5
Miami, Florida	260.5	1,951	7.3	30.4	28.1	17.2	68.8	37	2	3,889	0.3

Notes:

(1) Energy Information Administration, *State Energy Data 2000*, tabulated unit measures converted to Kilojoules
(2) U.S. Census Bureau, data calculated using both U.S. *Census 1990* and U.S. *Census 2000* .
(3) U.S. Environmental Protection Agency, data from www.epa.gov/airnow/

Table A-3. Summary of Indexed Values for Selected Indicators or Measures of Energy Use

Sunbelt City	Energy Use Per Capita Million Kilojoule/Yr.	Alternate or no Fuels for Heating	Single Occupant Households	Mean Travel Time to Work Min.	Alternative Transport	0 to 1 Vehicles per Household	AQI Index	Days Unhealthy & Unhealthy for sensitive groups	Population Density Per Sq KM	Energy Use Index
Los Angeles, California	0.98	0.59	0.52	0.00	0.74	0.62	0.10	0.33	0.72	4.60
Houston Texas	0.51	0.09	0.46	0.20	0.31	0.26	0.34	0.60	0.25	3.02
Phoenix, Arizona	1.00	0.09	0.68	0.32	0.21	0.37	0.15	0.77	0.17	3.76
San Diego, California	0.98	0.03	0.54	0.58	0.38	0.34	0.26	0.76	0.29	4.16
Dallas, Texas	0.51	0.04	0.29	0.25	0.17	0.63	0.39	0.73	0.24	3.25
San Antonio, Texas	0.51	0.04	0.69	0.53	0.17	0.41	0.53	0.98	0.22	4.08
Jacksonville, Florida	0.98	0.13	0.64	0.40	0.09	0.35	1.00	0.98	0.01	4.58
Austin, Texas	0.51	0.03	0.30	0.65	0.30	0.43	0.48	0.91	0.15	3.76
Memphis, Tennessee	0.82	0.03	0.41	0.60	0.08	0.67	0.33	0.79	0.18	3.91
Nashville/Davidson, Tennessee	0.82	0.06	0.24	0.57	0.11	0.18	0.33	0.83	0.03	3.17
El Paso, Texas	0.51	0.01	1.00	0.65	0.07	0.31	0.33	0.92	0.15	3.95
Charlotte, North Carolina	0.69	0.00	0.47	0.41	0.15	0.33	0.31	0.77	0.16	3.29
Fort Worth, Texas	0.51	0.03	0.51	0.45	0.00	0.40	0.43	0.85	0.09	3.27
Oklahoma City, Oklahoma	0.74	0.04	0.40	0.80	0.00	0.37	0.48	0.95	0.00	3.78
Tucson, Arizona	1.00	0.09	0.32	0.73	0.26	0.60	0.44	1.00	0.20	4.64
New Orleans, Louisiana	0.00	0.06	0.27	0.35	0.96	1.00	0.44	0.85	0.22	4.15
Las Vegas, Nevada	0.88	0.00	0.70	0.38	0.24	0.45	0.31	0.96	0.25	4.17
Long Beach, California	0.98	0.43	0.46	0.08	0.28	0.65	0.10	0.33	0.84	4.15
Albuquerque, New Mexico	0.89	0.04	0.41	0.84	0.16	0.32	0.34	0.98	0.23	4.21
Fresno, California	0.98	0.11	0.79	0.72	0.13	0.51	0.00	0.00	0.30	3.54
Virginia Beach, Virginia	0.86	0.07	0.94	0.52	0.01	0.00	0.50	1.00	0.09	3.99
Atlanta, Georgia	0.84	0.04	0.00	0.12	1.00	0.89	0.24	0.55	0.24	3.92
Mesa, Arizona	1.00	0.03	0.74	0.34	0.08	0.33	0.34	0.90	0.21	3.97
Tulsa, Oklahoma	0.74	0.00	0.24	1.00	0.13	0.47	0.44	0.92	0.14	4.08
Miami, Florida	0.98	1.00	0.42	0.14	0.70	0.98	0.54	0.98	1.00	6.74

Table A-4. Index Values for Significantn Indicators and Measures of Energy Use (For Sustainability Ranking of Sumbelt Cities)

Sunbelt City	Alternative Transport	Population Density Per Sq KM	Alternate or no Fuels for Heating	Days Unhealthy & Unhealthy for sensitive groups	0 to 1 Vehicles per Household	Total Index Score
Los Angeles, California	0.74	0.72	0.59	0.33	0.62	3.00
Houston Texas	0.31	0.25	0.09	0.60	0.26	1.51
Phoenix, Arizona	0.21	0.17	0.09	0.77	0.37	1.61
San Diego, California	0.38	0.29	0.03	0.73	0.34	1.77
Dallas, Texas	0.17	0.24	0.04	0.73	0.63	1.81
San Antonio, Texas	0.17	0.22	0.04	0.98	0.41	1.82
Jacksonville, Florida	0.09	0.01	0.13	0.98	0.35	1.56
Austin, Texas	0.30	0.15	0.03	0.91	0.43	1.82
Memphis, Tennessee	0.08	0.18	0.03	0.79	0.67	1.75
Nashville/Davidson, Tennessee	0.11	0.03	0.06	0.83	0.18	1.21
El Paso, Texas	0.07	0.15	0.01	0.92	0.31	1.46
Charlotte, North Carolina	0.15	0.16	0.00	0.77	0.33	1.41
Fort Worth, Texas	0.00	0.09	0.03	0.85	0.40	1.37
Oklahoma City, Oklahoma	0.00	0.00	0.04	0.95	0.37	1.36
Tucson, Arizona	0.26	0.20	0.09	1.00	0.60	2.15
New Orleans, Louisiana	0.96	0.22	0.06	0.85	1.00	3.09
Las Vegas, Nevada	0.24	0.25	0.00	0.96	0.45	1.90
Long Beach, California	0.28	0.84	0.43	0.33	0.65	2.53
Albuquerque, New Mexico	0.16	0.23	0.04	0.98	0.32	1.73
Fresno, California	0.13	0.30	0.11	0.00	0.51	1.05
Virginia Beach, Virginia	0.01	0.09	0.07	1.00	0.00	1.17
Atlanta, Georgia	1.00	0.24	0.04	0.55	0.89	2.72
Mesa, Arizona	0.08	0.21	0.03	0.90	0.33	1.55
Tulsa, Oklahoma	0.13	0.14	0.00	0.92	0.47	1.66
Miami, Florida	0.70	1.00	1.00	0.98	0.98	4.66

Table A-5. Changes in Energy Use 1990-2000

Sunbelt Cities	2000 City % of State Population (1)	1990 City % of State Population (2)	2000 Energy Use per capita Million Kj (3)	1990 Energy Use per capita Million Kj (3)	1990-2000 % Change in per capita Energy Use	2000 Energy Use of city Billion Kj	1990 Energy Use of city Billion Kj	1990-2000 % Change in total Energy Use
Type A								
Miami	2.3	2.8	260.5	264.3	-1.5	94,412	94,776	-0.4
Los Angeles	10.9	11.7	265.5	273.9	-3.1	980,874	954,512	2.8
San Diego	3.6	3.7	265.5	273.9	-3.1	324,779	304,135	6.8
Atlanta	5.1	6.1	357.2	366.8	-2.6	173,825	148,681	16.9
Tucson	9.5	7.9	250.1	192.9	29.7	121,740	78,183	55.7
Long Beach	1.4	1.4	265.5	273.9	-3.1	122,522	117,605	4.2
Austin	3.1	2.7	586.6	615.1	-4.6	385,164	286,391	34.5
Houston	9.4	9.6	586.6	615.1	-4.6	1,146,073	1,002,908	14.3
Las Vegas	23.9	21.5	334.3	361.5	-7.5	159,926	93,375	71.3
San Antonio	5.5	5.5	586.6	615.1	-4.6	671,492	575,667	16.6
Dallas	5.7	5.9	586.6	615.1	-4.6	697,266	619,302	12.6
Albuquerque	24.7	25.4	360.2	417.3	-13.7	412,278	390,591	5.6
Phoenix	25.7	26.8	250.1	270.0	-7.3	330,439	265,480	24.5
Average	10.1	10.1	381.2	396.5	-2.4	432,368.6	379,354.2	20.4
Type B								
El Paso	2.7	3.0	586.6	615.1	-4.6	330,665	316,973	4.3
New Orleans	10.8	11.8	936.6	906.4	3.3	453,931	450,443	0.8
Tulsa	11.4	11.7	428.4	461.8	-7.2	168,388	169,624	-0.7
Fort Worth	2.6	2.6	586.6	615.1	-4.6	313,672	275,318	13.9
Memphis	11.4	12.5	375.9	390.8	-3.8	244,356	238,496	2.5
Jacksonville	4.6	4.9	260.5	264.3	-1.5	191,606	167,912	14.1
Mesa	7.7	7.9	250.1	350.1	-28.6	99,147	100,871	-1.7
Nashville	10.0	10.5	375.9	390.8	-3.8	214,208	199,594	7.3
Fresno	1.3	1.2	265.5	273.9	-3.1	113,530	97,002	17.0
Virginia Beach	6.0	6.4	343.5	333.5	3.0	146,083	131,097	11.4
Oklahoma City	14.7	14.1	428.4	461.8	-7.2	216,835	205,376	5.6
Charlotte	6.7	6.0	373.1	352.0	6.0	201,776	138,373	45.8
Average	7.5	7.7	434.3	451.3	-4.3	224,516.4	207,589.8	10.0

(1) U.S. Census Bureau, data calculated using U.S. Census 2000.
(2) U.S. Census Bureau, data calculated using U.S. Census 1900.
(3) Energy Information Administration, State Energy Data 2000, tabulated unit measures converted to kilojoules.

Index